Algebra

Sandra Luna McCune, Ph.D.

Associate Professor and Math Specialist
Department of Elementary Education
Stephen F. Austin State University
Nacogdoches, Texas

E. Donice McCune, Ph.D.

Professor
Department of Mathematics and Statistics
Stephen F. Austin State University
Nacogdoches, Texas

June R. Turner, M.A.

Chairman
Department of Mathematics
The Rockwell School
Austin, Texas

Dedication
To our parents, Joe and Kathryne Luna, Ennis and Winnie McCune, and John and June Turner.

All inquiries should be addressed to:
Barron's Educational Series, Inc.
250 Wireless Boulevard
Hauppauge, New York 11788

Library of Congress Catalog Card No.: 96-54296

International Standard Book No. 0-8120-9746-7

Library of Congress Cataloging-in-Publication Data

McCune, Sandra L.
Algebra / Sandra Luna McCune, E. Donice McCune, June R. Turner.
p. cm. — (Barron's college review series)
Includes index.
ISBN 0-8120-9746-7
1. Algebra—Outlines, syllabi, etc. I. McCune, E. Donice. II. Turner, June R. III. Title. IV. Series: College review series.
QA155.M33 1997
512.9′0076—dc21 96-54296
CIP

PRINTED IN THE UNITED STATES OF AMERICA

987654321

CONTENTS

Preface . . . . **vi**
List of Symbols . . . . **vii**

1 • Basic Concepts **1**
1.1 Sets 1
1.2 Sets of Numbers 2
1.3 Variables 6
1.4 Equality of Numbers 6
1.5 The Real Number Line 7
1.6 The Order Relationship and Inequalities 8
1.7 Intervals and Interval Notation 9
1.8 Field Properties of the Real Number System 10
1.9 Absolute Value 17
1.10 Sums and Differences 21
1.11 Products and Quotients 22
1.12 Integer Exponents 24
1.13 Scientific Notation 31
1.14 Radicals and Fractional Exponents 32
1.15 Order of Operations 34
Chapter 1 Summary 35
Practice Problems for Chapter 1 38
Solutions to Practice Problems 39

2 • Polynomials and Rational Expressions **43**
2.1 Algebraic Terminology 43
2.2 Sums and Differences of Polynomials 44
2.3 Products of Polynomials 45
2.4 Simplifying Polynomial Expressions 48
2.5 Quotients of Polynomials By Monomials 49
2.6 Long Division of Polynomials 50
2.7 Factoring Polynomials 51
2.8 Reducing Rational Expressions 59
2.9 Products and Quotients of Rational Expressions 61
2.10 Sums and Differences of Rational Expressions 62
2.11 Simplifying Complex Fractions 64
Chapter 2 Summary 65
Practice Problems for Chapter 2 67
Solutions to Practice Problems 69

3 • Solving One-Variable Equations and Inequalities . . **73**
3.1 Solving One-Variable Linear Equations 73

3.2 Solving Formulas 78
3.3 Solving a Two-Variable Equation for One of the Variables . . . 80
3.4 Solving One-Variable Absolute Value Equations 82
3.5 Solving One-Variable Linear Inequalities 83
3.6 Solving One-Variable Absolute Value Inequalities 87
3.7 Solving One-Variable Quadratic Equations 89
3.8 Solving Other Equations in One-Variable 104
3.9 Solving One-Variable Quadratic Inequalities 111
Chapter 3 Summary 114
Practice Problems for Chapter 3 117
Solutions to Practice Problems 118

4 • Problem Solving 125
4.1 Translating into Mathematical Language 125
4.2 Working with Units of Measurement 126
4.3 Steps in Problem Solving 127
Chapter 4 Summary 158
Practice Problems for Chapter 4 159
Solutions to Practice Problems 160

5 • Functions, Relations, and Their Graphs 169
5.1 The Cartesian Coordinate Plane 169
5.2 Relations and Functions 178
5.3 Functional Notation 179
5.4 Sketching Graphs of Functions 185
5.5 Special Function Characteristics 195
5.6 The Arithmetic of Functions 200
5.7 Composition of Functions 202
5.8 Inverse of Functions 204
5.9 Functions and Their Graphs 207
5.10 Relations Defined By Conic Sections 234
5.11 Relations Defined By Inequalities in the Plane 239
5.12 Applications of Functions 243
Chapter 5 Summary 246
Practice Problems for Chapter 5 254
Solutions to Practice Problems 256

6 • Systems of Equations 261
6.1 Solving a System of Two Linear Equations in Two Variables . 261
6.2 Linear Systems in Three Variables 271
6.3 Solving Other Systems of Equations in Two Variables 274
6.4 Applications of Systems of Equations 275
Chapter 6 Summary 280
Practice Problems for Chapter 6 281
Solutions to Practice Problems 282

APPENDICES

A • Calculator Use 290

B • Measurement Units and Conversions 295

C • Common Formulas 297

D • Synthetic Division 306

E • Simplifying Radicals 308

Glossary 310

Index 323

PREFACE

Success in college-level math usually depends on a proficient understanding of college algebra. This book is intended to help students increase their ability to perform routine manipulative algebra and develop a fuller understanding of algebra concepts.

College Review: Algebra can be used by itself or as a supplement to other texts. It consists of six easy-to-read chapters on fundamental algebraic topics including the real number system; polynomials and rational expressions; equation solving; problem solving; functions, relations, and their graphs; and systems of equations. In addition, it contains five appendices that cover calculator use, measurement units and conversions, common formulas, synthetic division, and simplifying radicals. An understanding of this material will provide you with a strong background for subsequent courses.

Note the following features:

- The writing style is informal and easy to read. Explanations are clear, complete, and easy to follow.
- The material is mathematically rigorous yet nontechnical.
- Key terms are introduced in bold type.
- Important rules, procedures, and formulas are boxed for easy reference.
- Each section contains examples, each of which is worked out in a detailed, step-by-step process.
- Visual aids, such as illustrations and annotations, are provided when appropriate.
- Helpful messages warn students about common mistakes or misconceptions.
- The chapter on solving equations shows methods for solving equations that can be applied to many situations.
- The chapter on problem solving explains a general strategy that can be used in later math courses for solving many problems. It includes detailed, fully explained illustrated examples with step-by-step guidelines for setting up the equations. The examples contain a variety of problems from business, chemistry, geometry, and so forth.
- Chapter summaries highlight important terms, ideas, and procedures presented in the chapter.
- Practice exercises appear at the end of each chapter.
- A comprehensive glossary of important terms appears at the end of the book.

The authors are indebted to many people for their suggestions and encouragement. We would like to thank our friends, families, and students. We would like to express our appreciation to our editor, Linda Turner, the editorial staff at Barron's Educational Series, Inc., and the reviewers of this book.

LIST OF SYMBOLS

Symbol	Meaning
$=$	is equal to
$\neq$	is not equal to
$\cong$	is approximately equal to
$\approx$	is nearly equal to
$>$	is greater than
$\geq$	is greater than or equal to
$<$	is less than
$\leq$	is less than or equal to
$\not>$	is not greater than
$\not<$	is not less than
$+$	plus
$-$	minus
$\pm$	plus or minus
$\mp$	minus or plus
$\{\ \}$	set notation
$a \cdot b$, ab, or $(a)(b)$	a times b
$a \div b$ or $\frac{a}{b}$	a divided by b
$\cap$	intersection
$\cup$	union
$\subseteq$	is a subset of

Symbol	Meaning
$\not\subseteq$	is not a subset of
$\subset$	is a proper subset of
$\in$	is an element of
$\notin$	is not an element of
Σ	the sum of
$f \circ g$	the composite of f and g
$\lvert x \rvert$	the absolute value of x
$[x]$	the greatest integer less than or equal to x
π	approximately 3.14159
e	approximately 2.71828
$\angle$	angle
$\triangle$	triangle
$\llcorner$	right angle
$\overline{AB}$	line segment AB
$\overleftrightarrow{AB}$	line AB
$\perp$	is perpendicular to
$\parallel$	is parallel to
$\sim$	is similar to
$\cong$	is congruent to

1
BASIC CONCEPTS

1.1 SETS

When a collection of objects is described so that you can tell whether an object is or is not in the collection, the collection is said to be *well-defined* and is called a **set.** We often use capital letters to represent a set. The set can be described in words or, if it is convenient, you may use braces to enclose a list of its members or **elements.** For example,

$$A = \{1, 2, 3, x, y, z\}$$

denotes the set A with elements 1, 2, 3, x, y, and z.

To indicate that 3 is an element of $\boldsymbol{A}$, write $3 \in A$.
To indicate that r is not an element of A, write $r \notin A$.

The **universal set** is the set of all objects you are considering in a given situation. It is usually denoted by the letter U. For example, if you were considering sets composed of letters from the English alphabet, the universal set would contain the letters a through z.

$$U = \{a, b, c, \ldots, x, y, z\}$$

The **empty set,** or **null set,** is the set that has no elements. To indicate the empty set, write $\varnothing$ or { }.

A helpful message: Notice that the empty set has *no* elements—do not use $\{\varnothing\}$ or $\{0\}$ to indicate the empty set. The set $\{\varnothing\}$ is *not* empty because it contains one element: the set $\varnothing$. The set $\{0\}$ is *not* empty because it contains one element: the number 0.

When two sets A and B contain exactly the same elements, they are **equal.** For example, if B is the set of vowels in the English alphabet and $A = \{a, e, i, o, u\}$, then sets A and B have exactly the same elements, and $A = B$.
If each element of set C is also an element of set A, then set C is called a **subset** of set A. For example, $C = \{a, e\}$ is a subset of $A = \{a, e, i, o, u\}$.

To indicate that set C is a subset of set A, write $C \subseteq A$.

The symbol $\subset$ can also be used to indicate that C is a subset of A: $C \subset A$. The distinction is that $\subset$ is used only when each element of set C is contained in set A, but A contains other elements as well. For example, set $A = \{a, e, i, o, u\}$ contains the three elements i, o, and u besides the two elements a and e that comprise the set $C = \{a, e\}$. When this happens, set C is called a **proper subset** of set A: $C \subset A$.

To indicate that set A is not a subset of set C, write $A \not\subseteq C$.

The **union** of two sets A and B is the set containing all the elements in either or both of sets A and B. To indicate the union of sets A and B, write $A \cup B$. For example, if $A = \{1, 2, 3, x, y, z\}$ and $B = \{2, 4, 5, r, x, z\}$, then

$$A \cup B = \{1, 2, 3, 4, 5, r, x, y, z\}$$

The **intersection** of two sets A and B is the set containing all the elements common to both sets A and B. To indicate the intersection of sets A and B, write $A \cap B$. For example, if $A = \{1, 2, 3, x, y, z\}$ and $B = \{2, 4, 5, m, x, z\}$, then

$$A \cap B = \{2, x, z\}$$

The **complement** of a set A is the set containing all the elements of the universal set that are *not* in A. To denote the complement of A, write A^c or A'. For example, if set $A = \{a, e, i, o, u\}$ and $u =$ the set of all letters from the English alphabet, then

$$A^c = \{b, c, d, f, g, h, j, k, l, m, n, p, q, r, s, t, v, w, x, y, z\}$$

1.2 SETS OF NUMBERS

Usually the sets you will be concerned with in algebra are sets of numbers. Particularly, the set of **real numbers** will frequently be the universal set under consideration. Let's begin by looking at some important subsets of the real numbers.

The set of **natural numbers** consists of the counting numbers; that is,

$$\text{natural numbers} = \{1, 2, 3, \ldots\}$$

We frequently subdivide the set of natural numbers into two subsets: **prime** numbers and **composite** numbers. A **prime** number is a natural number greater than 1 that is divisible only by itself and by 1 (without a remainder). Thus,

$$\text{primes} = \{2, 3, 5, 7, 11, 13, 17, 19, \ldots\}$$

The natural numbers, other than 1, that are *not* prime are called the **composite** numbers. Thus,

$$\text{composites} = \{4, 6, 8, 9, 10, 12, 14, 15, \ldots\}$$

The number 1 is neither prime nor composite.

A set of numbers, such as the natural numbers, which continues on forever is called an **infinite set.** If a set stops and does not continue on forever, it is called a **finite set.**

If you include the number 0 with the set of natural numbers you obtain the set of **whole numbers;** that is,

$$\text{whole numbers} = \{0\} \cup \{1, 2, 3, \ldots\} = \{0, 1, 2, 3, \ldots\}$$

Two important subsets of the whole numbers are the set of *perfect squares* and the set of *perfect cubes*. A whole number is a **perfect square** if it is the product of a whole number multiplied by itself. For example, 4 is a perfect square because $2 \cdot 2 = 4$. Thus,

$$\text{perfect squares} = \{1, 4, 9, 16, \ldots\}$$

A whole number is a **perfect cube** if it is the product of a whole number multiplied by itself twice. For example, 8 is a perfect cube because $2 \cdot 2 \cdot 2 = 8$. Thus,

$$\text{perfect cubes} = \{1, 8, 27, 64, \ldots\}$$

If we expand the set of whole numbers to include the negatives of the natural numbers, we obtain the set of **integers.** The negatives of the natural numbers are denoted using the $-$ symbol. For example, the negative of 2 is written -2 and is read as "negative 2." Thus,

$$\begin{aligned}\text{integers} &= \{\ldots, -3, -2, -1\} \cup \{0, 1, 2, 3, \ldots\} \\ &= \{\ldots, -3, -2, -1, 0, 1, 2, 3, \ldots\}\end{aligned}$$

A helpful message: We have adopted the common practice of omitting the + sign on positive numbers. If no sign is written with a number, the number is understood to be positive.

Besides designating the integers as positive, negative, or zero, the integers can also be subdivided as either *even* integers or *odd* integers. Integers that are divisible by 2 (without a remainder) are called **even** integers. Integers that are *not* divisible by 2 (without a remainder) are called **odd** integers.

A helpful message: The integer 0 is even, because $0 \div 2 = 0$ (and there is no remainder).

The set of **rational numbers** consists of the set of numbers that can be expressed as a quotient of an integer divided by an integer other than zero. Thus,

$$\text{rational numbers} = \left\{\frac{p}{q}, \text{ where } p \text{ and } q \text{ are integers and } q \neq 0\right\}$$

Rational numbers can be expressed as fractions or decimals. The decimal representation of a rational number is obtained by performing the indicated division. For example, $\frac{3}{4}$ can be interpreted as $3 \div 4$:

$$4\overline{)3.00}^{\,0.75}$$

Thus, the decimal form of $\frac{3}{4}$ is 0.75. In this instance, the decimal **terminates** (eventually has a zero remainder). For some rational numbers, the decimal representation contains a block of one or more digits that repeats endlessly. These decimals are called **repeating decimals.** An example is $\frac{5}{6}$.

$$6\overline{)5.000...}\quad 0.8333...$$

No matter how many zeros you put after the 5, the 3s in the quotient continue without end. This repetition is indicated by a bar over the repeating digit (or digits when more than one digit repeats). Thus, $\frac{5}{6} = 0.8\overline{3}$.

The natural numbers, whole numbers, and integers are rational numbers because they can be written as fractions. For example, 3 can be expressed as $\frac{3}{1}$, 0 as $\frac{0}{1}$, -5 as $\frac{-5}{1}$, and so forth.

A helpful message: Don't forget—*division by zero is undefined;* that is, $\frac{n}{0}$ and $n \div 0$ have no meaning no matter what number you put in the place of n. Consider the quotient $\frac{5}{0}$. To be defined, the answer must be a number that multiplies times 0 to give 5; but, there is no such number! When n is zero, the quotient $\frac{0}{0}$ is still meaningless because *any* number—0, -5, $\frac{5}{6}$, $\sqrt{3}$, and so on—can be used as the quotient, since any number times 0 gives 0. So, the answer cannot be defined to be a specific number. A trick for remembering that $\frac{n}{0}$ is undefined is to note that when 0 is being under the fraction bar, the expression is undefined.

The **irrational numbers** are the real numbers that cannot be expressed as $\frac{p}{q}$, where p and q are integers and $q \neq 0$. The decimal representations of irrational numbers neither terminate nor repeat. There are infinitely many decimals that are neither terminating nor repeating decimals. For example, 0.343344333444 . . . is such a decimal. Even though you can predict that four 3's followed by four 4's likely come next in the decimal representation, the number is not a repeating decimal because no *block of digits* repeats. Therefore, 0.343344333444 . . . is an irrational number. Other examples of irrational numbers are $\sqrt{6}$ (the square root of 6) and $\sqrt[3]{10}$ (the cube root of 10), e (Euler's constant), and $-\pi$ (negative pi). With these numbers, there is no discernible repeating block of digits in the decimal representation. Thus, we write

$$\sqrt{6} = 2.449489\ldots,\ \sqrt[3]{-10} = -2.154434\ldots,\ e = 2.718281\ldots,\ -\pi = -3.141592\ldots$$

A helpful message: Some roots of rational numbers are rational and others are not. For example, $\sqrt{6}$ is irrational, but $\sqrt{4}$ is 2, a rational number. Also, $\sqrt[3]{-10}$ is irrational, but $\sqrt[3]{-8}$ is -2, a rational number. Furthermore, be cautious with *even* roots of rational numbers. When working with real numbers, the square root, fourth root, eighth root, and all even roots of a *negative* number are *not* real numbers.

For computational purposes we can only approximate irrational numbers. For example, if you want to use $\sqrt{6}$, $\sqrt[3]{-10}$, e, or $-\pi$ in computations, you can obtain an approximate value for each using a scientific calculator that can be programmed to display a preselected number of decimal places. For example, the decimal representations of these numbers to three places are as follows:

$$\sqrt{6} \cong 2.449, \sqrt[3]{-10} \cong -2.154, e \cong 2.718, -\pi \cong -3.142$$

Note: The symbol $\cong$ is read "is approximately equal to."

A helpful message: It is important to remember that if the *exact* value of an irrational root is desired, the radical symbol must be retained. For instance, if the area of a square is 6 cm^2, then the exact length of each side of the square is $\sqrt{6}$ cm.

The union of the set of rational numbers and the set of irrational numbers is the set of **real numbers;** that is,

$$\text{real numbers} = \text{rational numbers} \cup \text{irrational numbers}$$

The relationship of the subsets of the real numbers is illustrated in Figure 1.1. Each set in the figure contains those sets below it to which it is connected; that is, natural numbers $\subset$ whole numbers $\subset$ integers $\subset$ rationals $\subset$ real numbers, and irrational numbers $\subset$ real numbers:

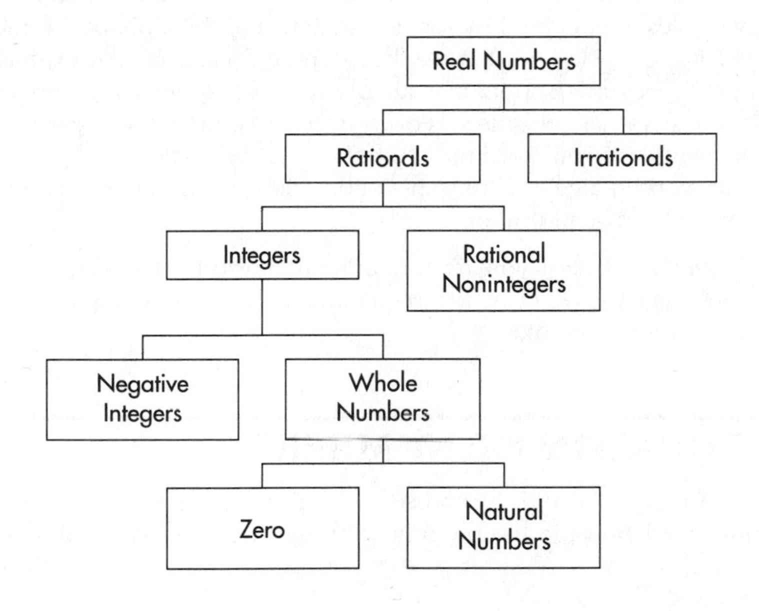

FIGURE 1.1 *The Real Numbers*

The set of **complex numbers** consists of all those numbers that can be written in the form $a + bi$, where a and b are real numbers and $i^2 = -1$. The coefficients a and b are called the **real part** and **imaginary part,** respectively, of a complex number. When a is zero, the resulting set of numbers is called the **pure imaginary** numbers. When b is zero, the resulting set of numbers consists of the **real numbers.** Thus, the real numbers are a subset of the complex numbers.

1.3 VARIABLES

The language of algebra is symbolic. In algebra, we use two kinds of numerical quantities: *constants* and *variables.*

A **constant** is a number whose value does not change. For example, all of the real numbers, including the rationals, irrationals, integers, whole numbers, and natural numbers, are constants. Each has a fixed, definite value. Consequently, when you use letters to name constants, the letter represents one fixed value. For example, the Greek letter π stands for a number whose value is approximately 3.14159. Similarly, the letter e stands for a number whose value is approximately 2.71828. The numbers π and e always have the same fixed values.

A **variable** is a symbol used to represent an unspecified element from a set, usually the set of real numbers. A variable may assume many different values. Often, variables are denoted by lowercase letters of the alphabet. Sometimes the first letter of a key word is used to denote a variable; for example, let l = length. To expand the pool of available symbols, uppercase letters and letters with subscripts are also used. For example, to denote the ages of a group of five students you could use the symbols x_1, x_2, x_3, x_4, x_5.

For our work in algebra, constants and variables will represent *real* numbers unless otherwise indicated.

A helpful message: Unless you are certain no confusion will occur, you should avoid using the letters *o*, *O*, *e*, or *i* (remember, in the set of complex numbers, $i^2 = -1$) to denote variables.

1.4 EQUALITY OF NUMBERS

If two variables a and b represent the same number we write $a = b$. The equality relationship has the following properties for real numbers a, b, and c:

THE PROPERTIES OF EQUALITY

The reflexive property: $a = a$
The symmetric property: If $a = b$, then $b = a$.
The transitive property: If $a = b$ and $b = c$, then $a = c$.

The equality properties allow you to substitute a quantity for one that is equal to it in any mathematical expression without altering the meaning of the expression. For instance,

Substitute $100 - x$ for y in xy

If $xy = 1600$ and $y = 100 - x$, then $x(100 - x) = 1600$.

The slash (/) is used to indicate "not equal." For instance, $5 \neq 7$ is read "5 is not equal to 7."

1.5 THE REAL NUMBER LINE

To give a pictorial representation of the real numbers, draw a straight line and label a point as 0. Then select a convenient length for one unit and plot on the line the points corresponding to the integers. It should look similar to Figure 1.2.

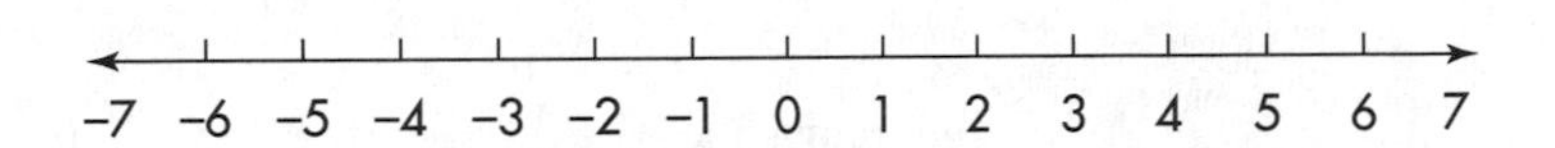

FIGURE 1.2 *Number Line*

Use arrows on the ends of the number line to show that it continues on forever in both directions. The point that is labeled 0 on the number line is called the **origin.** The **positive numbers** are located to the *right* of the origin and the **negative numbers** are located to the *left* of the origin. Numbers other than integers can also be indicated on the number line. For example,

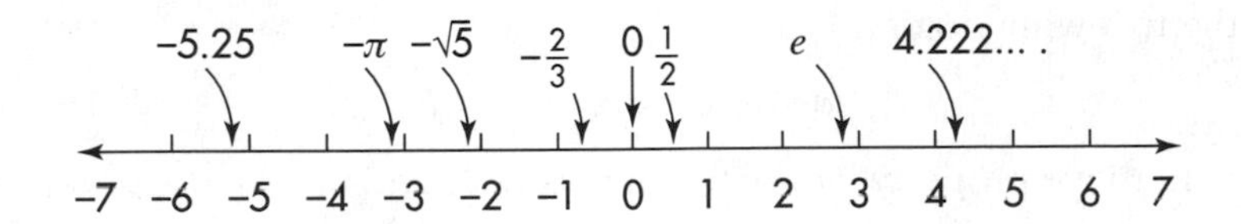

FIGURE 1.3

In fact, there is a one-to-one matching between the set of real numbers and the points on the number line. That is,

Every real number, rational or irrational, corresponds to exactly one point on the number line; and, conversely, every point on the number line corresponds to exactly one real number.

1.6 THE ORDER RELATIONSHIP AND INEQUALITIES

If you graph two numbers on the number line, exactly one of three situations occurs:

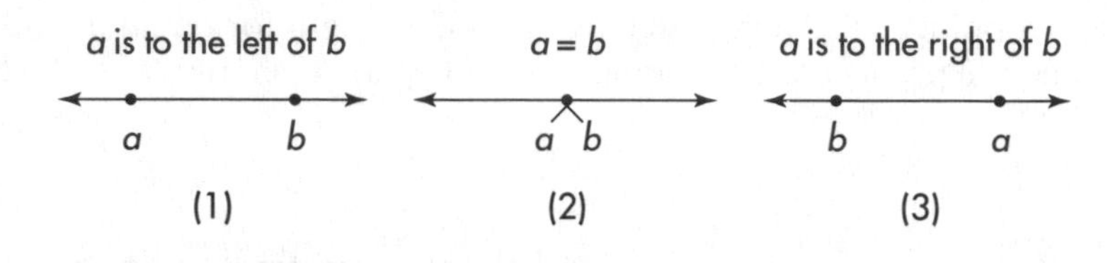

FIGURE 1.4

When a is to the left of b, as depicted in Figure 1.4 (1), we say a is **less than** b, denoted $a < b$. Similarly, when a is to the right of b, as depicted in Figure 1.4 (3), we say a is **greater than** b, denoted $a > b$. This is called an **order relationship** for the real numbers. Two important properties concerning the order of the real numbers follow:

ORDER OF THE REAL NUMBERS

The trichotomy property: For any two real numbers a and b, *exactly one* of the following holds:

$$a < b,\ a = b, \text{ or } a > b$$

The transitive property: For real numbers a, b, and c,

If $a < b$, and $b < c$, then $a < c$; and, similarly,

if $c > b$ and $b > a$, then $c > a$.

The statements $a < b$ and $a > b$ are called **inequalities.** Other inequality statements can be defined for the real numbers. Table 1.1 summarizes commonly used inequalities:

TABLE 1.1. INEQUALITIES

Inequality	Read As	Means
$x < b$	"x is less than b"	x is to the left of b on the number line
$x > a$	"x is greater than a"	x is to the right of a on the number line
$x \leq b$	"x is less than or equal to b"	either $x < b$ or $x = b$, but not both
$x \geq a$	"x is greater than or equal to a"	either $x > a$ or $x = a$, but not both
$a < x < b$	"x is greater than a and less than b"	x is between a and b, but not equal to either; that is, $a < x$ and $x < b$
$a \leq x < b$	"x is greater than or equal to a and less than b"	x is between a and b, and possibly equal to a, but not b; that is, $a \leq x$ and $x < b$
$a < x \leq b$	"x is greater than a and less than or equal to b"	x is between a and b, and possibly equal to b, but not a; that is, $a < x$ and $x \leq b$
$a \leq x \leq b$	"x is greater than or equal to a and less than or equal to b"	x is between a and b, and possibly equal to either; that is, $a \leq x$ and $x \leq b$

As you move from left to right on the number line, the numbers increase; and, conversely, as you move from right to left, the numbers decrease. Notice that all positive numbers are greater than zero and all negative numbers are less than zero. Furthermore, if a number a is positive or 0, we say it is **nonnegative,** denoted $a \geq 0$. If a number b is negative or 0, we say it is **nonpositive,** denoted $b \leq 0$.

1.7 INTERVALS AND INTERVAL NOTATION

Inequalities can be depicted as portions of the number line called **intervals.** Intervals that include neither endpoint are called **open intervals.** Those that include both endpoints are called **closed intervals.** If only one endpoint is included, the interval is called a **half-open** (or **half-closed**) **interval.** Intervals that can be represented as finite segments on the number line are called **bounded intervals;** while those that extend indefinitely to the right or left on the number line are called **unbounded intervals. Interval notation** can be used to describe the intervals. In interval notation, parentheses or brackets are placed around the endpoints of an interval. If the endpoint is included in the interval, a bracket is used; if the endpoint is not included, a parenthesis is used. If the interval is unbounded, the symbols ∞ and $-\infty$ are used in the notation; however, these symbols do not represent real numbers, but are used to convey the idea of extending indefinitely to the right and indefinitely to the left, respectively. A representation of the interval on the number line is called its **graph.** Shading of the number line is used to represent the numbers included in the interval. An open circle is used to indicate an excluded endpoint and a solid circle is used to indicate an included endpoint. The following table summarizes common inequalities and their corresponding intervals:

TABLE 1.2. INEQUALITIES AND THEIR INTERVALS

Inequality	Interval Notation and Type	Graph
$x < b$	$(-\infty, b)$ unbounded, open	b
$x > a$	(a, ∞) unbounded, open	a
$x \leq b$	$(-\infty, b]$ unbounded, half-open	b
$x \geq a$	$[a, \infty)$ unbounded, half-open	a
$a < x < b$	(a, b) bounded, open	a b
$a \leq x < b$	$[a, b,)$ bounded, half-open	a b
$a < x \leq b$	$(a, b,]$ bounded, half-open	a b
$a \leq x \leq b$	$[a, b]$ bounded, closed	a b

1.8 FIELD PROPERTIES OF THE REAL NUMBER SYSTEM

For the most part, in algebra you will be concerned with the set of real numbers along with two operations, called **addition,** denoted by the plus symbol (+), and **multiplication,** denoted by the raised (·). The parts of an addition problem are

addend + addend = sum

The parts of a multiplication problem are

factor · factor = product

When no confusion will occur, the notation ab, $a(b)$, $(a)b$, or $(a)(b)$ will also be used to denote $a \cdot b$. Throughout this section, we will let a, b, and c represent any numbers that are members of the set of real numbers. The following properties, known as **field properties,** hold:

CLOSURE PROPERTY:

$a + b$ and ab are real numbers.

The closure property means the sum or product of any two real numbers is also a real number.

COMMUTATIVE PROPERTY:

$a + b = b + a$ and $ab = ba$

The commutative property means you can switch the order of the two numbers when you add or multiply without changing the answer. Thus,

same results

$$2 + 5 = 7 \text{ and } 5 + 2 = 7$$

Similarly,

same results

$$2 \cdot 5 = 10 \text{ and } 5 \cdot 2 = 10$$

ASSOCIATIVE PROPERTY:

$(a + b) + c = a + (b + c)$ and

$(ab)c = a(bc)$

The associative property tells you how to work with sums with more than two addends and products with more than two factors. It tells you the way the addends or factors are grouped does not affect the sum or product. Thus, same results

$$(3 + 1) + 5 = 4 + 5 = 9 \text{ and } 3 + (1 + 5) = 3 + 6 = 9$$

Likewise,

same results

$$(3 \cdot 2) \cdot 5 = 6 \cdot 5 = 30 \text{ and } 3 \cdot (2 \cdot 5) = 3 \cdot 10 = 30$$

DISTRIBUTIVE PROPERTY:

$a(b + c) = ab + ac$ and $(b + c)a = ba + ca$

The distributive property means there are two ways of evaluating certain expressions that involve both addition and multiplication. You can either add first and then multiply, or multiply first and then add. Either way the answer works out to be the same. Thus,

same results

$$4(3 + 5) = 4 \cdot 8 = 32 \text{ and } 4(3 + 5) = 4 \cdot 3 + 4 \cdot 5 = 12 + 20 = 32$$

add 1st then multiply multiply 1st then add

This property provides a connection between addition and multiplication; it is used frequently in algebra. For instance, it can be used to change the form of an expression from a product to a sum:

$$5(20 + 8) = 5 \cdot 20 + 5 \cdot 8$$

And it can also be used, in an opposite fashion, to express a sum as a product:

$$3a + 3b = 3(a + b)$$

Because when you add the real number 0 to a number it leaves the number unchanged, 0 is called the **additive identity.**

ADDITIVE IDENTITY PROPERTY:

There exists a real number, denoted 0, such that $a + 0 = a$ and $0 + a = a$.

This property ensures that 0 is a real number and that its sum with any number is the number. Zero is the only number that works this way. Thus,

$$3 + 0 = 3 \text{ and } 0 + 3 = 3$$

Similarly, because when you multiply the number 1 times a number, it leaves the number unchanged, 1 is called the **multiplicative identity.**

MULTIPLICATIVE IDENTITY PROPERTY:

There exists a real number, denoted 1, such that $a \cdot 1 = a$ and $1 \cdot a = a$.

This property ensures that 1 is a real number and that its product with any number is the number. One is the only number for which this is true. Thus,

$$3 \cdot 1 = 3 \text{ and } 1 \cdot 3 = 3$$

There are two additional properties of numbers that you will find useful in your work with algebra. Examine the number line in Figure 1.5. Notice that the points $\frac{1}{2}$ and $-\frac{1}{2}$, the points 2 and -2, the points π and $-\pi$, and the points 5 and -5 are the same distance from zero, but on opposite sides from each other.

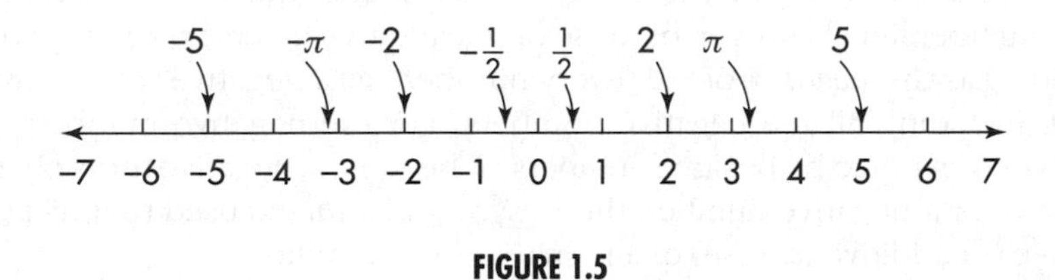

FIGURE 1.5

Such pairs of numbers are called **additive inverses** (or more commonly, **negatives**) of each other. (You may also have learned to call them "opposites" of each other.). Thus, $\frac{1}{2}$ and $-\frac{1}{2}$ are negatives of each other, 2 and -2 are negatives of each other, and so on. In general, if two real numbers are the same distance from zero, but on opposite sides of zero, they are negatives of each other. Moreover, the negative of zero is zero. Because every number has an additive inverse (or negative), we have the following property:

ADDITIVE INVERSE PROPERTY:

For every real number a, there exists a real number, denoted, $-a$, such that $a + (-a) = 0$ and $(-a) + a = 0$.

The additive inverse property means that for every number there is another number, opposite to it in sign, which when added to the number gives 0. For example,

$$2 + -2 = 0 \text{ and } -2 + 2 = 0$$

A helpful message: Our experience tells us that many students make mistakes when the word *negative* is used as a synonym for *additive inverse,* especially when they see the negative symbol, $-$, indicating the additive inverse. In the example, -2 and 2 are *negatives.* In other words, -2 is the "negative" of 2, and 2 is the "negative" of -2. The latter statement in the previous sentence is what may confuse some students: The number 2 is a *positive* number—how can it be a "negative"? The explanation is that in mathematics, the word "negative" and its symbol, $-$, are used in more than one way. You already know one instance: when they are used to indicate that a real number lies to the left of zero on the number line. In that case, the $-$ symbol is placed on the immediate left of the leftmost digit and is considered to be part of the number, called the **sign** of the real number. And we say the number is a negative number. For example, in Figure 1.5, -2 is a *negative* number because it is located to the *left* of zero on the number line. Its negative (or additive inverse), namely 2, is a *positive* number because it is located to the *right* of zero on the number line. Only numbers to the left of zero on the number line are properly called negative numbers. In Figure 1.5, 2 is *not* a negative number, because it does *not* lie to the left of zero on the number line; however, because of the additive inverse property, we can describe 2 as the *negative* of -2. Every number has a negative (or additive inverse), including all the negative numbers. For the negative numbers, their negatives turn out to be positive numbers. Whether the negative turns out to be a positive or a negative number, the $-$ symbol is *always* used to indicate the negative (or additive inverse) of a number. For example,

$$-(2) = \text{"the negative of 2"} = -2, \text{ and}$$

$$-(-2) = \text{"the negative of } -2\text{"} = 2$$

In general, $-x$ represent the number that is the negative (or additive inverse) of x. In this case, think of the $-$ symbol as telling you to change the sign of whatever number x represents. If x represents a *positive* number, its negative will be a *negative* number. On the other hand, if x represents a *negative* number, its negative will be a *positive* number. You don't know which it is unless you find out something about x. For instance, if you replace x with the number 10, then $-x = -(10) =$ "the negative of 10" $= -10$; on the other hand, if you replace x with the number -8, $-x = -(-8) =$ "negative of -8" $= 8$. So, what does this mean to you? In a nutshell, NEVER assume $-x$ is a negative number just because there is a $-$ symbol in front of it!

From your work with fractions in arithmetic, you may recall fractional pairs such as $\frac{1}{2}$ and $\frac{2}{1}$, $\frac{3}{4}$ and $\frac{4}{3}$, and so forth whose product is 1. Such pairs of numbers are called **multiplicative inverses** (or more commonly, **reciprocals**) of each other. In general, if the product of two numbers is 1, they are reciprocals of each other. Moreover, the reciprocal of 1 is 1 and *zero has no reciprocal* (since there is no number that will multiply times zero to give 1). Because every number *except zero* has a multiplicative inverse (or reciprocal) we have the following property:

MULTIPLICATIVE INVERSE PROPERTY:

For every nonzero real number a, there exists a real number, denoted a^{-1} (or $\frac{1}{a}$), such that

$$a \cdot a^{-1} = 1 \text{ and } a^{-1} \cdot a = 1,$$

or, equivalently,

$$a \cdot \frac{1}{a} = 1 \text{ and } \frac{1}{a} \cdot a = 1$$

The multiplicative inverse property means for every real number, *excluding the number zero,* there is another real number which when multiplied times the number gives 1. Thus,

$$2 \cdot \frac{1}{2} = 1 \text{ and } \frac{1}{2} \cdot 2 = 1$$

A helpful message: In algebra, the reciprocal of a number is designated by writing a small raised "-1" immediately to the right of the number. You can read a^{-1} as "a inverse" or as "a to the negative one." Either is correct. Be careful—do not make the mistake of thinking the – symbol used in the $^{-1}$ notation is telling you to change the sign of the reciprocal. The $^{-1}$ is just a shorthand way to tell you to find the reciprocal. It is a convenient notation and, *except when it is attached to zero,* makes sense in other settings. If the number is written in fractional form, then all you need do is switch the numerator and denominator like this:

$$\left(\frac{x}{y}\right)^{-1} = \frac{y}{x}, \text{ since } \frac{x}{y} \cdot \frac{y}{x} = 1$$

After doing that, you can decide whether you should express the resulting fraction in decimal form. For instance,

$$\left(\frac{2}{3}\right)^{-1} = \frac{3}{2} = 1.5$$

If the number is not written in fractional form, then you divide 1 by the number:

$$(-0.25)^{-1} = \frac{1}{-0.25} = 1 \div -0.25 = -4$$

Since the reciprocal of any nonzero number x is $\frac{1}{x}$ (because $x \cdot \frac{1}{x} = 1$), depending on the situation, you may just want to indicate the division as a fraction whose numerator is 1 and whose denominator is the number. For instance,

$$\pi^{-1} = \frac{1}{\pi}$$

You can always check to see if the number you obtain is the reciprocal of the number you started with by multiplying them together. If the product is 1, they are reciprocals. For example,

$$\left(\frac{3}{4}\right)^{-1} = \frac{4}{3} \text{ because } \frac{3}{4} \cdot \frac{4}{3} = 1$$

Finally, if a number is *positive,* its reciprocal is *positive;* and if a number is *negative,* its reciprocal is *negative.* In other words, a number and its reciprocal *always* have the same sign.

Since every number has an additive inverse, **subtraction** of real numbers is defined as follows:

SUBTRACTION:

$$a - b = a + (-b)$$

This definition means that when you want to subtract one number from another number, change the sign of the number being subtracted (the number on the right) and then work the problem as an addition problem. Thus,

$$6 - (-8) = 6 + 8 = 14$$

A helpful message: In the above definition, the − symbol is used to indicate subtraction. Thus far, the − symbol has been used in four different situations:

(1) To indicate *negative numbers:* -8 is read "negative 8" and lies 8 units to the left of zero on the number line.
(2) To indicate the *additive inverse:* $-(-8)$ = "the negative of -8" = 8
(3) As part of the *reciprocal* symbol $^{-1}$: 2^{-1} = "the reciprocal of 2" = $\frac{1}{2}$
(4) To indicate *subtraction:* $6 - (-8) = 6 + 8$

Since every number *except zero* has a multiplicative inverse, **division** of real numbers is defined as follows:

DIVISION:

$$a \div b = \frac{a}{b} = a \cdot b^{-1} = a \cdot \frac{1}{b}$$

This definition means that when you want to divide one number by another number, you replace the divisor with its reciprocal and then work the problem as a multiplication problem. Thus,

$$18 \div 6 = 18 \cdot \frac{1}{6} = 3$$

A helpful message: Have you ever wondered why subtraction and division are given *new* definitions in algebra? Why not do it the old way—like you learned in arithmetic? The simple reason is that subtraction is neither commutative nor associative, and neither is division. (Make up some examples to convince yourself about this.) That's why changing to addition or multiplication, whichever one is applicable, is better—because these two operations have *all* the useful properties we've discussed in this section.

1.9 ABSOLUTE VALUE

The **absolute value** of a real number is its distance from zero on the number line. Distance always has a *nonnegative* (positive or zero) value. For example, as Figure 1.6 shows, both 5 and −5 have an absolute value of 5 because each is 5 units from zero on the number line.

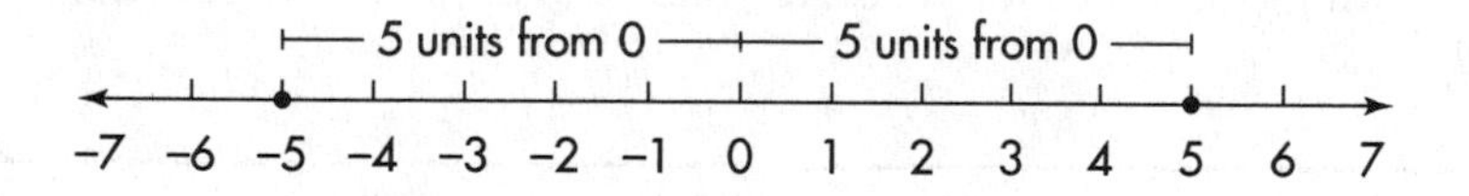

FIGURE 1.6

The absolute value is indicated by two vertical bars, one on either side of the number:

$$|5| = |-5| = 5$$

Formally, absolute value is defined as follows:

> For any real number x, the absolute value of x is
>
> $$|x| = \begin{cases} x, \text{ if } x \geq 0 \\ -x, \text{ if } x < 0 \end{cases}$$

Thus, for every real number x, its absolute value, denoted $|x|$, is either x or $-x$, whichever is a *nonnegative* number (that is, whichever one is further to the right on the number line).

A helpful message: Remember, $-x$ can be a positive number. Don't be confused by the $-$ symbol to the left of x. The $-$ symbol to the immediate left of x tells you to change the sign of x. When x itself is *nonnegative* (positive or zero), $|x|$ is the nonnegative number x; but when x is *negative,* $|x|$ is the *positive* number $-x$.

Figure 1.7 illustrates the absolute value definition graphically.

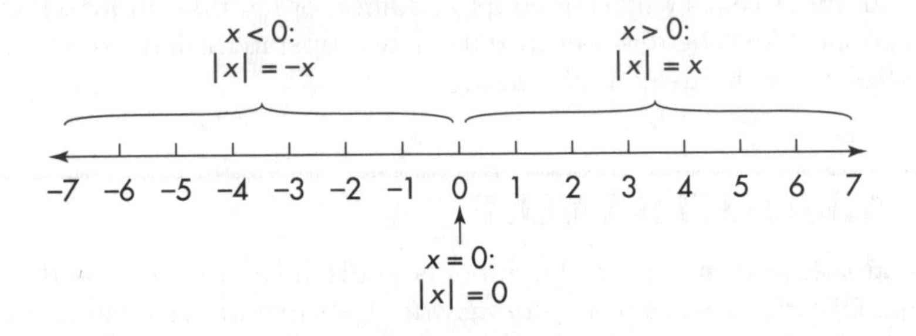

FIGURE 1.7

The following properties for absolute value will apply for all real numbers x, y, and c.

$$|x| = |-x|$$

This property means a number and its negative have the same absolute value. Thus,

$$|5| = |-5| = 5$$

$$|xy| = |x||y|$$

This property means the absolute value of a product is the same as the product of the absolute values of the factors. Therefore, you can multiply first, then find the absolute value of the product; or, if you prefer, you can find the absolute values first, then multiply. Thus, same results

$$|7 \cdot 2| = |14| = 14 \text{ and } |7||2| = 7 \cdot 2 = 14$$

$$\left|\frac{x}{y}\right| = \frac{|x|}{|y|}, \text{ provided } y \neq 0$$

This property means the absolute value of a quotient is the same as the quotient of the absolute values. Therefore, you can divide first, then find the absolute value of the quotient; or, if you prefer, you can find the absolute values first, then divide. Thus,

same results

$$\left|\frac{12}{4}\right| = |3| = 3 \text{ and } \frac{|12|}{|4|} = \frac{12}{4} = 3$$

$$|x + y| \leq |x| + |y|$$

This property means the absolute value of a sum is always less than or equal to the sum of the absolute values. For example,

$$|10 + 2| \leq |10| + |2| \text{ because}$$
$$|10 + 2| = |12| = 12 \text{ and } |10| + |2| = 10 + 2 = 12$$

In this example, the two sides are equal. But, remember, x and y could be replaced with *any* two real numbers, not necessarily two positive numbers as shown here. In the next section, you will see that when x and y are replaced with numbers that have opposite signs, their sum will be a number whose absolute value is *less than* the sum of the absolute values of the two numbers. For example,

$$|10 + (-2)| \leq |10| + |-2| \text{ because}$$
$$|10 + (-2)| = |8| = 8 \text{ and } |10| + |-2| = 10 + 2 = 12$$

In addition, you do not have a choice whether you add first or take the absolute value first: you should add first before you find the absolute value of the sum.

If c is any positive number,

$$|x| = c \Leftrightarrow \text{either } x = c \text{ or } x = -c$$

A helpful message: The symbol $\Leftrightarrow$ is read "if and only if."

Thus, the equation $|x| = 5$ means x is 5 units from 0 on the number line; that is,

$$|x| = 5 \Leftrightarrow x = -5 \text{ or } x = 5$$

This is illustrated in Figure 1.8.

|x| = 5 if and only if
x is 5 units from 0:

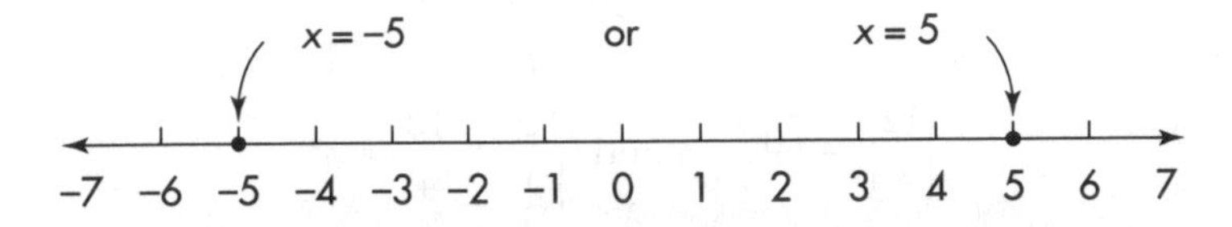

FIGURE 1.8

If c is any positive number,

$$|x| < c \Leftrightarrow -c < x < c$$

Because $-c < x < c$ is a connected segment of the number line, we say $|x| < c$ results in a **conjunction.** The equation $|x| < 5$ means x is less than 5 units from 0 on the number line; that is,

$$|x| = 5 \Leftrightarrow -5 < x < 5$$

This conjunction is illustrated in Figure 1.9.

|x| < 5 if and only if
x is less than 5 units
from 0:

−5 < x < 5

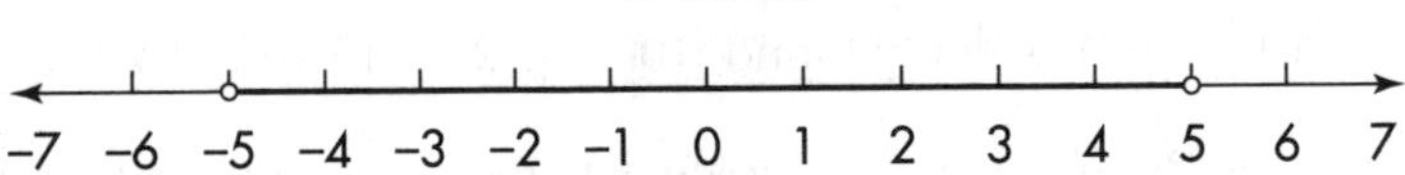

FIGURE 1.9

If c is any positive number,

$$|x| > c \Leftrightarrow x < -c \text{ or } x > c$$

Because $-c < x$ and $x < c$ are two unconnected segments of the number line, we say $|x| > c$ results in a **disjunction.** The equation $|x| > 5$ means x is more than 5 units from 0 on the number line; that is,

$$|x| > 5 \text{ if and only if } x < -5 \text{ or } x > 5$$

This disjunction is illustrated in Figure 1.10.

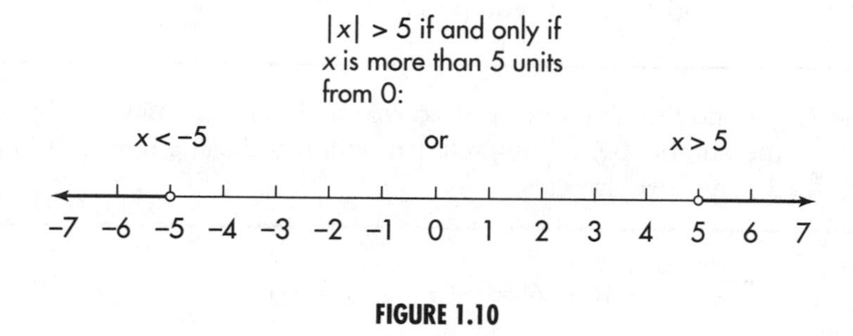

FIGURE 1.10

Furthermore, these properties for absolute value hold if $<$ is replaced everywhere with $\leq$ and/or $>$ is replaced with $\geq$.

1.10 SUMS AND DIFFERENCES

Computations using real numbers are performed by using the absolute values (which are always positive or zero) of the numbers.

Rule 1. To find the sum of two real numbers that have the same sign, add their absolute values and attach the common sign to the sum.

- $5 + 9 = 14$
- $-5 + -9 = -14$

Rule 2. To find the sum of two real numbers that have opposite signs, subtract the smaller absolute value from the larger absolute value and indicate the result has the same sign as the number with the larger absolute value; if both numbers have the same absolute value, the sum is 0.

- $5 + -9 = -4$
- $-5 + 9 = 4$
- $5 + -5 = 0$

The two rules for addition tell you how to add two real numbers, but often you will want to add more than two numbers. To do this, add in pairs, keeping track of the appropriate sign as you proceed.

Rule 3. To find the difference of two real numbers, change the sign of the number being subtracted and then add using Rule 1 or Rule 2, whichever applies.

- $-10 - 6 = -10 + -6 = -16$
- $-10 - (-6) = -10 + 6 = -4$

1.11 PRODUCTS AND QUOTIENTS

Rule 1. To find the product of two numbers that have the same sign, multiply their absolute values and indicate the product is positive (no sign is necessary).

- $5 \cdot 9 = 45$
- $(-5)(-9) = 45$

Rule 2. To find the product of two numbers that have opposite signs multiply their absolute values and use a $-$ sign with the result to indicate the product is negative.

- $(-5)(9) = -45$
- $(5)(-9) = -45$

A helpful message: Be careful not to confuse this rule with the similar rule for addition. In addition, the sum has the same sign as the number with the *larger* absolute value. For example,

$$-5 + 9 = 4, \text{ whereas } 5 + -9 = -4$$

But when multiplying, the product is negative, *regardless* which of the numbers has the larger absolute value.

The two rules for multiplication tell you how to multiply two numbers, but often you will want to find the product of more than two numbers. To do this, multiply in pairs. You can keep track of the appropriate sign as you proceed, or you can use the following:

> **Rule 3.** When *zero* is one of the factors, the product is always zero; otherwise, products involving an *even* number of negative factors are positive, whereas those involving an *odd* number of negative factors are negative.

- $(-10)(2)(0)(-5) = 0$
- $(-10)(2)(3)(-5) = 300$
- $(-10)(2)(3)(-5)(-1) = -300$

> **Rule 4.** To find the quotient of two numbers that have the same sign, divide their absolute values and indicate the quotient is positive (no sign is necessary).

- $\dfrac{45}{9} = 5$
- $\dfrac{-45}{-9} = 5$

> **Rule 5.** To find the quotient of two numbers that have opposite signs, divide their absolute values and use a $-$ sign with the result to indicate the product is negative.

- $\frac{-45}{9} = -5$
- $\frac{45}{-9} = -5$

1.12 INTEGER EXPONENTS

In mathematical expressions, **exponentiation** of a quantity is indicated by a small raised number, called the **exponent,** written to the upper right of the quantity.

Positive integers are used as exponents to indicate repeated multiplication. For example, to indicate 2 is to be used as a factor 5 times, a small raised 5 is attached to the upper right of 2. Thus,

$$2^5 = \underbrace{2 \cdot 2 \cdot 2 \cdot 2 \cdot 2}_{\text{5 factors of 2}} = 32$$

Here 2^5 (read "2 to the fifth power" or simply "2 to the fifth") is called the **exponential expression;** 2 is the **base;** the small raised 5 is the **exponent;** and 32 is the fifth **power** of 2. That is,

$$\underset{\text{Base}}{2}^{\overset{\text{Exponent}}{5}} = \underset{\text{Fifth power of 2}}{32}$$

This leads to the following definition:

POSITIVE INTEGER EXPONENTS

If x is any real number and n is a positive integer,

$$x^n = \underbrace{x \cdot x \cdot x \cdot \ldots \cdot x}_{n \text{ factors of } x}$$

where x^n is read as "x to the nth power" or as "x to the n."

For example,

$$x^1 = x$$
$$x^2 = x \cdot x$$
$$x^3 = x \cdot x \cdot x$$

$$\vdots$$

$$x^8 = x \cdot x \cdot x \cdot x \cdot x \cdot x \cdot x \cdot x$$

$$\vdots$$

The following is a list of powers that are encountered so frequently in algebra and higher level math courses that it will be to your advantage to memorize them:

$2^2 = 4$	$8^2 = 64$	$14^2 = 196$	$2^3 = 8$	$2^4 = 16$
$3^2 = 9$	$9^2 = 81$	$15^2 = 225$	$3^3 = 27$	$3^4 = 81$
$4^2 = 16$	$10^2 = 100$	$16^2 = 256$	$4^3 = 64$	$4^4 = 256$
$5^2 = 25$	$11^2 = 121$	$17^2 = 289$	$5^3 = 125$	$5^4 = 625$
$6^2 = 36$	$12^2 = 144$	$20^2 = 400$		
$7^2 = 49$	$13^2 = 169$	$25^2 = 625$		

A helpful message: The exponent 2 on a number is usually read "squared" rather than as "to the second power." Likewise, the exponent 3 is usually read "cubed" rather than as "to the third power."

Because, for instance, $x^2 = x \cdot x$ and $x^3 = x \cdot x \cdot x$, the product can be written as follows:

$$x^2 \cdot x^3 = x \cdot x \cdot x \cdot x \cdot x = x^5$$

In general,

$$x^m \cdot x^n = \underbrace{x \cdot x \cdot x \cdot \ldots \cdot x}_{m \text{ factors of } x} \cdot \underbrace{x \cdot x \cdot x \cdot \ldots \cdot x}_{n \text{ factors of } x} = \underbrace{x \cdot x \cdot x \cdot x \cdot \ldots \cdot x}_{m + n \text{ factors of } x} = x^{m+n}$$

This suggests the following:

PRODUCT RULE FOR EXPONENTIAL EXPRESSIONS

For any real number x, and positive integers m and n,

$$x^m \cdot x^n = x^{m+n}$$

This rule means that to multiply exponential expressions that have the *same base,* add the exponents and keep the same base. Thus,

$$2^2 \cdot 2^3 = 2^{2+3} = 2^5$$

Consider $(xy)^3 = xy \cdot xy \cdot xy$. Using the commutative and associative properties, the factors can be rearranged as follows:

$$(xy)^3 = \underbrace{xy \cdot xy \cdot xy}_{\text{3 factors of } xy} = \underbrace{x \cdot x \cdot x}_{\text{3 factors of } x} \cdot \underbrace{y \cdot y \cdot y}_{\text{3 factors of } y} = x^3 \cdot y^3$$

In general,

$$(xy)^p = \underbrace{xy \cdot xy \cdot xy \cdot \ldots \cdot xy}_{p \text{ factors of } xy} = \underbrace{x \cdot x \cdot x \cdot \ldots \cdot x}_{p \text{ factors of } x} \cdot \underbrace{y \cdot y \cdot y \cdot \ldots \cdot y}_{p \text{ factors of } y} = x^p \cdot y^p$$

This suggests the following:

POWER OF A PRODUCT

For any real numbers x and y, and positive integer p,

$$(xy)^p = x^p y^p$$

This rule means a product raised to a power is the product of the factors each raised to that power. Thus,

$$(ab)^7 = a^7 b^7$$

Similarly,

POWER OF A QUOTIENT

For any real numbers x and y, $y \neq 0$, and positive integer p,

$$\left(\frac{x}{y}\right)^p = \frac{x^p}{y^p}$$

This rule means a quotient raised to a power is the quotient of the dividend and divisor each raised to that power. For example,

$$\left(\frac{x}{y}\right)^7 = \frac{x^7}{y^7}$$

Next, observe what happens when you find a power of an expression that is itself a power:

$$(x^2)^3 = \underbrace{x^2 \cdot x^2 \cdot x^2}_{\text{3 factors of } x^2} = x^{2+2+2} = x^{2 \cdot 3} = x^6$$

In general,

$$(x^n)^p = \underbrace{x^n \cdot x^n \cdot x^n \cdot \ldots \cdot x^n}_{p \text{ factors of } x^n} = x^{\overbrace{}^{}\underbrace{n+n+n+\ldots+n}_{p \text{ times}}} = x^{np}$$

This suggests the following rule:

POWER OF A POWER

For any real number x, and positive integers n and p,

$$(x^n)^p = x^n \cdot x^n \cdot x^n \cdot \ldots \cdot x^n = x^{np}$$

This rules means that to raise to a power a number that is itself raised to a power, you multiply the powers and keep the same base. For example,

$$(a^5)^4 = a^{20}$$

This rule can be combined with the power of a product and the power of a quotient rules, yielding, for instance,

$$(a^5b)^4 = a^{20}b^4, \text{ and}$$

$$\left(\frac{x^3}{y^5}\right)^7 = \frac{x^{21}}{y^{35}}$$

We can extend the definition of exponents to include exponents other than the positive integers. First, you know that according to the product rule for exponents,

$$x^0 \cdot x^5 = x^{0+5} = x^5$$

This means that x^0 is a number that multiplies times x^5 and gives x^5 back. According to the multiplicative identity property for real numbers, there is only one number for which this is true: the number 1. Thus, if you want things to work the way they are supposed to, the following must be true:

$$x^0 = 1$$

Therefore, we have the following definition:

ZERO EXPONENT

For any real number x (except 0),

$$x^0 = 1$$

This definition means that any number (except 0) to the zero power will be 1. For example,

- $6^0 = 1$
- $(a^5b)^0 = 1$
- $(-7)^0 = 1$
- $\left(\frac{x^3}{y^5}\right)^0 = 1$
- $\pi^0 = 1$
- 0^0 is undefined.

Now consider that, according to the product rule,

$$x^5 \cdot x^{-5} = x^{5+-5} = x^0 = 1 \text{ (by definition)}$$

This means x^{-5} is a number that multiplies times x^5 and gives 1 as the product. According to the multiplicative inverse property for real numbers, this means x^{-5} is the reciprocal of x^5; that is,

$$x^{-5} = \frac{1}{x^5}, \text{ provided } x \neq 0$$

In general,

$$x^n \cdot x^{-n} = x^{n+-n} = x^0 = 1 \text{ and } x^{-n} \cdot x^n = x^{-n+n} = x^0 = 1$$

This means that x^{-n} and x^n are reciprocals of each other; that is,

$$x^{-n} = \frac{1}{x^n} \text{ and } x^n = \frac{1}{x^{-n}} \text{ provided } x \neq 0$$

Therefore, we have the following definition:

NEGATIVE INTEGER EXPONENTS

If x is any real number (except zero) and $-n$ is a negative integer,

$$x^{-n} = \frac{1}{x^n} \text{ and } \frac{1}{x^{-n}} = x^n$$

This definition means an exponential expression whose exponent is negative equals the reciprocal of the corresponding positive power. For example,

- $x^{-7} = \frac{1}{x^7}$
- $2^{-3} = \frac{1}{2^3} = \frac{1}{8}$
- $\frac{1}{3^{-4}} = \frac{3^4}{1} = 81$

Also, it follows that an exponential expression whose exponent is positive equals the reciprocal of the corresponding negative power:

- $x^7 = \dfrac{1}{x^{-7}}$

- $2^3 = \dfrac{1}{2^{-3}}$

- $\dfrac{1}{3^4} = \dfrac{3^{-4}}{1} = 3^{-4}$

Furthermore, factors whose exponents are negative in the numerator of a fraction assume the corresponding positive exponents when moved down to the denominator; and factors whose exponents are negative in the denominator assume the corresponding positive exponents when moved up to the numerator. For example,

- $\dfrac{1}{x^{-7}} = x^7$

- $\dfrac{ab^{-1}c^{-3}}{x^{-7}y^4} = \dfrac{ax^7}{bc^3y^4}$

- $\dfrac{1 + a^{-1}}{b^{-1}} = b\left(1 + \dfrac{1}{a}\right)$

As a result of the two previous definitions, all of the rules for positive integer exponents hold for all integer exponents—positive, negative, or zero. When you have zero or negative exponents in an expression, as a general rule, you should rewrite it so that it no longer has zero or negative exponents. As you examine the following examples, keep in mind that division by zero must *never* occur:

- $2^0x^{-3} = 1 \cdot \dfrac{1}{x^3} = \dfrac{1}{x^3}$

- $3a^{-4}b^2c^{-3} = \dfrac{3b^2}{a^4c^3}$

- $\dfrac{x^{-1}}{y^{-1}} = \dfrac{y}{x}$

Using the power of a quotient rule and the definition for negative exponents, we have these useful results

- $\left(\dfrac{x}{y}\right)^{-1} = \dfrac{x^{-1}}{y^{-1}} = \dfrac{y}{x}$

- $\left(\dfrac{x}{y}\right)^{-n} = \dfrac{x^{-n}}{y^{-n}} = \dfrac{y^n}{x^n} = \left(\dfrac{y}{x}\right)^n$

If we have the quotient of two exponential expressions, we can use the definition of division of real numbers, the definition for negative exponents, and the product rule for exponents to proceed as follows:

$$\frac{x^m}{x^n} = x^m\left(\frac{1}{x^n}\right) = x^m x^{-n} = x^{m+(-n)} = x^{m-n}$$

This leads to the following rule:

QUOTIENT RULE FOR EXPONENTIAL EXPRESSIONS

For any real number x (except zero), and positive integers m and n,

$$\frac{x^m}{x^n} = x^{m-n}$$

This rule means that to divide exponential expressions that have the *same base,* subtract the exponent of the divisor from the exponent of the dividend and keep the same base. Thus,

- $\frac{x^5}{x^2} = x^{5-2} = x^3$
- $\frac{x^2}{x^5} = x^{2-5} = x^{-3} = \frac{1}{x^3}$

Problems involving exponents may require you to use an extension or combination of the rules for exponents. For instance,

$$\frac{x^7x^3y^{-3}z^5}{x^4y^{-2}z^{-6}z^4} = x^{7+3-4}y^{-3-(-2)}z^{5-(-6)-4} = x^6y^{-1}z^7 = \frac{x^6z^7}{y}$$

A helpful message: Many students have difficulty working with exponents. Here are some things to remember:

The product and quotient rules for exponential expressions can be used only when the exponential expressions have exactly the same base:

$a^2 \cdot b^3$ and $\frac{x^4}{y^3}$ cannot be simplified further.

Exponentiation is not commutative:

$$2^5 \neq 5^2$$

Exponentiation does not distribute over addition (or subtraction):

$$(a + b)^5 \neq a^5 + b^5$$

An exponent applies only to the factor it is attached to:

$$ab^5 \neq a^5b^5$$
$$-2^2 = -(2^2) = -4$$

Use parentheses around the factors for which the exponent applies:

$$(ab)^5 = a^5b^5$$

A negative number raised to an even power yields a positive product:

$$(-2)^4 = (-2)(-2)(-2)(-2) = 16$$

A negative number raised to an odd power yields a negative product:

$$(-2)^5 = (-2)(-2)(-2)(-2)(-2) = -32$$

A nonzero number or mathematical expression raised to the zero power is 1:

$$\begin{pmatrix}\text{nonzero number}\\ \text{or mathematical}\\ \text{expression}\end{pmatrix}^0 = 1 \quad \text{ALWAYS!}$$

Only exponential expressions that are <u>*factors*</u> *in the numerator or denominator of a fraction can be moved simply by changing the sign of the exponent*:

$$\frac{1}{x^{-1}y^{-1}} = \frac{xy}{1} = xy, \text{ but } \frac{1}{x^{-1} + y^{-1}} = \frac{1}{\frac{1}{x} + \frac{1}{y}} = \frac{1}{\frac{x + y}{xy}} = \frac{xy}{x + y} \neq x + y$$

1.13 SCIENTIFIC NOTATION

A number expressed in scientific notation is written as a number greater than or equal to 1 but less than 10 multiplied times an integral power of 10. For instance,

- $2.73 \cdot 10^8$
- $6.5 \cdot 10^{-6}$

To change the number to ordinary notation, perform the indicated multiplication. Thus,

- $2.73 \cdot 10^8 = 2.73 \cdot 100{,}000{,}000 = 273{,}000{,}000$
- $6.5 \cdot 10^{-6} = 6.5 \cdot \frac{1}{1{,}000{,}000} = 6.5 \cdot 0.000001 = 0.0000065$

To write a number in scientific notation, move the decimal point to the immediate right of the first nonzero digit. Then indicate multiplication by the proper power of 10. If you moved the decimal point to the *left,* the number of places you moved it is the value of the exponent for 10. Thus,

$$273{,}000{,}000 = 2.73000000. \cdot 10^8 = 2.73 \cdot 10^8$$

Moved decimal point 8 places to *left.*

If you moved the exponent to the *right,* the *negative* of the number of places you moved it is the value of exponent for 10. Thus,

$$0.0000065 = 0.000006.5 \cdot 10^{-6} = 6.5 \cdot 10^{-6}$$

Moved decimal point 6 places to *right.*

The powers of 10 in products and quotients of numbers expressed in scientific notation follow the rules for exponents. For example,

- $(2.73 \cdot 10^8)(1.4 \cdot 10^5) = (2.73)(1.4)(10^{8+5}) = 3.822 \cdot 10^{13}$
- $(6.5 \cdot 10^{-6})(5.0 \cdot 10^6) = (6.5)(5.0)(10^{-6+6}) = 32.5 \cdot 10^0 = 32.5$
- $\dfrac{2.73 \cdot 10^8}{1.3 \cdot 10^6} = \left(\dfrac{2.73}{1.3}\right)(10^{8-6}) = 2.1 \cdot 10^2$
- $\dfrac{(5.5 \cdot 10^{23})(4.3 \cdot 10^{20})(1.2 \cdot 10^{-19})}{(3.2 \cdot 10^{15})(4.5 \cdot 10^{35})} = \dfrac{(5.5)(4.3)(1.2)(10^{23+20-19})}{(3.2)(4.5)(10^{15+35})} =$
  $\dfrac{(28.38)(10^{24})}{(14.4)(10^{50})} \cong 1.97 \cdot 10^{24-50} = 1.97 \cdot 10^{-26}$

1.14 RADICALS AND FRACTIONAL EXPONENTS

A number a such that $a^2 = x$ is called a **square root** of x. Finding the square root of a number is the reverse of squaring a number. The square root of zero is zero. Every positive number has two square roots that are equal in absolute value and opposite in sign. For example,

Since $(5)^2 = 25$ and $(-5)^2 = 25$, 5 and -5 are square roots of 25.

Negative numbers do not have real number square roots.

The **principal square root** of x, when $x \geq 0$, is its nonnegative square root, denoted $\sqrt{x}$. Thus, $\sqrt{0} = 0$, $\sqrt{25} = 5$, $\sqrt{144} = 12$, and so forth. The negative root of a positive number x is indicated by $-\sqrt{x}$. Thus, $-\sqrt{25} = -5$, $-\sqrt{144} = -12$, and so forth. The two square roots of a positive number are indicated by $\pm\sqrt{x}$. Thus, $\pm\sqrt{25} = \pm 5$, $\pm\sqrt{144} = \pm 12$, and so forth. Finally, the $\sqrt{x}$ *always* denotes the nonnegative root so that $\sqrt{(-5)^2} = \sqrt{25} = 5$, not -5.

A number a such that $a^3 = x$ is called a **cube root** of x. Finding the cube root of a number is the reverse of cubing a number. Every real number has exactly one real cube root, which is called its **principal cube root.** For example,

Since $(-5)^3 = -125$, -5 is the principal cube root of -125.

In general, if $a^n = x$, a is called an ***n*th root** of x, where n is a natural number. We denote the principal nth root of x as

$$\sqrt[n]{x}$$

provided $x = 0$ when n is even.

The $\sqrt{}$ is called a **radical,** x is called the **radicand,** and n is called the **index** and indicates which root is desired. If no index is written, it is understood to be 2:

$$\sqrt[2]{25} = \sqrt{25} = 5$$

As a rule, a *positive* real number x has exactly one real positive nth root whether n is even or odd; and *every* real number x, whether positive or negative, has exactly one real nth root, which has the same sign as x, when n is odd. Negative numbers do not have real nth roots when n is even. Thus,

- $\sqrt{36} = 6$
- $\sqrt[3]{125} = 5$
- $\sqrt[4]{16} = 2$

Positive numbers have positive real roots.

- $\sqrt[5]{32} = 2$
- $\sqrt[3]{-64} = -4$

Odd roots of positive numbers are positive.
Odd roots of negative numbers are negative.

- $\sqrt{-64}$ is not a real number.
- $\sqrt[4]{-16}$ is not a real number.

Even roots of negative numbers are not real numbers.

Finally, if n is a natural number, the nth root of zero is zero, whether n is even or odd.

$$\sqrt[n]{0} = 0 \text{ (always)}$$

Now, we can define rational exponents:

RATIONAL EXPONENTS

If x is a real number and m and n are natural numbers,

$$x^{\frac{1}{n}} = \sqrt[n]{x};$$

$$x^{\frac{m}{n}} = (\sqrt[n]{x})^m \text{ or } x^{\frac{m}{n}} = \sqrt[n]{x^m}; \text{ and}$$

$$x^{-\frac{m}{n}} = \frac{1}{x^{\frac{m}{n}}} \ (x \neq 0),$$

provided, in all cases, that $x \geq 0$ when n is even.

This definition extends the idea of exponents to include all rational numbers. For example,

- $25^{\frac{1}{2}} = \sqrt{25} = 5$
- $-25^{\frac{1}{2}} = -\sqrt{25} = -5$
- $25^{\frac{3}{2}} = (\sqrt{25})^3 = 5^3 = 125$
- $(-8)^{\frac{2}{3}} = (\sqrt[3]{-8})^2 = (-2)^2 = 4$

} Usually, it is better to find the root before raising to the power.

- $25^{-\frac{1}{2}} = \dfrac{1}{\sqrt{25}} = \dfrac{1}{5}$
- $25^{-\frac{3}{2}} = \dfrac{1}{25^{\frac{3}{2}}} = \dfrac{1}{(\sqrt{25})^3} = \dfrac{1}{5^3} = \dfrac{1}{125}$
- $0^n = 0$, provided $n > 0$; 0^0 is not defined.

All the rules for exponents hold for rational exponents. For instance,

- $x^{\frac{1}{2}} x^{\frac{1}{2}} = x^{\frac{1}{2}+\frac{1}{2}} = x^1 = x$
- $(a^{-3}b^6)^{\frac{1}{3}} = a^{-\frac{3}{3}} b^{\frac{6}{3}} = a^{-1}b^2 = \dfrac{b^2}{a}$

1.15 ORDER OF OPERATIONS

Mathematicians have agreed upon the following order of operations:

1. When grouping symbols are present, perform all computations within grouping symbols—parentheses ( ), brackets [ ], or braces { }—inside absolute value bars | |, under $\sqrt{}$ symbols, and above and below all fraction bars, starting with the innermost grouping symbol and working out.
2. In an evaluation, always proceed as follows:

   First: Perform any exponentiation—powers and roots—as they occur from left to right.
   Second: Perform all multiplications and divisions in the order in which they occur from left to right.
   Third: Perform all additions and subtractions in the order in which they occur from left to right.

A commonly used mnemonic is the sentence, "Please excuse my dear Aunt Sally." The first letters of the words remind you of the following:

ORDER OF OPERATIONS

Operations enclosed in Parentheses
(or other grouping symbol, if present)
Exponentiation
Multiplication and Division from left to right, *whichever comes first.*
Addition and Subtraction from left to right, *whichever comes first.*

For example,

- $-15[-12 - (-7 + 4)] = -15[-12 - (-3)] = -15[-12 + 3]$
  $= -15[-9] = 135$
- $1 + 72 \div 9 - 2 = 1 + 8 - 2 = 9 - 2 = 7$
- $6 \cdot 3 - 4(5 + 2) = 6 \cdot 3 - 4(7) = 18 - 28 = -10$
- $20 + 10(3 + 2^3) = 20 + 10(3 + 8) = 20 + 10(11) = 20 + 110 = 130$
- $-4\{[-6 - (9 + 12)] + 17\} = -4\{[-6 - (21)] + 17\} = -4\{[-6 - 21] + 17\}$
  $= -4\{-27 + 17\} = -4\{-10\} = 40$
- $25 - 36^{\frac{1}{2}} + 7^0 = 25 - 6 + 1 = 20$
- $\dfrac{4^2 - 20}{3 \cdot 2 - 8} = \dfrac{16 - 20}{6 - 8} = \dfrac{-4}{-2} = 2$
- $2|-10 + 3| = 2|-7| = 2(7) = 14$
- $-|-5| - 4^2 + 5 = -5 - 16 + 5 = -16$
- $\sqrt{4^2 + 3^2} = \sqrt{16 + 9} = \sqrt{25} = 5$

A helpful message: $-\sqrt{-25} \neq \sqrt{25} = 5$ because $\sqrt{-25}$ is not a real number.

CHAPTER 1 SUMMARY

Chapter 1 covers the fundamental properties of and operations on numbers that should be mastered for an understanding of algebra. Included are the following:

1.1 A well-defined **set** is one that is described so that you can tell whether an **element** is or is not in the set. The **universal set** is the set of all objects under consideration. The **empty set** has no elements. **Equal** sets have exactly the same elements. If every element of a set A is also an element of set B, set A is a **subset** of set B. The **union** of two sets is a set containing all the elements in either or both sets. The **intersection** of two sets is a set containing the elements common to both. The **complement** of a set is the set containing all the elements in the universal set that are not in the set.

1.2 Important sets of numbers in algebra follow:

natural numbers = $\{1, 2, 3, \ldots\}$
whole numbers = $\{0, 1, 2, 3, \ldots\}$
integers = $\{\ldots -3, -2, -1, 0, 1, 2, 3, \ldots\}$
rationals = $\{\frac{p}{q}$, where p and q are integers and $q \neq 0\}$
irrationals = {nonterminating, nonrepeating decimals}
reals = rationals $\cup$ irrationals
complex numbers = $\{a + bi$, where a and b are real numbers and $i^2 = -1\}$

The natural numbers, whole numbers, integers, rationals, and irrationals are subsets of the reals. The reals are a subset of the complex numbers.

1.3 A **constant** is a quantity whose value does not change. A **variable** represents an unspecified element from a set and may assume many different values.

1.4 Equality has the following properties:

Reflexive: $a = a$

Symmetric: If $a = b$, then $b = a$.

Transitive: If $a = b$ and $b = c$, then $a = c$.

Substitution: If $a = b$, then a can be replaced by b or b replaced by a in any statement without changing the meaning of the statement.

1.5 The real numbers can be represented graphically on a **number line.** Zero is called the **origin.** The **positive** numbers are located to the right of zero; the **negative** numbers, to the left of zero.

1.6 **Less than** ($<$) and **greater than** ($>$) determine an **order relationship** for the real numbers. The **trichotomy** property states that for any two real numbers a and b, exactly one of the following is true:

$$a < b,\ a = b, \text{ or } a > b$$

1.7 **Intervals** can be depicted on the number line. Intervals are **open** (no endpoints included), **closed** (both endpoints included), or **half-open** (only one endpoint included), and are either **bounded** or **unbounded.** Interval notation can be used to describe the intervals.

1.8 The set of real numbers has the following **field properties** under addition and multiplication:

Closure Property: $a + b$ and ab are real numbers.

Commutative Property: $a + b = b + a$ and $ab = ba$

Associative Property: $(a + b) + c = a + (b + c)$ and $(ab)c = a(bc)$

Distributive Property: $a(b + c) = ab + ac$ and $(b + c)a = ba + ca$

Additive Identity Property: There exists a real number, denoted 0, such that $a + 0 = a$ and $0 + a = a$.

Multiplicative Identity Property: There exists a real number, denoted 1, such that $a \cdot 1 = a$ and $1 \cdot a = a$.

Additive Inverse Property: For every real number a, there exists a real number, denoted $-a$, such that $a + (-a) = 0$ and $(-a) + a = 0$.

Multiplicative Inverse Property: For every nonzero real number a, there exists a real number, denoted a^{-1}, such that $a \cdot a^{-1} = 1$ and $a^{-1} \cdot a = 1$.

Subtraction is defined as follows: $a - b = a + (-b)$

Division is defined as follows: $a \div b = \frac{a}{b} = a \cdot b^{-1} = a \cdot \frac{1}{b}$

1.9 The **absolute value** of a number is its distance from zero on the number line. The absolute value of x is either x or $-x$, whichever of these is **nonnegative.** A number and its additive inverse have equal absolute values. Absolute value has the following properties for all real numbers x, y, and c.

1. $|x| = |-x|$
2. $|xy| = |x||y|$
3. $\frac{|x|}{|y|} = \frac{|x|}{|y|}$, provided $y \neq 0$.
4. $|x + y| \leq |x| + |y|$
5. If c is any positive number: $|x| = c \Leftrightarrow$ either $x = c$ or $x = -c$.
6. If c is any positive number: $|x| < c \Leftrightarrow -c < x < c$. **(conjunction)**
7. If c is any positive number: $|x| > c \Leftrightarrow x < -c$ or $x > c$. **(disjunction)**

1.10 The **sum of two positive numbers** is the positive sum of their absolute values; the **sum of two negative numbers** is the negative of the sum of their absolute values; and the **sum of a positive and negative number** is the difference between their absolute values and has the same sign as the addend with the greater absolute value.

1.11 The **product** or **quotient** of two numbers that have the **same sign** is positive; and the **product** or **quotient** of two numbers that have **different signs** is negative. A quotient is **undefined** when the denominator is zero.

1.12 **Exponentiation** is indicated by a small raised number, called the **exponent,** written to the upper right of a quantity. A **positive integer exponent** indicates repeated multiplication. Any number raised to the zero power is 1. A **negative integer exponent** indicates the reciprocal of the corresponding positive power. The following **rules for exponents** hold: For real numbers x and y and integers m, n, and p:

$x^1 = x$ $\qquad x^0 = 1$ $\qquad 0^0$ is undefined.

$$x^{-n} = \frac{1}{x^n} \qquad \left(\frac{x}{y}\right)^{-1} = \frac{y}{x} \qquad \left(\frac{x}{y}\right)^{-n} = \left(\frac{y}{x}\right)^n$$

$$(x^n)^p = x^{np} \qquad \left(\frac{x}{y}\right)^p = \frac{x^p}{y^p} \qquad (xy)^p = x^p y^p$$

$$x^m x^n = x^{m+n} \qquad \frac{x^m}{x^n} = x^{m-n}$$

provided that division by zero does not occur, and that $x > 0$ when n is even.

1.13 A number expressed in **scientific notation** is written as a number greater than or equal to 1, but less than 10, multiplied times a power of 10. The powers of 10 in products and quotients of numbers expressed in scientific notation follow the rules for exponents.

1.14 Every positive number has two square roots that are equal in absolute value and opposite in sign. The *positive* square root is called the **principal square root** of the number. Every real number has exactly one real cube root. If $a^n = x$, a is called an ***n*th root** of x, where n is a natural number. A *positive* real number x has exactly one real positive nth root whether n is even or odd; and *every* real number x has exactly one real nth root that has the same sign as x when n is odd. Negative numbers do not have real nth roots when n is even. If n is a natural number, the nth root of zero is zero. A **fractional** or **rational exponent** indicates a root. We write $x^{\frac{1}{n}} = \sqrt[n]{x}$. The $\sqrt{}$ is called a **radical,** x is called the **radicand,** and n is called the **index** and indicates which root is desired. If no index is written, the radical expression indicates the *principal* square root of the radicand. Rational exponents follow the rules for exponents given in Section 1.11.

1.15 Mathematicians have agreed upon the following order of operations:

First, operations enclosed within grouping symbols (parentheses, brackets, braces, etc.), starting within the innermost grouping symbol.

Next, exponentiation—powers and roots—from left to right.

Next, multiplication and division from left to right, *whichever comes first.*

Last, addition and subtraction from left to right, *whichever comes first.*

PRACTICE PROBLEMS FOR CHAPTER 1

Mark each statement as True or False. Explain the error in each false statement.

1. $\varnothing = \{0\}$
2. $\frac{5}{0} = 0$
3. $a(bc) = (ab)(ac)$
4. $-x < 0$ for all real numbers x
5. $3^{-2} = -\frac{1}{9}$
6. $-|-x| = |x|$
7. $(20)(-3) = 60$
8. $a^m + a^n = a^{m+n}$
9. $x^2 + y^2 = (x + y)^2$
10. $(2 + 3)^2 = 2^2 + 3^2 = 4 + 9 = 13$
11. $\left(\frac{b^{12}}{b^3}\right) = b^4$
12. $-3^2 = 9$
13. $\frac{2^{-1} + 20}{5} = \frac{20}{10} = 2$
14. $\sqrt{49} = \pm 7$
15. $-\sqrt{-49} = 7$

16. $\sqrt{5^2 + 12^2} = 5 + 12 = 17$

Tell which property of the real numbers justifies each statement:

17. $2 + 9 = 9 + 2$
18. $78 + 0 = 78$
19. $(13)\left(\frac{1}{13}\right) = 1$
20. $4(5 + 2) = 4 \cdot 5 + 4 \cdot 2$
21. $4(2 \cdot 3) = (4 \cdot 2)3$
22. $7 + (-7) = 0$
23. $(1)(5) = 5$

Write in scientific notation:

24. 45,000,000
25. 0.0098

Simplify:

26. $a^7 a^2$
27. $\left(\frac{b^8}{b^2}\right)$
28. $(a^5 b^2)^3$
29. $\frac{xy^{-2}z^4}{x^{-7}y^4}$
30. $\frac{(3.5 \cdot 10^{21})(5.3 \cdot 10^{-15})(2.2 \cdot 10^5)}{(5.7 \cdot 10^4)(1.5 \cdot 10^{-25})}$

Perform the indicated operations:

31. $-12[-15 - (-5 + 3)] =$
32. $5 + 56 \div 8 - 6 =$
33. $8 \cdot 3 - 5(7 - 2) =$
34. $30 + 2(6 + 4^3) =$
35. $-5\{[-4 - (5 + 11)] + 8\} =$
36. $16 - 49^{\frac{1}{2}} + 8^0 =$
37. $\frac{24 - 8^2}{2 \cdot 5 - 2} =$
38. $5|-6 - 4| =$
39. $-|-7| - 3^2 - 8 =$
40. $\sqrt{8^2 + 6^2} =$

SOLUTIONS TO PRACTICE PROBLEMS

1. False. $\varnothing = \{\}$; The set $\{0\}$ is not empty because it contains the element 0.
2. False. $\frac{5}{0}$ is undefined because division by zero has no meaning.

3. False. $a(bc) = (ab)(c)$; Do not confuse this problem with those involving the distributive property, in which $a(b + c) = ab + bc$.
4. False. $-x < 0$ *only* when x is a positive number; if x is negative, $-x > 0$.
5. False. $3^{-2} = \frac{1}{3^2} = \frac{1}{9}$
6. False. $-|-x| = -|x|$ because $|-x| = |x|$ for all real numbers.
7. False. $(20)(-3) = -60$; When multiplying, the product is negative when the signs are different.
8. False. $a^m + a^n$ cannot be simplified further. Do not confuse this problem with problems involving the product rule for exponents, where $a^m a^n = a^{m+n}$.
9. False. $x^2 + y^2$ cannot be simplified further. Exponents do not distribute over addition.
10. False. $(2 + 3)^2 = (5)^2 = 25$. First, perform the addition in parentheses, then square the sum.
11. False. $\left(\frac{b^{12}}{b^3}\right) = b^{12-3} = b^9$; When dividing exponential expressions that have the same base, subtract (not divide) the exponents.
12. False. $-3^2 = -9$; Since the $-$ symbol is not enclosed in parentheses, the exponent does not apply to it. First, square 3 to obtain 9, then write its negative.
13. False. $\frac{2^{-1} + 20}{5} = \frac{\frac{1}{2} + 20}{5} = \frac{20.5}{5} = 4.1$
14. False. $\sqrt{49} = 7$; The $\sqrt{}$ symbol *always* indicates the one nonnegative square root of a number.
15. False. $-\sqrt{-49}$ is not a real number. Since, by the order of operations, you must find $\sqrt{-49}$ *before* you find the negative, the answer is not a real number because $\sqrt{-49}$ is not a real number.
16. False. $\sqrt{5^2 + 12^2} = \sqrt{25 + 144} = \sqrt{169} = 13$. By the order of operations, you must first evaluate the expression under the $\sqrt{}$, doing the exponentiation before the addition, *then* take the square root. You cannot do the square root first.

Tell which property of the real numbers justifies each statement:

17. Commutative property (for addition)
18. Additive identity property
19. Multiplicative inverse property
20. Distributive property
21. Associative property (for multiplication)
22. Additive inverse property
23. Multiplicative identity property

Write in scientific notation:

24. $45{,}000{,}000 = 4.5 \cdot 10^7$
25. $0.0098 = 9.8 \cdot 10^{-3}$

Simplify:

26. $a^7a^2 = a^{7+2} = a^9$
27. $\left(\dfrac{b^8}{b^2}\right) = b^{8-2} = b^6$
28. $(a^5b^2)^3 = a^{(5)(3)}\ b^{(2)(3)} = a^{15}b^6$
29. $\dfrac{xy^{-2}z^4}{x^{-7}y^4} = x^{1+7}y^{-2-4}z^4 = x^8y^{-6}z^4 = \dfrac{x^8z^4}{y^6}$
30. $\dfrac{(3.5 \cdot 10^{21})(5.3 \cdot 10^{-15})(2.2 \cdot 10^5)}{(5.7 \cdot 10^4)(1.5 \cdot 10^{-25})} = \dfrac{(3.5)(5.3)(2.2)(10^{21})(10^{-15})(10^5)}{(5.7)(1.5)(10^4)(10^{-25})} =$

    $\dfrac{(40.81)(10^{21-15+5})}{(8.55)(10^{4-25})} = \dfrac{(40.81)(10^{11})}{(8.55)(10^{-21})} \cong 4.77 \cdot 10^{11+21} = 4.77 \cdot 10^{32}$

Perform the indicated operations:

31. $-12[-15 - (-5 + 3)] = -12[-15 - (-2)] = -12[-15 + 2] =$
    $-12[-13] = 156$
32. $5 + 56 \div 8 - 6 = 5 + 7 - 6 = 6$
33. $8 \cdot 3 - 5(7 - 2) = 8 \cdot 3 - 5(5) = 24 - 25 = -1$
34. $30 + 2(6 + 4^3) = 30 + 2(6 + 64) = 30 + 2(70) = 30 + 140 = 170$
35. $-5\{[-4 - (5 + 11)] + 8\} = -5\{[-4 - (16)] + 8\} =$
    $-5\{[-4 - 16] + 8\} = -5\{[-20] + 8\} = -5\{-20 + 8\} = -5\{-12\} = 60$
36. $16 - 49^{\frac{1}{2}} + 8^0 = 16 - 7 + 1 = 10$
37. $\dfrac{24 - 8^2}{2 \cdot 5 - 2} = \dfrac{24 - 64}{10 - 2} = \dfrac{-40}{8} = -5$
38. $5|-6 - 4| = 5|-10| = 5(10) = 50$
39. $-|-7| - 3^2 - 8 = -(7) - 9 - 8 = -7 - 9 - 8 = -24$
40. $\sqrt{8^2 + 6^2} = \sqrt{64 + 36} = \sqrt{100} = 10$

2
POLYNOMIALS AND RATIONAL EXPRESSIONS

2.1 ALGEBRAIC TERMINOLOGY

An **algebraic expression** is any symbol or combination of symbols that represents a number. It consists of one or more variables joined by one or more operations with or without numbers included. For example, $3x$, xy, $-y$, $5x + 4$, $3x^4 - 4x^2 + 4$, $9(x + 1) - 3$, $7abc$, $y(2x + 3)$, $\frac{3xy}{2(x+3z)}$, and $\frac{2}{x+1} + \frac{1}{x-1}$ are algebraic expressions.

A **term** is a constant, variable, or any product of constants or variables. For instance, x, $2xy$, $-6y$, $7abc$, $y(-2x)$, and 10 are terms. The expression $5x + 4y - 7$ has three terms. The terms $5x$, $4y$, and 7 are separated by + and − signs. Note that quantities enclosed within grouping symbols are considered single terms even though they may contain + or − signs. Thus, $y(-2x) + 9$ has 2 terms: $y(-2x)$ and 9. In a term with two or more factors, the **coefficient** of each factor is the product of all the other factors in that term. The numerical factor of a term is called the **numerical coefficient.** If no numerical coefficient is written, it is understood that the numerical coefficient is 1:

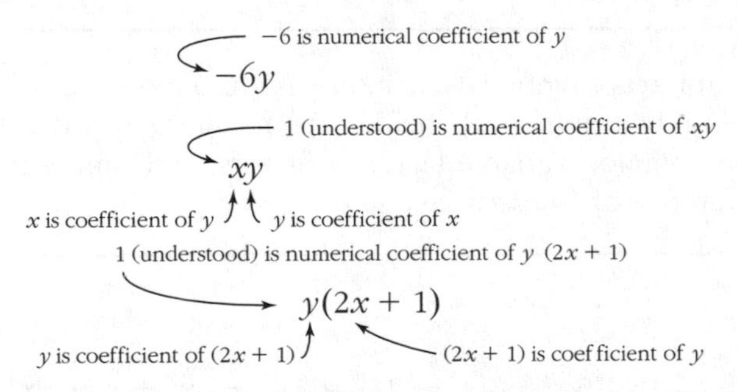

An expression such as $4x^2$ consisting of a numerical coefficient times one or more variables each raised to a nonnegative power is called a **monomial.** The **degree** of a monomial is the sum of the exponents of its variables. The degree of any nonzero constant c is zero, since $c = cx^0$ for any number c. The degree of the monomial 0 is undefined. For example,

- The degree of x is 1 (since $x = x^1$).
- The degree of xy is 2 (since $xy = x^1y^1$).
- The degree of $3x^4$ is 4.
- The degree of 10 is zero (since $10 = 10x^0$).

Monomials that differ only in their numerical coefficients are called **like terms;** that is, their variable factors are the same (same letters with the same respective exponents). For instance, $2x^2y$ and $-3x^2y$ are like terms; however, $2x^2y$ and $-3xy$ are **unlike terms.** All constants are like terms.

A **polynomial** is a sum of monomials. A polynomial such as $5x + 4$, consisting of two unlike terms, is called a **binomial.** A polynomial such as $3x^4 - 4x^2 + 4$, consisting of three unlike terms, is called a **trinomial.** A polynomial is **simplified** when it contains no like terms. The **degree** of a polynomial is the same as the greatest of the degrees of its monomial terms after the polynomial has been simplified.

2.2 SUMS AND DIFFERENCES OF POLYNOMIALS

The distributive property allows us to add like monomial terms by adding or subtracting their numerical coefficients. Addition or subtraction of unlike terms can only be indicated. For example,

- $2x + 5x = (2 + 5)x = 7x$.
- $2x^2y - 3x^2y = (2 - 3)x^2y = -1x^2y = -x^2y$.
- $2x^2y - 3xy$ cannot be further simplified.

Therefore, we have the following:

> To add or subtract polynomials, combine *like* monomial terms by adding or subtracting their numerical coefficients, using the result as the coefficient of the common variable factor or factors, and simply indicate the sum or difference of *unlike* terms.

For example,

- $(3x^4 - 4x^2 + 4) + (5x^4 - 2x^2 - 24) = 3x^4 - 4x^2 + 4 + 5x^4 - 2x^2 - 24$
  $= 8x^4 - 6x^2 - 20$
- $(3x^4 - 4x^2 + 4) - (5x^4 - 2x^2 - 24) = 3x^4 - 4x^2 + 4 - 5x^4 + 2x^2 + 24$
  $= -2x^4 - 2x^2 + 28$
- $(6x^2 - 5) + (3x^4 - 2x^2 - 7) = 6x^2 - 5 + 3x^4 - 2x^2 - 7$
  $= 3x^4 + 4x^2 - 12$
- $(12xy + 5) - (3xy - 4) = 12xy + 5 - 3xy + 4 = 9xy + 9$

These examples illustrate the following principles:

When simple parentheses (or brackets or braces) are immediately preceded by a + symbol, they can be removed without changing the signs of the terms within; if the parentheses are immediately preceded by a − symbol, the sign of *every* term within the grouping must be changed when the parentheses are removed.

- $(2x^5 - 7) + (5x^5 - x^3 + 9) = 7x^5 - x^3 + 2$
- $(2x^5 - 7) - (5x^5 - x^3 + 9) = 2x^5 - 7 - 5x^5 + x^3 - 9$
  $= -3x^5 + x^3 - 16$

A helpful message: Exercise caution when a − symbol appears before a radical symbol, absolute value bars, parentheses to which an exponent is attached, or other symbolism for an operation that should be performed *before* doing subtraction or finding the negative of a quantity. In general,

- $-\sqrt{-x} \neq \sqrt{x}$
- $-|-x| \neq |x|$
- $-(-x)^2 \neq x^2$

2.3 PRODUCTS OF POLYNOMIALS

To multiply a monomial by a monomial, multiply both the numerical coefficients and the variable factors:

- $(4a)(3a) = 12a^2$
- $4(-5x) = -20x$
- $(2x^2y)(-3xy) = -6x^3y^2$
- $(-2x^2yz^5)(-3xy) = 6x^3y^2z^5$

A helpful message: Notice that both like terms and unlike terms can be multiplied together.

To multiply a monomial by a polynomial of two or more terms, use the distributive property to multiply *each* term of the polynomial by the monomial:

- $2x(3x + 1) = \overbrace{(2x)(3x) + (2x)(1)}^{\text{Do this step mentally:}} = 6x^2 + 2x$
- $-3a(4a + 8b) = \overbrace{(-3a)(4a) + (-3a)(8b)}^{\text{Do this step mentally:}} = -12a^2 - 24ab$

When a term is subtracted, mentally insert a + sign and think of the – symbol as belonging to the term that follows:

- $-3a(4a - 8b) = \overbrace{(-3a)(4a) + (-3a)(-8b)}^{\text{Do this step mentally:}} = -12a^2 + 24ab$

We now have the following:

To multiply polynomials, multiply each term in the second polynomial by each term of the first polynomial, then combine like terms.

- $(2x+4)(3x+1) = \overbrace{(2x)(3x) + (2x)(1) + (4)(3x) + (4)(1)}^{\text{Do this step mentally:}} =$
  $6x^2 + 2x + 12x + 4 = 6x^2 + 14x + 4$

- $(x+2)(x-2) = \overbrace{(x)(x) + (x)(-2) + (2)(x) + (2)(-2)}^{\text{Do this step mentally:}} =$
  $x^2 - 2x + 2x - 4 = x^2 - 4$

The product of two binomials can be obtained quickly by using the **FOIL** method. The acronym **FOIL** is a mnemonic that reminds you to

- multiply the two **F**irst terms;
- multiply the two **O**uter terms;
- multiply the two **I**nner terms;
- multiply the two **L**ast terms;
- then simplify the results, if possible.

(First + Last)(First + Last) **(F)(F) (O)(O) (I)(I) (L)(L)**

$$(a + b)(c + d) = ac + ad + bc + bd$$

(Outer + Inner)(Inner + Outer) **F O I L**

For example,

- $(2x + 5)(3x + 1) = 6x^2 + 2x + 15x + 5 = 6x^2 + 17x + 5$
- $(5x - 1)(7x + 2) = 35x^2 + 10x - 7x - 2 = 35x^2 + 3x - 2$
- $(3x + 4y)(3x - 4y) = 9x^2 - 12xy + 12xy - 16y^2 = 9x^2 - 16y^2$

Notice, if the first terms are like terms and the last terms are like terms, the products of the outer and inner terms combine into a single term that can be computed mentally:

Compute this mentally:

- $(3x - 1)(5x + 2) = 15x^2 + \overbrace{6x - 5x} - 2 = 15x^2 + x - 2$

Compute this mentally:

- $(x - y)(x + y) = x^2 + \overbrace{xy - xy} - y^2 = x^2 - y^2$

For larger polynomials, you may find it more convenient to use a vertical format. For example, $(4x^2 + 3x - 5)(2x^2 - x - 3)$ can be computed as follows:

$$\begin{array}{lllll}
4x^2 & + 3x & - 5 & & \\
\underline{2x^2} & \underline{- x} & \underline{- 3} & & \\
8x^4 & + 6x^3 & - 10x^2 & & \leftarrow 2x^2(4x^2 + 3x - 5) \\
 & - 4x^3 & - 3x^2 & + 5x & \leftarrow -x(4x^2 + 3x - 5) \\
 & & - 12x^2 & - 9x + 15 & \leftarrow -3(4x^2 + 3x - 5) \\
\hline
8x^4 & + 2x^3 & - 25x^2 & - 4x + 15 &
\end{array}$$

When several polynomial factors are involved, multiply two at a time:

$$\begin{aligned}
(3x - 1)(5x + 2)(x - 1) &= (15x^2 + x - 2)(x - 1) \\
&= 15x^3 - 15x^2 + x^2 - x - 2x + 2 \\
&= 15x^3 - 14x^2 - 3x + 2
\end{aligned}$$

Some special products you will need to know are the following:

perfect trinomial squares:

- $(x + y)^2 = (x + y)(x + y) = \mathbf{x^2 + 2xy + y^2}$
- $(x - y)^2 = (x - y)(x - y) = \mathbf{x^2 - 2xy + y^2}$

difference of two squares:

$$(x + y)(x - y) = \mathbf{x^2 - y^2}$$

sum of two cubes:

$$(x + y)(x^2 - xy + y^2) = x^3 - x^2y + xy^2 + x^2y - xy^2 + y^3 = \mathbf{x^3 + y^3}$$

difference of two cubes:

$$(x - y)(x^2 + xy + y^2) = x^3 + x^2y + xy^2 - x^2y - xy^2 - y^3 = \mathbf{x^3 - y^3}$$

perfect cubes:

$$\begin{aligned}(x+y)^3 &= (x+y)(x+y)(x+y)\\ &= (x+y)(x^2+2xy+y^2)\\ &= \mathbf{x^3+3x^2y+3xy^2+y^3}\\ (x-y)^3 &= (x-y)(x-y)(x-y)\\ &= (x-y)(x^2-2xy+y^2)\\ &= \mathbf{x^3-3x^2y+3xy^2-y^3}\end{aligned}$$

A helpful message: Notice that the middle term in a perfect trinomial square is plus or minus *twice* the product of the square roots of the first and last terms. If this is not the case the trinomial is *not* a perfect trinomial square. For instance,

- $4x^2 - 4xy + y^2$ is a perfect square, but $4x^2 - 2xy + y^2$ is *not*;
- $9x^2 - 24xy + 16y^2$ is a perfect square, but $9x^2 - 12xy + 16y^2$ is *not*;
- $25x^2 + 90xy + 81y^2$ is a perfect square, but $25x^2 + 81y^2$ is *not.*

When multiplying polynomials, if possible, arrange the two factors in descending or ascending powers of a common variable:

$$\begin{aligned}&(5x^4+3x^2-x+4)(x^6-2x-1)\\ &\quad = 5x^{10}-10x^5-5x^4+3x^8-6x^3-3x^2-x^7+2x^2+x+4x^6-8x-4\\ &\quad = 5x^{10}+3x^8-x^7+4x^6-10x^5-5x^4-6x^3-x^2-7x-4\end{aligned}$$

2.4 SIMPLIFYING POLYNOMIAL EXPRESSIONS

To simplify a polynomial expression:

1. When grouping symbols are present, perform all operations within grouping symbols, starting with the innermost grouping symbol and working outward;
2. If powers or products of polynomials are involved, perform all indicated multiplication, enclosing the product in parentheses if it is to be multiplied by an additional factor;
3. Remove all remaining parentheses and combine like terms.

Some examples follow:

- $2 + 4(x + 3y - 2) = 2 + 4x + 12y - 8$
  $= 4x + 12y - 6$
- $2a - 5a(4x - 3a) + 8ax = 2a - 20ax + 15a^2 + 8ax$
  $= 15a^2 + 2a - 12ax$
- $(2x - 5)(4x - 5) - (x - 3)(x + 3) = 8x^2 - 30x + 25 - (x^2 - 9)$
  $= 8x^2 - 30x + 25 - x^2 + 9$
  $= 7x^2 - 30x + 34$
- $5x^2 - (2x - 3)^2 + 9 = 5x^2 - (4x^2 - 12x + 9) + 9$
  $= 5x^2 - 4x^2 + 12x - 9 + 9 = x^2 + 12x$
- $x - [3x - (2x - 5)] = x - [3x - 2x + 5]$
  $= x - [x + 5]$
  $= x - x - 5$
  $= -5$
- $-2\{3x - [2x^2 - 5x(3x + 1) + 2] - 6\} = -2\{3x - [2x^2 - 15x^2 - 5x + 2] - 6\}$
  $= -2\{3x - [-13x^2 - 5x + 2] - 6\}$
  $= -2\{3x + 13x^2 + 5x - 2 - 6\}$
  $= -2\{13x^2 + 8x - 8\}$
  $= -26x^2 - 16x + 16$

2.5 QUOTIENTS OF POLYNOMIALS BY MONOMIALS

To divide a monomial by a monomial, divide both the numerical coefficients and the variable factors.

- $\dfrac{12a^2}{4a} = 3a$
- $\dfrac{-20x}{-5x} = 4$
- $\dfrac{-6x^3y^2}{2x^2y} = -3xy$
- $\dfrac{6x^3y^2z^5}{-3xy} = -2x^2yz^5$

A helpful message: Notice that both like terms and unlike terms can be divided. We now have the following:

> To divide a polynomial of two or more terms by a monomial, divide each term of the polynomial by the monomial.

- $\dfrac{-12x^6 + 6x^3}{-3x^2} = \dfrac{-12x^6}{-3x^2} + \dfrac{6x^3}{-3x^2} = 4x^4 - 2x$
- $\dfrac{25a^7b^4c^3 + 10a^6b^2c - 15a^3bc}{-5a^2bc} = \dfrac{25a^7b^4c^3}{-5a^2bc} + \dfrac{10a^6b^2c}{-5a^2bc} + \dfrac{-15a^3bc}{-5a^2bc}$
  $= -5a^5b^3c^2 - 2a^4b + 3a$

A helpful message: This process of dividing the polynomial by the monomial is related to the distributive property. Notice that

$$\frac{-12x^6 + 6x^3}{-3x^2} = \frac{1}{-3x^2}(-12x^6 + 6x^3)$$

If you think of this relationship, it will help you avoid the common mistake of trying to divide a monomial by a polynomial, for example,

$$\frac{8x^2}{16x^6 + 24x^3} \neq 2x^4 + 3x$$

2.6 LONG DIVISION OF POLYNOMIALS

Long division of polynomials is accomplished in a manner analogous to long division in arithmetic.

Example

Find the quotient

$$\underbrace{(-2x^2 + 4 + 3x^4)}_{\text{dividend}} \div \underbrace{(x^2 - 2 + x)}_{\text{divisor}}$$

Step 1: Write both dividend and divisor in descending powers of x, using a coefficient of 0 when a power of x is missing:

$$x^2 + x - 2 \overline{)3x^4 + 0x^3 - 2x^2 + 0x + 4}$$

Step 2: Divide the first term in the dividend by the first term in the divisor:

$$\begin{array}{r} 3x^2 \\ x^2 + x - 2 \overline{)3x^4 + 0x^3 - 2x^2 + 0x + 4} \end{array}$$

Step 3: Multiply each term in the divisor by $3x^2$, subtract, and bring down the next term from the dividend:

$$\begin{array}{r} 3x^2 \\ x^2 + x - 2 \overline{)3x^4 + 0x^3 - 2x^2 + 0x + 4} \\ \underline{3x^4 + 3x^3 - 6x^2} \\ -3x^3 + 4x^2 + 0x \end{array}$$

Step 4: Repeat steps 2 and 3. Continue until the degree of the remainder is less than the degree of the divisor:

$$\begin{array}{rl} 3x^2 - 3x + 7 & \leftarrow \text{quotient} \\ x^2 + x - 2 \overline{)3x^4 + 0x^3 - 2x^2 + 0x + 4} & \\ \underline{3x^4 + 3x^3 - 6x^2} & \text{multiply} \\ -3x^3 + 4x^2 + 0x & \text{subtract and bring down} \\ \underline{-3x^3 + 3x^2 + 6x} & \text{multiply} \\ 7x^2 - 6x + 4 & \text{subtract and bring down} \\ \underline{7x^2 + 7x - 14} & \text{multiply} \\ \text{remainder} \rightarrow \quad -13x + 18 & \text{subtract} \end{array}$$

Solution

$$\underbrace{(-2x^2 + 4 + 3x^4)}_{\text{dividend}} \div \underbrace{(x^2 - 2 + x)}_{\text{divisor}} = \underbrace{3x^2 - 3x + 7}_{\text{quotient}} \quad \text{R: } \underbrace{-13x + 18}_{\text{remainder}}$$

It is customary to write the results by forming a fraction with the remainder:

$$(-2x^2 + 4 + 3x^4) \div (x^2 - 2 + x) = \frac{3x^4 - 2x^2 + 4}{x^2 + x - 2}$$

$$= 3x^2 - 3x + 7 + \frac{-13x + 18}{x^2 + x - 2}$$

2.7 FACTORING POLYNOMIALS

Factoring is the direct reverse of multiplication. In multiplication, you are given the factors and are required to find the product. In factoring, the product is given and you are required to find the factors. Here are examples from arithmetic using only prime numbers as factors:

- $15 = 3 \cdot 5$
- $100 = 2 \cdot 2 \cdot 5 \cdot 5 = 2^2 \cdot 5^2$

Factoring a polynomial means to find two or more polynomials whose product is the original polynomial. For example,

$$6x^2 + 17x + 5 = (2x + 5)(3x + 1)$$

This is the **factorization** of $6x^2 + 17x + 5$ because $(2x + 5)(3x + 1) = 6x^2 + 17x + 5$. The polynomials $(2x + 5)$ and $(3x + 1)$ are **factors** of $6x^2 + 17x + 5$.

Before you can say that a polynomial factor is a **prime factor** of a polynomial, you must specify the set of numbers from which the coefficients of the polynomial factors are to be chosen. Unless you are directed otherwise, a common rule to use is that if the original polynomial has *integer* coefficients, then use only *integer* coefficients in the polynomial factors; and if the original polynomial has *rational* coefficients, then use only *rational* coefficients in the polynomial factors. For example,

- $x^2 - 4 = (x + 2)(x - 2)$ is factorable, since $(x + 2)(x - 2) = x^2 - 4$.
- $x^2 - 8$ is prime, since it is not factorable using the integers or rationals (i.e., fractions) (even though $(x + \sqrt{8})(x - \sqrt{8}) = x^2 - 8$).
- $x^2 - \frac{1}{4} = \left(x + \frac{1}{2}\right)\left(x - \frac{1}{2}\right)$ is factorable using the rationals, since $\left(x + \frac{1}{2}\right)\left(x - \frac{1}{2}\right) = x^2 - \frac{1}{4}$.

For the remainder of this section, factoring will be done using the set of integers, which means all numerical coefficients must be integers. Every polynomial with integer coefficients can be factored as itself and 1. When a polynomial cannot be factored in any other way using the set of integers, it is called **prime.** A polynomial is **factored completely** using the set of the integers when it is written as the product of prime polynomials.

Factoring Out Common Monomial Factors

The **greatest common monomial factor** for a polynomial is the monomial of highest degree and greatest coefficient that is a factor of each term of the polynomial.

> Use the distributive property to factor a polynomial whose terms have a common monomial factor as the product of two factors: the first factor is the greatest common monomial factor of the polynomial; the second factor is the quotient of the polynomial divided by the greatest common monomial factor.

- $6x^3 - 12x^6 = \underbrace{(6x^3)}(1) - \underbrace{(6x^3)}(2x^3) = 6x^3(1 - 2x^3)$

  greatest common monomial factor

- $25a^7b^4c^3 + 10a^6b^2c - 15a^3bc = \underbrace{(5a^3bc)}(5a^4b^3c^2) + \underbrace{(5a^3bc)}(2a^3b) - \underbrace{(5a^3bc)}(3)$

  greatest common monomial factor

  $= 5a^3bc(5a^4b^3c^2 + 2a^3b - 3)$

A helpful message: You should check your work by mentally multiplying the two factors using the distributive property. For the above examples,

- $6x^3(1 - 2x^3) = (6x^3)(1) - (6x^3)(2x^3) = 6x^3 - 12x^6$
- $5a^3bc(5a^4b^3c^2 + 2a^3b - 3) = (5a^3bc)(5a^4b^3c^2) + (5a^3bc)(2a^3b) - (5a^3bc)(3)$
  $= 25a^7b^4c^3 + 10a^6b^2c - 15a^3bc$

Consider these examples where the common monomial factor is -1:

- $-x - y = -1(x + y) = -(x + y)$
- $-x + y = -1(x - y) = -(x - y)$

Notice that -1 can be factored from a polynomial by changing the sign of *each* term of the polynomial and showing the results in parentheses prefixed by a $-$ symbol. Naturally, you can extend this idea to factoring out any negative common term:

$$-12x^6 + 6x^3 = -6x^3(2x^3 - 1)$$

Sometimes a common monomial factor is a quantity enclosed in parentheses:

$$3x\underbrace{(x - 2)} - 5\underbrace{(x - 2)} = (x - 2)(3x - 5)$$

common monomial factor

You may have to organize the terms of the polynomial into smaller groups to be able to factor out a common monomial factor:

Because you are factoring out a -1, this sign must be $-$:

$$\begin{aligned} 3x^2 - 6x - 5x + 10 &= (3x^2 - 6x) - (5x - 10) \\ &= 3x(x - 2) - 5(x - 2) \\ &= (x - 2)(3x - 5) \end{aligned}$$

Factoring Binomials

difference of two squares:

$$x^2 - y^2 = (x + y)(x - y)$$

sum of two squares:

$x^2 + y^2$ is prime over the real numbers

sum of two cubes:

$$x^3 + y^3 = (x + y)(x^2 - xy + y^2)$$

difference of two cubes:

$$x^3 - y^3 = (x - y)(x^2 + xy + y^2)$$

To factor a **difference of two squares,** find the principal square roots of the two squares and form two binomial factors: the sum of the two principal square roots and the difference of the two principal square roots.

- $x^2 - y^2 = (x + y)(x - y)$
- $a^2x^2 - b^2y^2 = (ax + by)(ax - by)$
- $9x^2 - 16y^2 = (3x + 4y)(3x - 4y)$
- $a^4 - y^2 = (a^2 + y)(a^2 - y)$

The **sum of two squares** like $x^2 + y^2$ is not factorable using the set of real numbers. Note that $x^2 + y^2 \neq (x + y)^2$ because $(x + y)^2 = x^2 + 2xy + y^2$.

To factor a **sum of two cubes,** form the product of a binomial factor and a trinomial factor. The binomial factor is the sum of the cube roots of the two cubes. The first and last terms of the trinomial factor are the squares of the two cube roots, and the middle term is the negative product of the two cube roots.

- $a^3 + b^3 = \underbrace{(\overset{\sqrt[3]{a^3}}{a} + \overset{\sqrt[3]{b^3}}{b})}_{\text{binomial}}\underbrace{(a^2 - ab + b^2)}_{\text{trinomial}}$
- $$\begin{aligned} 64x^3 + 27y^3 &= (\overset{\sqrt[3]{64x^3}}{4x} + \overset{\sqrt[3]{27y^3}}{3y})[(4x)^2 - (4x)(3y) + (3y)^2)] \\ &= \underbrace{(4x + 3y)}_{\text{binomial}}\underbrace{(16x^2 - 12xy + 9y^2)}_{\text{trinomial}} \end{aligned}$$

To factor a **difference of two cubes,** form the product of a binomial factor and a trinomial factor. The binomial factor is the difference of the cube roots of the two cubes. The first and last terms of the trinomial factor are the squares of the two cube roots, and the middle term is the positive product of the two cube roots.

- $a^3 - b^3 = \underbrace{(\overset{\sqrt[3]{a^3}}{a} - \overset{\sqrt[3]{b^3}}{b})}_{\text{binomial}}\underbrace{(a^2 + ab + b^2)}_{\text{trinomial}}$
- $$\begin{aligned} 64x^3 - 27y^3 &= (\overset{\sqrt[3]{64x^3}}{4x} - \overset{\sqrt[3]{27y^3}}{3y})[(4x)^2 + (4x)(3y) + (3y)^2)] \\ &= \underbrace{(4x - 3y)}_{\text{binomial}}\underbrace{(16x^2 + 12xy + 9y^2)}_{\text{trinomial}} \end{aligned}$$

Factoring Trinomials

general quadratic trinomials:

$$x^2 + (a + b)x + ab = (\boldsymbol{x} + \boldsymbol{a})(\boldsymbol{x} + \boldsymbol{b})$$

$$acx^2 + (ad + bc)x + bd = (\boldsymbol{ax} + \boldsymbol{b})(\boldsymbol{cx} + \boldsymbol{d})$$

perfect trinomial squares:

$$a^2x^2 + 2abxy + b^2y^2 = (\boldsymbol{ax} + \boldsymbol{by})^2$$

$$a^2x^2 - 2abxy + b^2y^2 = (\boldsymbol{ax} - \boldsymbol{by})^2$$

To factor $x^2 + (a + b)x + ab$ using **trial and error,** look for two binomial factors:

First, find two terms whose product is x^2 and use them as the first terms of the two trial binomial factors: $(x \quad)(x \quad)$

Second, find two terms whose product is ab and whose sum is $a + b$. If ab is *positive,* the two terms will have the *same* sign and it will be the same as the sign of the middle term; if ab is *negative,* the two terms will have *different* signs.

These terms are the last terms of the trial binomial factors.

- To factor $z^2 - 10z + 21$:

First, find two terms whose product is z^2 and use them as the first terms of the two trial binomial factors: $(z \quad)(z \quad)$

Second, find two terms whose product is 21 and whose sum is -10 and use them as the last terms of the trial binomial factors: $(z - 3)(z - 7)$

Thus, $z^2 - 10z + 21 = (z - 3)(z - 7)$.

To factor $acx^2 + (ad + bc)x + bd$ using **trial and error,** look for two binomial factors:

First, find two terms whose product is acx^2. These terms are the first terms of the two trial binomial factors.

Second, find two terms whose product is bd. If bd is *positive,* the two terms will have the *same* sign and it will be the same as the sign of the middle term; if bd is *negative,* the two terms will have *different* signs.

These terms are the last terms of the trial binomial factors.

Third, combine the products of the outer and inner terms of the trial binomial factors and check the result against the middle term of the original expression. If the result equals the middle term, the trial binomial factors are the factors of the quadratic expression; if not, modify. If no such factors can be determined, the polynomial is not factorable.

- To factor $3x^2 - 14x - 5$:

First, find two terms whose product is $3x^2$ and use them as the first terms of the two trial binomial factors: $(x \quad)(3x \quad)$
Second, find two terms whose product is -5 and which have different signs and use them as the last terms of the trial binomial factors: $(x - 1)(3x + 5)$
Third, combine the products of the outer and inner terms of the trial binomial factors and check the result against the middle term, $-14x$:

$$5x - 3x = 2x \neq -14x$$

Modify, by interchanging the last terms: $(x + 5)(3x - 1)$
Check again: $-x + 15x = 14x \neq -14x$
Modify, by interchanging the signs of the last terms: $(x - 5)(3x + 1)$
Check again: $x - 15x = -14x =$ middle term
Since the middle term is correct, $3x^2 - 14x - 5 = (x - 5)(3x + 1)$.

- To factor $9x^2 + 24x + 16$:

First, find two terms whose product is $9x^2$ and use them as the first terms of the two trial binomial factors: $(3x \quad)(3x \quad)$
Second, find two terms whose product is 16 and which have the same sign as $24x$ and use them as the last terms of the trial binomial factors: $(3x + 4)(3x + 4)$
Third, combine the products of the outer and inner terms of the trial binomial factors and check the result against the middle term, $24x$:

$$12x + 12x = 24x$$

Since the middle term is correct, $9x^2 + 24x + 16 = (3x + 4)(3x + 4) = (3x + 4)^2$.

Quadratic expressions like $9x^2 + 24x + 16$ are called **perfect trinomial squares,** since they are the result of squaring a binomial.

> The middle term of a perfect trinomial square is either plus or minus *twice* the product of the square roots of the first and last terms.

You should always check to see this is the case *before* you attempt to factor the expression as a perfect trinomial square.

To factor a **perfect trinomial square** that has a *positive* middle term, find the square roots of the first and last terms, then express the trinomial as the *sum* of the square roots squared.

- $x^2 + 2xy + y^2 = (x + y)(x + y) = (x + y)^2$
- $9x^2 + 24x + 16 = (3x + 4)(3x + 4) = (3x + 4)^2$

- $a^2x^2 + 2abxy + b^2y^2 = (ax + by)(ax + by) = (ax + by)^2$
- $x^4 + 2x^2y^2 + y^4 = (x^2 + y^2)(x^2 + y^2) = (x^2 + y^2)^2$

To factor a perfect trinomial square that has a *negative* middle term, find the square roots of the first and last terms, then express the trinomial as the *difference* of the square roots squared.

- $x^2 - 2xy + y^2 = (x - y)(x - y) = (x - y)^2$
- $9x^2 - 24x + 16 = (3x - 4)(3x - 4) = (3x - 4)^2$
- $a^2x^2 - 2abxy + b^2y^2 = (ax - by)(ax - by) = (ax - by)^2$
- $x^4 - 2x^2y^2 + y^4 = (x^2 - y^2)(x^2 - y^2) = (x^2 - y^2)^2$

Factoring Polynomials of Four Terms

Sometimes, a **polynomial of four terms** can be factored by grouping some of the terms together and factoring separately first.

- $$\begin{aligned} 9x^2 - a^2 + 2ab - b^2 &= 9x^2 - (a^2 - 2ab + b^2) \\ &= (3x)^2 - (a - b)^2 \\ &= [3x + (a - b)][3x - (a - b)] \\ &= (3x + a - b)(3x - a + b) \end{aligned}$$
- $$\begin{aligned} ax + bx + ay + by &= (ax + bx) + (ay + by) \\ &= x(a + b) + y(a + b) \\ &= (x + y)(a + b) \end{aligned}$$

A helpful message: The expression $x(a + b) + y(a + b)$ is *not* in factored form. An expression is in factored form when it is written as a *product* of two or more factors. The expression $x(a + b) + y(a + b)$ is not a product; it is the *sum* of two products. The factored form of this expression is $(x + y)(a + b)$.

You may have learned to rewrite quadratic trinomials of the form $ax^2 + bx + c$ as polynomials of four terms for the purposes of factoring by grouping. This can be an efficient method for factoring trinomials.

To factor $ax^2 + bx + c$ by **grouping:**

First, find the product ac.
Second, find two numbers whose product is ac and whose sum is b.
Third, rewrite the trinomial, replacing bx with the sum of two terms whose coefficients are the numbers just determined.
Fourth, factor by grouping.

- To factor $3x^2 - 14x - 5$ by grouping:

First, find the product $3(-5) = -15$.
Second, find two numbers whose product is -15 and whose sum is -14: 1 and -15

Third, rewrite $3x^2 - 14x - 5$, replacing $-14x$ with the sum of two terms whose coefficients are 1 and -15: $3x^2 \underbrace{- 14x} - 5 = 3x^2 \underbrace{- 15x + 1x} - 5$

Fourth, factor by gr ouping:

$$\begin{aligned} 3x^2 - 15x + 1x - 5 &= (3x^2 - 15x) + (1x - 5) \\ &= 3x\,\underbrace{(x - 5)} + 1\underbrace{(x - 5)} \end{aligned}$$

common monomial factor

$$= (x - 5)(3x + 1)$$

To **factor a polynomial completely** over the set of integers means to find the prime factors of the polynomial. Your ability to factor polynomials depends on your skill at recognizing the special products and procedures you have seen thus far. The following procedure for factoring a polynomial completely summarizes what you have learned:

> To factor a polynomial completely over the set of integers:
>
> 1. Look for a greatest common monomial factor. If there is one, factor the polynomial. Then continue factoring until all factors except monomial factors are prime.
> 2. If a factor is a binomial, see if it is a difference of two squares, a sum of two cubes, or the difference of two cubes. If it is one of these, factor it.
> 3. If a factor is a trinomial, see if it is a perfect square or a factorable quadratic expression. If it is one of these, factor it.
> 4. If a factor has four terms, look for a grouping arrangement that will work for factoring by grouping. If this can be done, factor it.
> 5. Write the original polynomial as the product of all the factors obtained. Check to make sure all polynomial factors except monomial factors are prime.
> 6. Check by multiplying the factors to obtain the original polynomial.

Some examples follow:

- $16a^2x^2 - 4a^2 = 4a^2(4x^2 - 1) = 4a^2(2x + 1)(2x - 1)$
- $27z^2 + 36yz + 12y^2 = 3(9z^2 + 12yx + 4y^2) = 3(3z + 2y)^2$
- $640x^3y - 270y^4 = 10y(64x^3 - 27y^3) = 10y(4x - 3y)(16x^2 + 12xy + 9y^2)$
- $$\begin{aligned} 2x^4y + 2x^3y - 18x^2y - 18xy &= 2xy(x^3 + x^2 - 9x - 9) \\ &= 2xy[(x^3 + x^2) - (9x + 9)] \\ &= 2xy[x^2(x + 1) - 9(x + 1) \\ &= 2xy[(x + 1)(x^2 - 9)] \end{aligned}$$

$$= 2xy[(x + 1)(x + 3)(x - 3)]$$
$$= 2xy(x + 1)(x + 3)(x - 3)$$

- $16x^2 - 4a^2 - 8ab - 4b^2 = 4(4x^2 - a^2 - 2ab - b^2)$
$$= 4[4x^2 - (a^2 + 2ab + b^2)]$$
$$= 4[(2x)^2 - (a + b)^2]$$
$$= 4[2x + (a + b)][2x - (a + b)]$$
$$= 4(2x + a + b)(2x - a - b)$$

2.8 REDUCING RATIONAL EXPRESSIONS

A **rational expression** is an algebraic fraction in which both the numerator and denominator are polynomials. Since division by zero is undefined, any value of the variable (or variables) in the denominator polynomial that makes the denominator zero must be excluded. To find an excluded value, factor the denominator polynomial completely, and then determine any replacements for the variable (or variables) that cause a factor to evaluate to zero. For instance,

- $\dfrac{x}{5}$ is a rational expression; it has no excluded value.
- $\dfrac{3}{x}$ is a rational expression where $x \neq 0$.
- $x^4y^{-2} = \dfrac{x^4}{y^2}$ is a rational expression where $y \neq 0$.
- $\dfrac{2}{x - 1}$ is a rational expression where $x \neq 1$.
- $\dfrac{x^2 - 4}{x^2 - 3x - 4} = \dfrac{(x + 2)(x - 2)}{(x - 4)(x + 1)}$ is a rational expression where $x \neq 4$ and $x \neq -1$.
- $2x^4 + 3x^3 - 18x - 1$ is a rational expression, since it can be written as $\dfrac{2x^4 + 3x^3 - 18x - 1}{1}$; it has no excluded value.

From the last example, it should be clear that all polynomials are rational expressions that have no excluded values since any polynomial can be written as an algebraic fraction whose denominator is 1.

Hereafter, whenever a rational expression is written, it will be understood that any values for which the expression is undefined are excluded.

The principles used in the study of arithmetic fractions are generalized in work with rational expressions. Recall that every fraction has three signs associated with it:

$$\text{Sign of fraction} \rightarrow + \frac{-3}{+4}$$

Sign of numerator

Sign of denominator

If any two of the three signs of a fraction are changed at the same time, the value of the fraction is unchanged:

$$+\frac{-3}{+4} = +\frac{+3}{-4} = -\frac{+3}{+4} = -\frac{-3}{-4} = -0.75$$

This rule of signs is helpful in simplifying rational expressions. It can be used to find an equivalent rational expression as in the following:

- $-\dfrac{-2x}{x+2} = \dfrac{2x}{x+2}$
- $\dfrac{3x-5}{-x} = -\dfrac{3x-5}{x}$
- $\dfrac{-1}{3-x} = \dfrac{1}{-(3-x)} = \dfrac{1}{-3+x} = \dfrac{1}{x-3}$
- $-\dfrac{2x-5}{x+1} = \dfrac{-2x+5}{x+1}$

Notice that if a sign change is applied to a quantity of more than one term, the sign of *each* term is changed. If, however, a sign change is applied to a product, the sign of only *one* factor is changed. Which factor is to receive the sign change is a matter of choice. For example,

$$-\frac{2x}{(3-x)(x-2)} = \frac{2x}{(x-3)(x-2)} = \frac{2x}{(3-x)(2-x)}$$

Also, to reduce an arithmetic fraction, we divide out common factors in the numerator and denominator. This process is commonly called "canceling":

$$\frac{15}{10} = \frac{\not{5} \cdot 3}{\not{5} \cdot 2} = \frac{3}{2}$$

> To reduce a rational expression to lowest terms, factor the numerator and denominator completely, then cancel common factors.

- $\dfrac{15x^3y^2}{10xy^2} = \dfrac{\cancel{5}\cancel{x}\cancel{y^2} \cdot 3x^2}{\cancel{5}\cancel{x}\cancel{y^2} \cdot 2} = \dfrac{3x^2}{2}$
- $\dfrac{8x}{4x+20} = \dfrac{{}^{2}\cancel{8}x}{\cancel{4}(x+5)} = \dfrac{2x}{x+5}$
- $\dfrac{a^2+4a+4}{a^2-4} = \dfrac{\cancel{(a+2)}(a+2)}{\cancel{(a+2)}(a-2)} = \dfrac{a+2}{a-2}$
- $\dfrac{x-3}{3-x} = -\dfrac{x-3}{x-3} = -1$
- $\dfrac{x+2}{4}$ cannot be reduced further.
- $\dfrac{x^2+5}{x+10}$ cannot be reduced further.

A helpful message: $\frac{x+2}{4} \neq \frac{x+1}{2}$; The number 2 is *not* a common factor of the numerator and denominator of the fraction $\frac{x+2}{4}$. similarly, $\frac{x^2+5}{x+10} \neq \frac{x+1}{2}$ because neither x nor 5 are common factors of the numerator and denominator of the fraction $\frac{x^2+5}{x+10}$. The important thing to remember is that you must *factor first before you reduce.* Then you will know whether a quantity is a common factor that can be divided out.

2.9 PRODUCTS AND QUOTIENTS OF RATIONAL EXPRESSIONS

To multiply rational expressions, factor all numerators and denominators completely, then cancel common factors (as in reducing). The product of the remaining numerator factors is the numerator of the answer, and the product of the remaining denominator factors is the denominator of the answer.

- $\dfrac{a^2+4a+4}{a^2+a-2} \cdot \dfrac{a^2-2a+1}{a^2-4} = \dfrac{\cancel{(a+2)}\cancel{(a+2)}}{\cancel{(a+2)}\cancel{(a-1)}} \cdot \dfrac{\cancel{(a-1)}(a-1)}{\cancel{(a+2)}(a-2)} = \dfrac{a-1}{a-2}$
- $$\begin{aligned}\dfrac{4y^2-4x^2}{y^2-2xy-3x^2} \cdot \dfrac{3x-y}{x^3-y^3} &= \dfrac{4(y^2-x^2)}{(y-3x)(y+x)} \cdot \dfrac{(3x-y)}{(x-y)(x^2+xy+y^2)} \\ &= \dfrac{4\cancel{(y+x)}\cancel{(y-x)}}{\cancel{(y-3x)}\cancel{(y+x)}} \cdot \dfrac{-\cancel{(y-3x)}}{-\cancel{(y-x)}(x^2+xy+y^2)} \\ &= \dfrac{4}{x^2+xy+y^2}\end{aligned}$$

> To divide two rational expressions, multiply the dividend by the reciprocal of the divisor.

- $$\frac{a^2 + 4a + 4}{a^2 + a - 2} \div \frac{a^2 - 4}{a^2 - 2a + 1} = \frac{a^2 + 4a + 4}{a^2 + a - 2} \cdot \frac{a^2 - 2a + 1}{a^2 - 4}$$
$$= \frac{\cancel{(a+2)}\cancel{(a+2)}}{\cancel{(a+2)}\cancel{(a-1)}} \cdot \frac{\cancel{(a-1)}(a-1)}{\cancel{(a+2)}(a-2)}$$
$$= \frac{a-1}{a-2}$$

- $$\frac{4y^2 - 4x^2}{y^2 - 2xy - 3x^2} \div \frac{x^3 - y^3}{3x - y} = \frac{4y^2 - 4x^2}{y^2 - 2xy - 3x^2} \cdot \frac{3x - y}{x^3 - y^3}$$
$$= \frac{4(y^2 - x^2)}{(y - 3x)(y + x)} \cdot \frac{(3x - y)}{(x - y)(x^2 + xy + y^2)}$$
$$= \frac{4\cancel{(y+x)}\cancel{(y-x)}}{\cancel{(y-3x)}\cancel{(y+x)}} \cdot \frac{-\cancel{(y-3x)}}{-\cancel{(y-x)}(x^2 + xy + y^2)}$$
$$= \frac{4}{x^2 + xy + y^2}$$

A helpful message: Notice that when multiplying or dividing rational expressions, the denominators do *not* have to be the same. For instance,

- $$\frac{x}{x+y} \cdot \frac{y}{x-y} = \frac{xy}{x^2 - y^2}$$
- $$\frac{x}{x+y} \div \frac{y}{x-y} = \frac{x}{x+y} \cdot \frac{x-y}{y} = \frac{x^2 - xy}{xy + y^2}$$

2.10 SUMS AND DIFFERENCES OF RATIONAL EXPRESSIONS

To add rational expressions that have the same denominator, add or subtract like terms of the numerators, put the result over the common denominator, then reduce, if possible.

- $$\frac{x+2}{x-3} + \frac{2x-11}{x-3} = \frac{x + 2 + 2x - 11}{x-3} = \frac{3x-9}{x-3} = \frac{3\cancel{(x-3)}}{\cancel{x-3}} = \frac{3}{1} = 3$$

Because the minus sign in front of the fraction applies to the *entire* numerator, this sign is −, not +:

- $\dfrac{5x^2}{3(x+1)} - \dfrac{4x^2+1}{3(x+1)} = \dfrac{5x^2-4x^2-1}{3(x+1)} = \dfrac{x^2-1}{3(x+1)} = \dfrac{\cancel{(x+1)}(x-1)}{3\cancel{(x+1)}} = \dfrac{x-1}{3}$

A helpful message: When subtracting a rational expression, change the sign of *each* term of the numerator before combining like terms.

Recall the fundamental principle that multiplying the numerator and denominator of a fraction by the same nonzero quantity results in an equivalent fraction:

$$\frac{3}{4} = \frac{3 \cdot 2}{4 \cdot 2} = \frac{6}{8}$$

This is true for rational expressions as well:

- $\dfrac{3}{x} = \dfrac{3(x+3)}{x(x+3)}$
- $\dfrac{x}{x+3} = \dfrac{x \cdot x}{x \cdot (x+3)} = \dfrac{x^2}{x(x+3)}$

In general,

To add or subtract rational expressions, follow these steps:

1. If the denominators are the same, add or subtract the numerators, put the result over the common denominator, then reduce.
2. If the denominators are different, factor each denominator completely, then form a product using each factor the *most* number of times it appears as a factor in any one denominator. This is the LCD.
3. Write each rational expression as an equivalent rational expression having the LCD as a denominator by multiplying the numerator and denominator by the factor needed to change that denominator to the LCD, then proceed as in Step 1.

- $\underbrace{\dfrac{3}{x} + \dfrac{x}{x+3}}_{\text{LCD} = x(x+3)} = \dfrac{3(x+3)}{x(x+3)} + \dfrac{x^2}{x(x+3)} = \dfrac{3x+9}{x(x+3)} + \dfrac{x^2}{x(x+3)} = \dfrac{x^2+3x+9}{x(x+3)}$

- $\dfrac{4x^2+28x-10}{x^2-25} + \dfrac{x}{x+5} - \dfrac{4x}{x-5} = \underbrace{\dfrac{4x^2+28x-10}{(x+5)(x-5)} + \dfrac{x}{x+5} - \dfrac{4x}{x-5}}_{\text{LCD} = (x+5)(x-5)}$

$$= \frac{4x^2 + 28x - 10}{(x+5)(x-5)} + \frac{x(x-5)}{(x+5)(x-5)} - \frac{4x(x+5)}{(x-5)(x+5)}$$

$$= \frac{4x^2 + 28x - 10}{(x+5)(x-5)} + \frac{x^2 - 5x}{(x+5)(x-5)} - \frac{4x^2 + 20x}{(x-5)(x+5)}$$

Notice this sign is −, not +:

$$= \frac{4x^2 + 28x - 10 + x^2 - 5x - 4x^2 - 20x}{(x+5)(x-5)}$$

$$= \frac{x^2 + 3x - 10}{(x+5)(x-5)} = \frac{\cancel{(x+5)}(x-2)}{\cancel{(x+5)}(x-5)} = \frac{x-2}{x-5}$$

2.11 SIMPLIFYING COMPLEX FRACTIONS

A complex fraction is a fraction that has one or more fractions in the numerator, or in the denominator, or possibly in both. For example,

$$\frac{\frac{1}{2} + \frac{1}{3}}{\frac{1}{2} - \frac{1}{3}}, \frac{\frac{1}{x} + \frac{1}{y}}{\frac{1}{x} - \frac{1}{y}}, \frac{a - \frac{1}{a}}{a - 1}, \frac{1}{\frac{1}{xy}}, \text{ and } \frac{\frac{1}{x} + y}{y - \frac{1}{x}}$$

Two methods for simplifying complex fractions are explained below. *Method 1:* Since a fraction bar is a grouping symbol, perform the indicated operations in the numerator and denominator of the complex fraction to obtain a single fraction in each, then divide the simplified numerator by the simplified denominator.

- $$\frac{\frac{1}{2} + \frac{1}{3}}{\frac{1}{2} - \frac{1}{3}} = \frac{\frac{3}{6} + \frac{2}{6}}{\frac{3}{6} - \frac{2}{6}} = \frac{\frac{5}{6}}{\frac{1}{6}} = \frac{5}{6} \cdot \frac{1}{6} = \frac{5}{\cancel{6}} \cdot \frac{\cancel{6}}{1} = \frac{5}{1} = 5$$

- $$\frac{\frac{1}{x} + \frac{1}{y}}{\frac{1}{x} - \frac{1}{y}} = \frac{\frac{y}{xy} + \frac{x}{xy}}{\frac{y}{xy} - \frac{x}{xy}} = \frac{\frac{y+x}{xy}}{\frac{y-x}{xy}} = \frac{y+x}{xy} \div \frac{y-x}{xy} = \frac{y+x}{\cancel{xy}} \cdot \frac{\cancel{xy}}{y-x} = \frac{y+x}{y-x}$$

- $$\frac{a - \frac{1}{a}}{a-1} = \frac{\frac{a^2}{a} - \frac{1}{a}}{a-1} = \frac{\frac{a^2-1}{a}}{a-1} = \frac{a^2-1}{a} \div (a-1) = \frac{a^2-1}{a} \div \frac{a-1}{1}$$
  $$= \frac{(a+1)\cancel{(a-1)}}{a} \cdot \frac{1}{\cancel{a-1}} = \frac{a+1}{a}$$

- $$\frac{1}{\frac{1}{xy}} = 1 \div \frac{1}{xy} = 1 \cdot \frac{xy}{1} = xy$$

- $$\frac{\frac{1}{x} + y}{y - \frac{1}{x}} = \frac{\frac{1}{x} + \frac{xy}{x}}{\frac{xy}{x} - \frac{1}{x}} = \frac{\frac{1+xy}{x}}{\frac{xy-1}{x}} = \frac{1+xy}{x} \div \frac{xy-1}{x}$$
  $$= \frac{1+xy}{\cancel{x}} \cdot \frac{\cancel{x}}{xy-1} = \frac{1+xy}{xy-1}$$

Method 2: Multiply the numerator and denominator of the complex fraction by the LCD for *all* the fractions *within* the numerator and denominator, then simplify.

- $$\frac{\frac{1}{2}+\frac{1}{3}}{\frac{1}{2}-\frac{1}{3}}=\frac{\frac{6}{1}\left(\frac{1}{2}+\frac{1}{3}\right)}{\frac{6}{1}\left(\frac{1}{2}-\frac{1}{3}\right)}=\frac{\frac{{}^{3}\cancel{6}}{1}\cdot\frac{1}{\cancel{2}}+\frac{{}^{2}\cancel{6}}{1}\cdot\frac{1}{\cancel{3}}}{\frac{{}^{3}\cancel{6}}{1}\cdot\frac{1}{\cancel{2}}-\frac{{}^{2}\cancel{6}}{1}\cdot\frac{1}{\cancel{3}}}=\frac{3+2}{3-2}=\frac{5}{1}=5$$

- $$\frac{\frac{1}{x}+\frac{1}{y}}{\frac{1}{x}-\frac{1}{y}}=\frac{\frac{xy}{1}\left(\frac{1}{x}+\frac{1}{y}\right)}{\frac{xy}{1}\left(\frac{1}{x}-\frac{1}{y}\right)}=\frac{\frac{\cancel{x}y}{1}\cdot\frac{1}{\cancel{x}}+\frac{x\cancel{y}}{1}\cdot\frac{1}{\cancel{y}}}{\frac{\cancel{x}y}{1}\cdot\frac{1}{\cancel{x}}-\frac{x\cancel{y}}{1}\cdot\frac{1}{\cancel{y}}}=\frac{y+x}{y-x}$$

- $$\frac{a-\frac{1}{a}}{a-1}=\frac{\frac{a}{1}\left(a-\frac{1}{a}\right)}{a(a-1)}=\frac{\frac{a}{1}\cdot a-\frac{\cancel{a}}{1}\cdot\frac{1}{\cancel{a}}}{a(a-1)}=\frac{a^2-1}{a(a-1)}$$
$$=\frac{(a+1)\cancel{(a-1)}}{a\cancel{(a-1)}}=\frac{a+1}{a}$$

- $$\frac{1}{\frac{1}{xy}}=\frac{xy\cdot 1}{\frac{\cancel{xy}}{1}\cdot\frac{1}{\cancel{xy}}}=\frac{xy}{1}=xy$$

- $$\frac{\frac{1}{x}+y}{y-\frac{1}{x}}=\frac{\frac{x}{1}\left(\frac{1}{x}+y\right)}{\frac{x}{1}\left(y-\frac{1}{x}\right)}=\frac{\frac{\cancel{x}}{1}\cdot\frac{1}{\cancel{x}}+\frac{x}{1}\cdot y}{\frac{x}{1}\cdot y-\frac{\cancel{x}}{1}\cdot\frac{1}{\cancel{x}}}=\frac{1+xy}{xy-1}$$

CHAPTER 2 SUMMARY

Chapter 2 covers the terms and operations of algebra necessary for understanding the similarities and differences between algebra and arithmetic. The areas of discussion follow:

2.1 An **algebraic expression** is any symbol or combination of symbols that represents a number. A **term** is a constant, variable, or any product of constants or variables. In a term, the **coefficient** of a factor is the product of all the other factors in the term. The numerical factor of a term is called the **numerical coefficient.** A **monomial** consists of one term. **Like monomials** differ only in their numerical coefficients. A **polynomial** is a sum of monomials. A **binomial** consists of two unlike terms. A **trinomial** consists of three unlike terms. The **degree of a monomial** is the sum of the exponents of its variables. The **degree of a polynomial** is the highest degree of its monomial terms.

2.2 Like monomials are **combined** by adding or subtracting their numerical coefficients and using the result as the coefficient of the common variable factors. Remove parentheses preceded by a minus sign by changing the sign of every term within the parentheses.

2.3 To **multiply** polynomials, multiply each term in the first polynomial by each term in the second polynomial. The FOIL method is useful for obtaining the product of two binomials. Some special products are the following:

- $(x + y)^2 = (x + y)(x + y) = x^2 + 2xy + y^2$
- $(x - y)^2 = (x - y)(x - y) = x^2 - 2xy + y^2$
- $(x + y)(x - y) = x^2 - y^2$
- $(x + y)(x^2 - xy + y^2) = x^3 + y^3$
- $(x - y)(x^2 + xy + y^2) = x^3 - y^3$
- $(x + y)^3 = x^3 + 3x^2y + 3xy^2 + y^3$
- $(x - y)^3 = x^3 - 3x^2y + 3xy^2 - y^3$

2.4 To **simplify** a polynomial expression:
First, perform all operations within grouping symbols, starting with the innermost grouping symbol and working outward.
Next, perform all indicated multiplication, including exponentiation.
Last, combine like terms.

2.5 To **divide** a polynomial by a monomial, divide each term of the polynomial by the monomial.

2.6 **Long division** of polynomials is accomplished in a manner analogous to long division in arithmetic. The result is usually written as a mixed expression: quotient + $\frac{\text{remainder}}{\text{divisor}}$.

2.7 **Factoring** a polynomial completely means writing it as a product of prime polynomials. The following summarizes factoring polynomials:

1. Look for a greatest common monomial factor.
2. If a factor is a binomial, check for:
   difference of two squares: $x^2 - y^2 = (x + y)(x - y)$
   sum of two cubes: $x^3 + y^3 = (x + y)(x^2 - xy + y^2)$
   difference of two cubes: $x^3 - y^3 = (x - y)(x^2 + xy + y^2)$
3. If a factor is a trinomial, check for:
   general factorable quadratic: $x^2 + (a + b)x + ab = (x + a)(x + b)$
   $acx^2 + (ad + bc)x + bd = (ax + b)(cx + d)$
   perfect trinomial square: $a^2x^2 + 2abxy + b^2y^2 = (ax + by)^2$
   $a^2x^2 - 2abxy + b^2y^2 = (ax - by)^2$
4. If a factor has four terms, look for a grouping arrangement that will work for factoring by grouping.
5. Write the original polynomial as the product of all the factors obtained. Check to make sure all polynomial factors except monomial factors are prime.
6. Check by multiplying the factors to obtain the original polynomial.

2.8 A **rational expression** is an algebraic fraction in which both the numerator and denominator are polynomials. Exclude values for which the denominator is zero. The principles used in the study of arithmetic fractions are generalized in work with rational expressions.

2.9 To **multiply** rational expressions: 1. factor completely, 2. cancel common factors, then 3. multiply the remaining numerator factors to obtain the numerator of the product and the remaining denominator factors to obtain the denominator of the product. Canceling is permitted for factors only—do not cancel terms! To **divide** rational expressions, multiply the dividend by the reciprocal of the divisor.

2.10 To **add** or **subtract** rational expressions: 1. find a common denominator, 2. express each term as an equivalent rational expression having the common denominator, 3. add or subtract numerators, 4. write the result over the common denominator, and 5. simplify.

2.11 A **complex fraction** is a fraction that contains fractions in its numerator or denominator or both. A complex fraction can be simplified by treating it as a division problem or by multiplying its numerator and denominator by the least common denominator of the fractions they contain, and then simplifying as necessary.

PRACTICE PROBLEMS FOR CHAPTER 2

Mark each statement as True or False. Explain the error in each false statement.

1. In the expression $6x(2z + 5)$, $6(2z + 5)$ is the coefficient of x.
2. The degree of the polynomial $4x^6y^3 + 27x^4y^3 + 2x^3y^2 + 7x^2y$ is 6.
3. $2x^5y^2 + 3x^5y^2 = 5x^{10}y^4$
4. $-2t + 2s = -2(t - s)$
5. $(3x - 4y)^2 = 9x^2 - 16y^2$
6. $(3x - 4y)^2 = 9x^2 - 12xy + 16y^2$
7. $\dfrac{20x^6 - 6x^3}{2x^2} = 10x^4 - 6x^3$
8. $125x^3 - 8y^3 = (5x - 2y)(25x^2 + 10xy + 4y^2)$
9. $25x^2 + 10xy + 4y^2 = (5x + 2y)^2$
10. $(a + b)^3 = a^3 + b^3$
11. $\dfrac{5 + x}{5 + y} = \dfrac{x}{y}$
12. $\dfrac{5x}{5y} = \dfrac{x}{y}$
13. $\dfrac{a}{b} + \dfrac{x}{y} = \dfrac{a + x}{b + y}$
14. $\dfrac{a}{x} + \dfrac{x}{y} = \dfrac{a}{y}$

15. $(a^{-1} + b^{-1})^{-1} = a + b$
16. $x^2 + 16$ is prime over the set of real numbers.
17. $3x(x - 2) - 5(x - 2)$ is in factored form.
18. $a^3 + 64 = (a + 4)^3$
19. $-3x + 3z = -3(x + z)$
20. $\frac{a}{x} - \frac{a-1}{x} = \frac{1}{x}$

Simplify:

21. $2x^2 - x + 5 + 3x^2 + x - 5$
22. $15 - 10z + 3z^2 + 8z^3 - (25 - 12z^2 + 3z^3 + 4z^5)$
23. $(-2x^2y^4z^3)(5x^3y^4z^2)$
24. $-4x^2y(25 - 12x^2 + 3y^3 + z^3)$
25. $(3x + 4)(5x - 1)$
26. $(2x - 1)(3x^2 + 3x + 1)$
27. $(3x^2 - 2x + 1)(5x^3 + 3x + 4)$
28. $(2x + 5)^3$
29. $2x + 5x(3y - 2)$
30. $(x - 5)(2x - 5) - (3x - 1)(x + 1)$
31. $14x^2 - (3x - 5)^2 + 3$
32. $2x - [3x - (4x - 1)]$
33. $\frac{36a^8b^5c^4 + 12a^7b^4c^3 - 18a^3bc}{-6a^2bc}$

Factor completely or write "Not factorable":

34. $80x^2y - 45y$
35. $2x + 2t + xt + t^2$
36. $3x^2y^3z^4 - 6x^3y^2z^3 + 15x^2y^3z^5$
37. $2x^2 + 9x - 5$
38. $a^2 - 2ab + b^2$
39. $a^2 + ab + b^2$
40. $16x^4 - 81y^4$
41. $25x^2 - a^2 - 2ab - b^2$

Perform the indicated operation:

42. $(4x^3 - 3x + 1) \div (x - 2) =$
43. $\frac{5x^2 + 14x + 8}{x^2 + 4x + 4} \cdot \frac{25x^2 - 16}{25x^2 + 40x + 16} =$
44. $\frac{8x^3 - 27}{x^3(x + 1)} \div \frac{6 - 4x}{2x^3 + 2} =$

45. $\dfrac{2x^2 - 5x - 1}{x^2 - 1} + \dfrac{3}{x + 1} - \dfrac{x}{x - 1} =$

46. $\dfrac{\dfrac{2}{a} + \dfrac{y}{b}}{\dfrac{x}{a} - \dfrac{3}{b}} =$

SOLUTIONS TO PRACTICE PROBLEMS

1. True. Each factor is the coefficient of the product of the remaining factors.
2. False. The degree is 9, the sum of the exponents of x and y in the term $4x^6y^3$ $(6 + 3 = 9)$. The degree of a polynomial is the highest degree of any of its terms. In the polynomial $4x^6y^3 + 27x^4y^3 + 2x^4y^2 + 7x^2y$, the term $4x^6y^3$ is the term of highest degree. The term $27x^4y^3$ has degree 7, $2x^3y^2$ has degree 5, and $7x^2y$ has degree 3.
3. False. $2x^5y^2 + 3x^5y^2 = 5x^5y^2$; do not add the exponents when adding similar terms; add only the coefficients.
4. True. $-2t + 2s = -2(t - s)$; the negative sign within the parentheses is necessary since -2 was removed as a factor from the term $2s$.
5. False. $(3x - 4y)^2 \neq 9x^2 - 16y^2$; do not forget the middle term when squaring a binomial: $(3x - 4y)^2 = 9x^2 - 24xy + 16y^2$
6. False. $(3x - 4y)^2 \neq 9x^2 - 12xy + 16y^2$; when squaring a binomial, remember that the middle term is *twice* the product of the two terms: $(3x - 4y)^2 = 9x^2 - 24xy + 16y^2$
7. False. $\dfrac{20x^6 - 6x^3}{2x^2} \neq 10x^4 - 6x^3$; *every* term in the dividend polynomial must be divided by the monomial divisor. Therefore, $\dfrac{20x^6 - 6x^3}{2x^2} = 10x^4 - 3x$.
8. True. $125x^3 - 8y^3 = (5x - 2y)(25x^2 + 10xy + 4y^2)$; this is the difference of two cubes.
9. False. $25x^2 + 10xy + 4y^2$ is not factorable over the integers.
10. False. $(a + b)^3 \neq a^3 + b^3$; exponents do not distribute over addition: $(a + b)^3 = (a + b)(a + b)(a + b) = (a + b)(a^2 + 2ab + b^2) = a^3 + 3a^2b + 3ab^2 + b^3$
11. False. $\dfrac{5 + x}{5 + y} \neq \dfrac{x}{y}$; you cannot cancel the 5's. The 5 is a *term* in both the numerator and denominator, but is not a *factor* of both.
12. True. $\dfrac{5x}{5y} = \dfrac{\cancel{5}x}{\cancel{5}y} = \dfrac{x}{y}$; you can cancel the 5's because 5 is a factor of both the numerator and denominator.

13. False. $\frac{a}{b} + \frac{x}{y} \neq \frac{a + x}{b + y}$; you do not add fractions by adding the numerators and adding the denominators. You must find a common denominator when adding or subtracting fractions: $\frac{a}{b} + \frac{x}{y} = \frac{ay}{by} + \frac{bx}{by} = \frac{ay + bx}{by}$

14. False. $\frac{a}{x} + \frac{x}{y} \neq \frac{a}{y}$; do not try to cancel the x's like you would do if the problem were multiplication: $\frac{a}{x} + \frac{x}{y} = \frac{ay}{xy} + \frac{x^2}{xy} = \frac{ay + x^2}{xy}$

15. False. $(a^{-1} + b^{-1})^{-1} \neq a + b$; exponents do not distribute over addition. Rewrite the problem using positive exponents, then simplify the resulting complex fraction:

$$(a^{-1} + b^{-1})^{-1} = \left(\frac{1}{a} + \frac{1}{b}\right)^{-1} = \frac{1}{\left(\frac{1}{a} + \frac{1}{b}\right)} = \frac{ab \cdot 1}{\frac{ab}{1}\left(\frac{1}{a} + \frac{1}{b}\right)}$$

$$= \frac{ab}{\frac{\cancel{a}b}{1} \cdot \frac{1}{\cancel{a}} + \frac{a\cancel{b}}{1} \cdot \frac{1}{\cancel{b}}} = \frac{ab}{b + a}$$

16. True. $x^2 + 16$ is the sum of two squares. It is not factorable over the set of real numbers.
17. False. $3x(x - 2) - 5(x - 2)$ is a sum, not a product.
18. False. $a^3 + 64 \neq (a + 4)^3$; $(a + 4)^3 = (a + 4)(a + 4)(a + 4) = a^3 + 12a^2 + 48a + 64$
19. False. $-3x + 3z \neq -3(x + z)$; when you factor out a negative term, change the sign of every term. Thus, $-3x + 3z = -3(x - z)$.
20. True. $\frac{a}{x} - \frac{a - 1}{x} = \frac{a - a + 1}{x} = \frac{1}{x}$

Simplify:

21. $2x^2 - x + 5 + 3x^2 + x - 5 = 5x^2$
22. $15 - 10z + 3z^2 + 8z^3 - (25 - 12z^2 + 3z^3 + 4z^5)$
$= 15 - 10z + 3z^2 + 8z^3 - 25 + 12z^2 - 3z^3 - 4z^5$
$= -4z^5 + 5z^3 + 15z^2 - 10z - 10)$
23. $(-2x^2y^4z^3)(5x^3y^4z^2) = -10x^5y^8z^5$
24. $-4x^2y(25 - 12x^2 + 3y^3 + z^3)$
$= -100x^2y + 48x^4y - 12x^2y^4 - 4x^2yz^3)$
25. $(3x + 4)(5x - 1) = 15x^2 + 17x - 4$
26. $(2x - 1)(3x^2 + 3x + 1) = 6x^3 + 6x^2 + 2x - 3x^2 - 3x - 1$
$= 6x^3 + 3x^2 - x - 1$

27. $(3x^2 - 2x + 1)(5x^3 + 3x + 4) = 15x^5 + 9x^3 + 12x^2 - 10x^4 - 6x^2 - 8x + 5x^3 + 3x + 4 = 15x^5 - 10x^4 + 14x^3 + 6x^2 - 5x + 4$
28. $(2x+5)^3 = (2x+5)(4x^2+20x+25) = 8x^3 + 40x^2 + 50x + 20x^2 + 100x + 125$
$= 8x^3 + 60x^2 + 150x + 125$
29. $2x + 5x(3x - 2) = 2x + 15xy - 10x = -8x + 15xy$; don't add $2x + 5x$ first—that is a mistake because it does not follow the order of operations.
30. $(x - 5)(2x - 5) - (3x - 1)(x + 1) = 2x^2 - 15x + 25 - (3x^2 + 2x - 1)$
$= 2x^2 - 15x + 25 - 3x^2 - 2x + 1$
$= -x^2 - 17x + 26$
31. $14x^2 - (3x - 5)^2 + 3 = 14x^2 - (9x^2 - 30x + 25) + 3$
$= 14x^2 - 9x^2 + 30x - 25 + 3$
$= 5x^2 + 30x - 22$
32. $2x - [3x - (4x - 1)] = 2x - [3x - 4x + 1]$
$= 2x - [-x + 1] = 2x + x - 1 = 3x - 1$
33. $\dfrac{36a^8b^5c^4 + 12a^7b^4c^3 - 18a^3bc}{-6a^2bc} = -6a^6b^4c^3 - 2a^5b^3c^2 + 3a$

Factor completely or write "Not factorable":

34. $80x^2y - 45y = 5y(16x^2 - 9) = 5y(4x + 3)(4x - 3)$
35. $2x + 2t + xt + t^2 = 2(x + t) + t(x + t) = (x + t)(2 + t)$
36. $3x^2y^3z^4 - 6x^3y^2z^3 + 15x^2y^3z^5 = 3x^2y^2z^3(yz - 2x + 5yz^2)$
37. $2x^2 + 9x - 5 = (x + 5)(2x - 1)$
38. $a^2 - 2ab + b^2 = (a - b)^2$
39. $a^2 + ab + b^2$ is not factorable over the integers.
40. $16x^4 - 81y^4 = (4x^2 + 9y^2)(4x^2 - 9y^2)$
$= (4x^2 + 9y^2)(2x + 3y)(2x - 3y)$
41. $25x^2 - a^2 - 2ab - b^2 = 25x^2 - (a^2 + 2ab + b^2) = (5x)^2 - (a + b)^2$
$= [5x + (a + b)][5x - (a + b)]$
$= (5x + a + b)(5x - a - b)$
42. $(4x^3 - 3x + 1) \div (x - 2) = 4x^2 + 8x + 13 + \dfrac{27}{4x^3 - 3x + 1}$;

This is shown by the following:

$$\begin{array}{r l}
 & 4x^2 + 8x + 13 \\
x - 2 & \overline{)4x^3 + 0x^2 - 3x + 1} \\
 & \underline{4x^3 - 8x^2} \\
 & 8x^2 - 3x \\
 & \underline{8x^2 - 16x} \\
 & 13x + 1 \\
 & \underline{13x - 26} \\
\text{remainder} \rightarrow & 27
\end{array}$$

43. $$\frac{5x^2+14x+8}{x^2+4x+4}\cdot\frac{25x^2-16}{25x^2+40x+16}=\frac{\cancel{(5x+4)}\cancel{(x+2)}}{(x+2)\cancel{(x+2)}}\cdot\frac{\cancel{(5x+4)}(5x-4)}{\cancel{(5x+4)}\cancel{(5x+4)}}$$
$$=\frac{5x-4}{x+2}$$

44. $$\frac{8x^3-27}{x^3(x+1)}\div\frac{6-4x}{2x^3+2}=\frac{8x^3-27}{x^3(x+1)}\cdot\frac{2x^3+2}{6-4x}$$
$$=\frac{(2x-3)(4x^2+6x+9)}{x^3(x+1)}\cdot\frac{\cancel{2}(x^3+1)}{\cancel{2}(3-2x)}$$
$$=\frac{\cancel{(2x-3)}(4x^2+6x+9)}{x^3\cancel{(x+1)}}\cdot\frac{\cancel{(x+1)}(x^2-x+1)}{-\cancel{(2x-3)}}$$
$$=-\frac{(4x^2+6x+9)(x^2-x+1)}{x^3}$$

45. $$\frac{2x^2-5x-1}{x^2-1}+\frac{3}{x+1}-\frac{x}{x-1}$$
$$=\frac{2x^2-5x-1}{(x+1)(x-1)}+\frac{3(x-1)}{(x+1)(x-1)}-\frac{x(x+1)}{(x-1)(x+1)}$$
$$=\frac{2x^2-5x-1+3x-3-x^2-x}{(x+1)(x-1)}$$
$$=\frac{x^2-3x-4}{(x+1)(x-1)}$$
$$=\frac{(x-4)\cancel{(x+1)}}{\cancel{(x+1)}(x-1)}$$
$$=\frac{x-4}{x-1}$$

46. $$\frac{\frac{2}{a}+\frac{y}{b}}{\frac{x}{a}-\frac{3}{b}}=\frac{\frac{ab}{1}\left(\frac{2}{a}+\frac{y}{b}\right)}{\frac{ab}{1}\left(\frac{x}{a}-\frac{3}{b}\right)}=\frac{\frac{\cancel{a}b}{1}\cdot\frac{2}{\cancel{a}}+\frac{a\cancel{b}}{1}\cdot\frac{y}{\cancel{b}}}{\frac{\cancel{a}b}{1}\cdot\frac{x}{\cancel{a}}-\frac{a\cancel{b}}{1}\cdot\frac{3}{\cancel{b}}}=\frac{2b+ay}{bx-3a}$$

3
SOLVING ONE-VARIABLE EQUATIONS AND INEQUALITIES

3.1 SOLVING ONE-VARIABLE LINEAR EQUATIONS

An **equation** is a statement that two mathematical expressions are equal. A variable or variables may hold the place for numbers in an equation. For example, $x + 2 = 10$ is an equation in which $x + 2$ is the left side of the equation and 10 is the right side of the equation. An equation may be true, or it may be false. For example, the equation $8 + 2 = 10$ is true, but the equation $-12 + 2 = 10$ is false.

A **one-variable linear equation** is an equation that can be written in the form $ax + b = 0$, where $a \neq 0$. Some examples follow:

- $5x + 1 = 0$
- $7x - 25 = 3(3x + 5)$
- $-2(1 - z) + 4z = 3z - 14$
- $\frac{2y}{3} - 45 = \frac{25}{2} - y$
- $0.04x = 140 + 0.06(2500 - x)$

A one-variable linear equation has only *one* variable, which (after terms have been simplified) occurs to the first power *only* (no squares, cubes, roots, radicals, and so on). The following are *not* one-variable linear equations:

- $2x^2 - 7x + 5 = 0$ (variable occurs to second power)
- $\frac{200}{t} + 25t = 60$ (when simplified, the variable will be raised to the second power)
- $0.05x = 200 - 0.08y$ (has two variables)
- $\sqrt{z} = 25$ (contains a radical)

A **solution,** or **root,** of an equation is a number that when substituted for the variable in the equation makes the equation true. To determine whether a number is a solution to an equation, replace the variable with the number and perform the operations indicated on each side of the equation. If this results in a true equation, the number is a solution. This process is called **checking** a solution. The set consisting of all the solutions of an equation is called its **solution set.** To **solve an equation** means to find its solution set. If the solution set consists of all real numbers, the equation is called an **identity.** If the solution set is empty, the equation has no solution. Equations that have the same solution set are called **equivalent.**

The process of solving a one-variable linear equation involves using a series of steps to produce an equivalent equation of the form $x =$ solution. Thus, an equation is solved when the variable has a coefficient of 1 and is by itself on one and only one side of the equation, the goal being to isolate the variable on one side so that the solution will be obvious. Two main tools are used to accomplish this:

Operations that result in equivalent equations:

1. Addition or subtraction of the same quantity on both sides of the equation
2. Multiplication or division by the same *nonzero* quantity on both sides of the equation

It will help to think of an equation as a balance scale, with the equal sign being in the middle. Thus, except for some special situations, if you *do the same thing to both sides,* it will still be balanced. What has been done to the variable determines the operation you should choose to perform on both sides of the equation. The relationship between the variable and a connected number is undone by using the inverse operation. The following procedure will aid in the decision making:

To solve a one-variable linear equation:

1. Remove grouping symbols, if any, using the distributive property; then simplify.
2. Eliminate fractions, if any, by multiplying both sides of the equation by the LCD of all the fractions in the equation; then simplify.
3. Undo indicated addition or subtraction to isolate the variable on one side; that is, get all terms containing the variable on one side and all other terms on the other side; then simplify.
4. If the side containing the variable consists of unlike terms, write that side as two factors, one of which is the variable.
5. Divide both sides by the coefficient of the variable; then simplify.
6. Check the solution in the original equation.

A helpful message: As you become skilled in solving equations, you will likely modify this procedure based on the particular equation you are attempting to solve. Sometimes, it is more efficient to change the order of the steps (e.g., eliminating fractions before removing parentheses).

- Solve $5x + 1 = 0$.

$$5x + 1 = 0$$

Isolate the variable by subtracting 1 from both sides:

$$5x + 1 - 1 = 0 - 1$$

Simplify:

$$5x = -1$$

Divide both sides by the coefficient 5:

$$\frac{5x}{5} = \frac{-1}{5}$$

Simplify:

$$x = -\frac{1}{5}$$

Check:

$$5\left(-\frac{1}{5}\right) + 1 \stackrel{?}{=} 0$$

$$-1 + 1 \stackrel{?}{=} 0$$

$$0 \stackrel{\checkmark}{=} 0$$

- Solve $7x - 25 = 3(3x + 5)$.

$$7x - 25 = 3(3x + 5)$$

Remove parentheses:

$$7x - 25 = 9x + 15$$

Isolate the variable by adding 25 and subtracting $9x$ on both sides:

$$7x - 25 + 25 - 9x = 9x + 15 + 25 - 9x$$

Simplify:

$$-2x = 40$$

Divide both sides by the coefficient -2:

$$\frac{-2x}{-2} = \frac{40}{-2}$$

Simplify:

$$x = -20$$

Check:

$$7(-20) - 25 \stackrel{?}{=} 3(3(-20) + 5)$$

$$-140 - 25 \stackrel{?}{=} 3(-55)$$

$$-165 \stackrel{\checkmark}{=} -165$$

Many students choose not to show all the details of the above procedure. Particularly, the algebraic procedure of eliminating a term from one side of the equation by adding or subtracting to obtain a sum or difference of zero can be accomplished mentally on the side in which the zero occurs, with the operation being indicated on the other side of the equation only. Also, dividing by the coefficient of the variable to obtain 1 as the final coefficient can be done mentally.

- Solve $-2(1 - z) + 4z = 3z - 14$.

$$-2(1 - z) + 4z = 3z - 14$$

Remove parentheses:

$$-2 + 2z + 4z = 3z - 14$$

Simplify:

$$-2 + 6z = 3z - 14$$

Isolate the variable by adding 2 and subtracting $3z$ on both sides:

$$\overbrace{-2 + 2}^{\text{Do this mentally.}} + \overbrace{6z - 3z}^{\text{Show this.}} = \overbrace{3z - 3z}^{\text{Do this mentally.}} - \overbrace{14 + 2}^{\text{Show this.}}$$

$$6z - 3z = -14 + 2$$

Simplify:

$$3z = -12$$

Divide both sides by the coefficient 3:

Do this mentally. $\left\{ \frac{3z}{3} = \frac{-12}{3} \right.$

Simplify:

$$z = -4$$

The streamlined version of the above problem looks like this:

$$-2(1 - z) + 4z = 3z - 14$$

Remove parentheses:

$$-2 + 2z + 4z = 3z - 14$$

Simplify:

$$-2 + 6z = 3z - 14$$

Isolate the variable by subtracting $3z$ and adding 2 on both sides:

$$6z - 3z = -14 + 2$$

$$3z = -12$$

Divide both sides by the coefficient 3:

$$z = -4$$

Check:

$$-2(1 - (-4)) + 4(-4) \stackrel{?}{=} 3(-4) - 14$$

$$-2(5) + 4(-4) \stackrel{?}{=} -12 - 14$$

$$-10 - 16 \stackrel{?}{=} -12 - 14$$

$$-26 \stackrel{\surd}{=} -26$$

- Solve $\frac{2y}{3} - 45 = \frac{25}{2} - y$.

$$\frac{2y}{3} - 45 = \frac{25}{2} - y$$

Eliminate fractions by multiplying *each* term on *both* sides by the LCD, $6 = \frac{6}{1}$:

$$\frac{{}^{2}\not{6}}{1}\left(\frac{2y}{\not{3}_{1}}\right) - 6(45) = \frac{{}^{3}\not{6}}{1}\left(\frac{25}{\not{2}_{1}}\right) - 6y$$

Simplify:

$$4y - 270 = 75 - 6y$$

Isolate the variable by adding $6y$ and 270 to both sides:

$$4y + 6y = 75 + 270$$

$$10y = 345$$

Divide both sides by the coefficient 10:

$$y = 34.5$$

Check:

$$\frac{2(34.5)}{3} - 45 \stackrel{?}{=} \frac{25}{2} - 34.5$$

$$23 - 45 \stackrel{?}{=} 12.5 - 34.5$$

$$-22 \stackrel{\surd}{=} -22.0$$

- Solve $0.04x = 140 + 0.06(2500 - x)$.

$$0.04x = 140 + 0.06(2500 - x)$$

Remove parentheses:

$$0.04x = 140 + 150 - 0.06x$$

Simplify:

$$0.04x = 290 - 0.06x$$

Isolate the variable by adding $0.06x$ to both sides:

$$0.04x + 0.06x = 290$$

$$0.10x = 290$$

Divide both sides by the coefficient 0.10:

$$x = 2900$$

Check:

$$0.04(2900) \stackrel{?}{=} 140 + 0.06(2500 - 2900)$$

$$116 \stackrel{?}{=} 140 + 0.06(-400)$$

$$116 \stackrel{?}{=} 140 - 24$$

$$116 \stackrel{\surd}{=} 116$$

3.2 SOLVING FORMULAS

An equation that expresses the relationship between two or more variables is called a **formula.** The procedure for solving one-variable linear equations can be used to solve a formula for a specific variable when the value(s) of the other variable(s) is (are) known. Simply substitute the given value(s) in the formula and solve the resulting one-variable equation. For example, the formula $C = \frac{5}{9}(F - 32)$ expresses the relationship between Celsius and Fahrenheit temperatures. To find the Fahrenheit temperature when the Celsius temperature is 30°:

Substitute 30 for C in the formula:

$$30 = \frac{5}{9}(F - 32)$$

Remove parentheses:

$$30 = \frac{5}{9}F - \frac{160}{9}$$

Eliminate fractions by multiplying each term on *both* sides by the LCD, $9 = \frac{9}{1}$.

$$9(30) = \frac{\not{9}}{1}\left(\frac{5}{\not{9}}F\right) - \frac{\not{9}}{1}\left(\frac{160}{\not{9}}\right)$$

$$270 = 5F - 160$$

Isolate the variable by adding 160 to both sides:

$$270 + 160 = 5F$$

$$430 = 5F$$

Divide both sides by the coefficient 5:

$$86 = F$$

Check:

$$30 \stackrel{?}{=} \frac{5}{9}(86 - 32)$$

$$30 \stackrel{?}{=} \frac{5}{9}(54)$$

$$30 \stackrel{?}{=} \frac{5}{{}_1\not{9}}(\not{54}^{6})$$

$$30 \stackrel{\checkmark}{=} 30$$

Thus, the Fahrenheit temperature is 86° when the Celsius temperature is 30°.

The procedure also can be used to solve a formula or **literal equation** (an equation with no numbers, only letters) for a specific variable in terms of the other variable(s). In general, isolate the specific variable and treat all other variable(s) as constants. This is called **changing the subject** of the formula or literal equation.

- To change the subject of the formula $C = \frac{5}{9}(F - 32)$ to F, solve for F:

$$C = \frac{5}{9}(F - 32)$$

Remove parentheses:

$$C = \frac{5}{9}F - \frac{160}{9}$$

Multiply each term on *both* sides by the LCD, $9 = \frac{9}{1}$:

$$9C = \frac{\not{9}}{1}\left(\frac{5}{\not{9}}F\right) - \frac{\not{9}}{1}\left(\frac{160}{\not{9}}\right)$$

$$9C = 5F - 160$$

Isolate the variable by adding 160 to both sides:

$$9C + 160 = 5F$$

Divide both sides by the coefficient 5:

$$\frac{9}{5}C + 32 = F$$

Applying the reflective property for equality yields:

$$F = \frac{9}{5}C + 32$$

- To change the subject of the formula $A = P + Prt$ to r, solve for r:

$$A = P + Prt$$

Isolate the variable by subtracting P from both sides:

$$A - P = Prt$$

Divide both sides by the coefficient Pt:

$$\frac{A - P}{Pt} = r$$

Applying the reflective property for equality yields:

$$r = \frac{A - P}{Pt}$$

- To change the subject of the formula $I = \dfrac{nE}{R + nr}$ to n, solve for n:

$$I = \frac{nE}{R + nr}$$

Multiply each term on *both* sides by the LCD, $(R + nr) = \dfrac{(R + nr)}{1}$:

$$I(R + nr) = \frac{(R + nr)}{1} \frac{nE}{(R + nr)}$$

$$IR + Inr = nE$$

Isolate the variable by subtracting nE and IR from both sides:

$$Inr - nE = -IR$$

Factor n from the left side to determine its coefficient:

$$n(Ir - E) = -IR$$

Divide both sides by the coefficient $(Ir - E)$:

$$n = \frac{-IR}{Ir - E}$$

3.3 SOLVING A TWO-VARIABLE EQUATION FOR ONE OF THE VARIABLES

The procedure for solving one-variable linear equations can be used to transform a two-variable equation such as $y = \frac{5x+6}{3}$ (where y is in terms of x) into an equivalent equation where x is in terms of y. For example:

To solve $y = \frac{5x+6}{3}$ for x:

$$y = \frac{5x + 6}{3}$$

Eliminate fractions by multiplying *each* term on *both* sides by the LCD, $3 = \frac{3}{1}$:

$$3y = \frac{\not{3}}{1}\left(\frac{5x + 6}{\not{3}}\right)$$

Simplify:

$$3y = 5x + 6$$

Isolate the variable x by subtracting 6 from both sides:

$$3y - 6 = 5x$$

Divide both sides by the coefficient 5:

$$\frac{3}{5}y - \frac{6}{5} = x$$

Another common use for the procedure for solving one-variable linear equations is to transform equations such as $3x - 2y = 6$ or $5(x - y) = -2(3y - 4x)$ into the form $y = mx + b$, where m and b are constants. You will encounter this form of a linear equation in the section on graphing in Chapter 5. For example:

- Express $3x - 2y = 6$ in the form $y = mx + b$:

$$3x - 2y = 6$$

Isolate the variable y by subtracting $3x$ from both sides:

$$-2y = -3x + 6$$

Divide both sides by the coefficient -2:

$$y = \frac{3}{2}x - 3$$

Thus, $3x - 2y = 6$ can be expressed as $y = \frac{3}{2}x - 3$, which is in the form $y = mx + b$, with $m = \frac{3}{2}$ and $b = -3$.

- Express $5(x - y) = -2(3y - 4x)$ in the form $y = mx + b$:

$$5(x - y) = -2(3y - 4x)$$

Remove parentheses:

$$5x - 5y = -6y + 8x$$

Isolate the variable y by adding $6y$ and subtracting $5x$ from both sides:

$$-5y + 6y = 8x - 5x$$

$$y = 3x$$

Thus, $5(x - y) = -2(3y - 4x)$ can be expressed as $y = 3x$, which is in the form $y = mx + b$, with $m = 3$ and $b = 0$.

3.4 SOLVING ONE-VARIABLE ABSOLUTE VALUE EQUATIONS

We know from the properties of absolute value given in Chapter 1, that

$$|x| = 0 \Leftrightarrow x = 0$$

More generally,

$$|ax + b| = 0 \Leftrightarrow ax + b = 0$$

Also, we have from Chapter 1:

If c is any positive number:

$$|x| = c \Leftrightarrow \text{either } x = c \text{ or } x = -c$$

And, more generally,

If c is any positive number:

$$|ax + b| = c \Leftrightarrow \text{either } ax + b = c \text{ or } ax + b = -c$$

We will use these properties and the procedure for solving one-variable linear equations to solve equations involving absolute value symbols. The solution set for the equations will be given in set notation. For example:

- If $|x| = 0$, then $x = 0$. The solution set is $\{0\}$.
- If $|x| = 9$, then $x = 9$ or $x = -9$. The solution set is $\{-9, 9\}$.
- To solve $|2x + 1| = 0$:

Apply the appropriate property:

$$|2x + 1| = 0 \Leftrightarrow 2x + 1 = 0$$

Solve the resulting equation:

$$2x + 1 = 0$$
$$2x = -1$$
$$x = -\frac{1}{2}$$

Check:

$$\left|2\left(-\frac{1}{2}\right) + 1\right| \stackrel{?}{=} 0$$
$$|-1 + 1| \stackrel{?}{=} 0$$
$$|0| \stackrel{?}{=} 0$$
$$0 \stackrel{\checkmark}{=} 0$$

The solution set for $|2x + 1| = 0$ is $\{-\frac{1}{2}\}$.

- To solve $|2x + 1| = 9$:

Apply the appropriate property:

$$|2x + 1| = 9 \Leftrightarrow 2x + 1 = 9 \quad \text{or} \quad 2x + 1 = -9$$

Solve each of the resulting equations:

$$2x + 1 = 9 \quad \text{or} \quad 2x + 1 = -9$$
$$2x = 9 - 1 \quad \text{or} \quad 2x = -9 - 1$$
$$2x = 8 \quad \text{or} \quad 2x = -10$$
$$x = 4 \quad \text{or} \quad x = -5$$

Check:

$$|2(4) + 1| \stackrel{?}{=} 9 \quad \text{or} \quad |2(-5) + 1| \stackrel{?}{=} 9$$
$$|8 + 1| \stackrel{?}{=} 9 \quad \text{or} \quad |-10 + 1| \stackrel{?}{=} 9$$
$$|9| \stackrel{?}{=} 9 \quad \text{or} \quad |-9| \stackrel{?}{=} 9$$
$$9 \stackrel{\checkmark}{=} 9 \quad \text{or} \quad 9 \stackrel{\checkmark}{=} 9$$

- The solution set for $|2x + 1| = 9$ is $\{-5, 4\}$.

3.5 SOLVING ONE-VARIABLE LINEAR INEQUALITIES

An **inequality** is a mathematical statement that two expressions are not equal. We use the symbol $>$ (greater than), $<$ (less than), $\geq$ (greater than or equal to), and $\leq$ (less than or equal to) to write one-variable linear inequalities. Like equations, inequalities may contain variables. For example, the inequality

$x > -2$ is true if x is any number to the right of -2 on the number line and false if x is -2 or any number to the left of -2 on the number line. This solution set can be illustrated on a number line as in Figure 3.1. This figure is called the **graph** of the inequality.

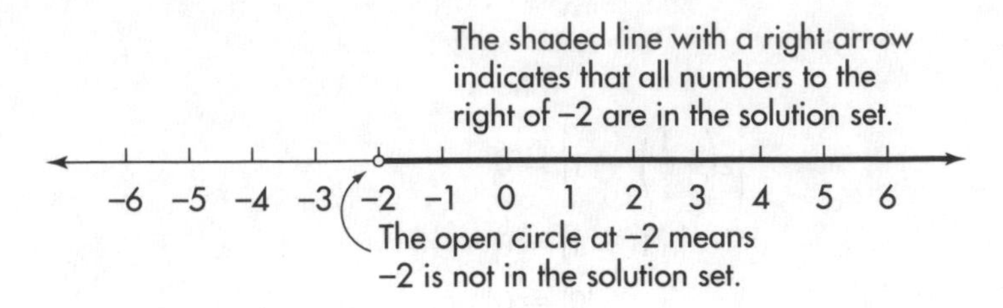

FIGURE 3.1

Inequalities, like equations, have two sides. It is usually easier to graph a one-variable linear inequality if the variable has the coefficient 1 and is on the *left* side of the inequality symbol. If the variable is a single term on the right, as in $-2 < x$, interchange the two sides and change the direction of the inequality symbol; that is, rewrite $-2 < x$ as $x > -2$ before graphing.

> When solving inequalities, treat them just like equations *except* for one difference: If you multiply or divide both sides of the inequality by a *negative* number, *reverse* the direction of the inequality.

Reversing the direction of the inequality applies *only* when you are *multiplying* or *dividing* both sides by a *negative* number. To see why this is so, look at a numerical example:

$$-2 < 6 \text{ is a true statement.}$$

If you multiply both sides by -5 and *do not reverse the direction of the inequality,* you obtain a false statement:

$$-2 < 6$$
$$-5(-2) < -5(6)$$
$$10 < -30$$

You must reverse the direction of the inequality to obtain a true statement:

$$-2 < 6$$
$$-5(-2) > -5(6)$$
$$10 > -30$$

On the other hand, when multiplying or dividing by a positive number, do *not* reverse the inequality symbol. For instance, if you multiply both sides of $-2 < 6$ by 5 without reversing the direction of the inequality, you obtain a true statement:

$$-2 < 6$$
$$5(-2) < 5(6)$$
$$-10 < 30$$

Also, if the operation is addition or subtraction, do *not* reverse the inequality:

$$-2 < 6$$
$$-2 - 5 < 6 - 5$$
$$-7 < 1$$

- Solve $2x + 17 > 9$.

$$2x + 17 > 9$$

Subtract 17 from both sides:

$$2x > 9 - 17$$
$$2x > -8$$

Divide both sides by 2 (don't reverse the inequality):

$$x > -4$$

The solution set for $2x + 17 > 9$ is the open, unbounded interval $(-4, \infty)$ illustrated in Figure 3.2:

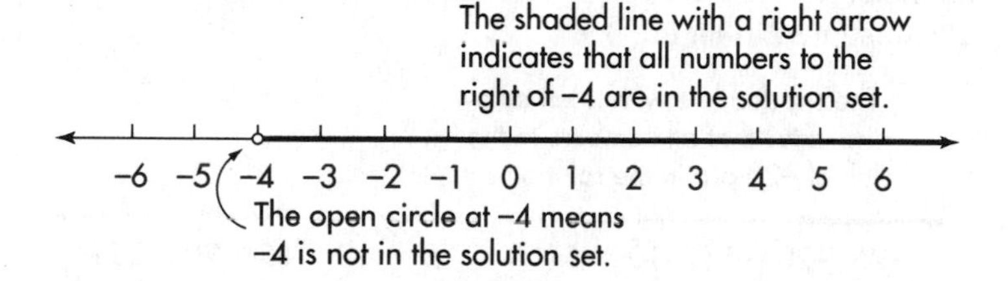

FIGURE 3.2

- Solve $-3x + 22 \geq 4$.

$$-3x + 22 \geq 4$$

Subtract 22 from both sides:

$$-3x \geq 4 - 22$$
$$-3x \geq -18$$

Divide both sides by -3 (reverse the inequality because -3 is negative):

$$x \leq 6$$

The solution set for $-3x + 22 \geq 4$ is the half-open, unbounded interval $(-\infty, 6]$ illustrated in Figure 3.3.

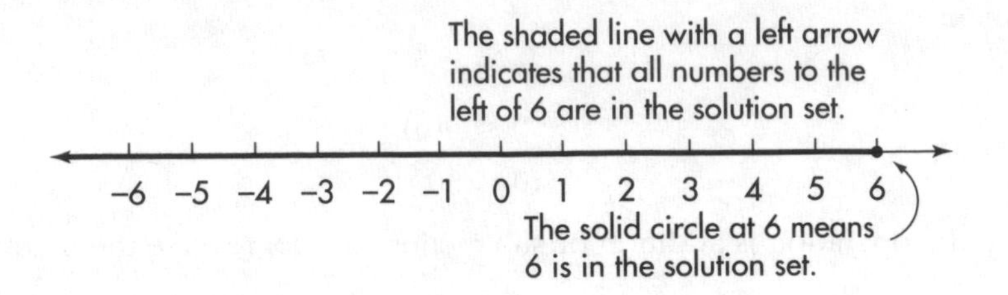

FIGURE 3.3

- Solve $7x - 25 \geq 3(3x + 5)$.

$$7x - 25 \geq 3(3x + 5)$$

Remove parentheses:

$$7x - 25 \geq 9x + 15$$

Subtract $9x$ from and add 25 to both sides:

$$7x - 9x \geq 15 + 25$$

$$-2x \geq 40$$

Divide both sides by -2 (reverse the inequality because -2 is negative):

$$x \geq -20$$

The solution set for $7x - 25 \geq 3(3x + 5)$ is the half-open, unbounded interval $(-\infty, -20]$ illustrated in Figure 3.4.

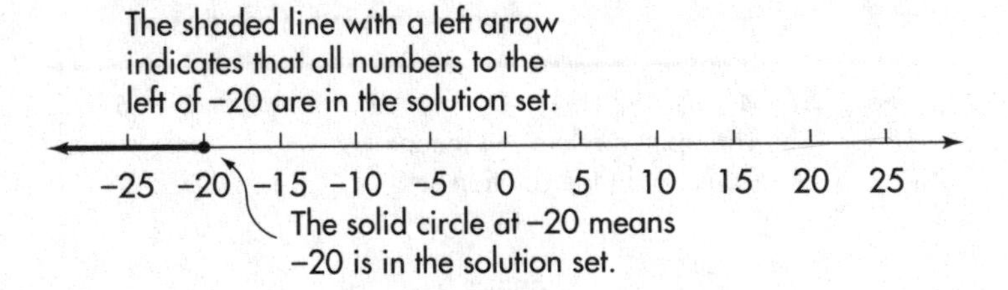

FIGURE 3.4

Sometimes two statements of inequality apply to a variable expression simultaneously and, thus, can be combined into a **double inequality.** For example, the two inequality statements

$$-12 < 5x + 3 \text{ and } 5x + 3 \leq 33$$

can be combined into the double inequality

$$-12 < 5x + 3 \leq 33$$

A double inequality can be solved by doing the same operations to the two combined statements simultaneously:

$$-12 < 5x + 3 \leq 33$$

Isolate the variable; subtract 3 from each of the three quantities:

$$-12 - 3 < \quad 5x \quad \leq 33 - 3$$

Simplify:

$$-15 < \quad 5x \quad \leq 30$$

Divide each of the three quantities by 5:

$$-3 < \quad x \quad \leq 6$$

Thus, the solution set for $-12 < 5x + 3 \leq 33$ is the half-open, bounded interval $(-3, 6]$ illustrated in Figure 3.5.

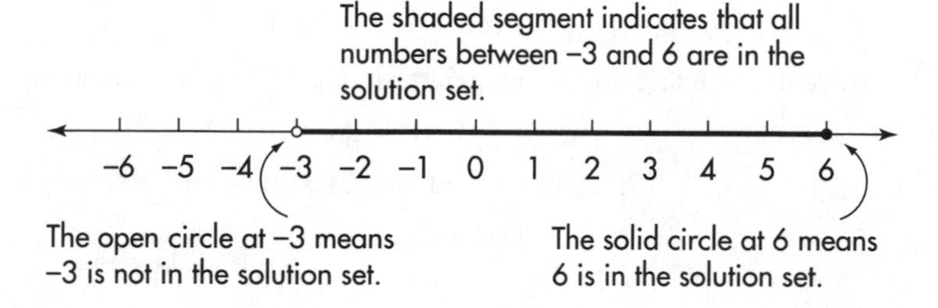

FIGURE 3.5

A helpful message: Be careful if you want to combine inequalities. Make sure *all* the resulting inequality statements indicated in the combination make sense. For instance,

$2 < x > 7$ does not make sense because $2 < 7$, *but* $2 \not> 7$.

3.6 SOLVING ONE-VARIABLE ABSOLUTE VALUE INEQUALITIES

The following properties for absolute value were given in Chapter 1:

> If c is any positive number:
>
> 1. $|x| < c \Leftrightarrow -c < x < c$ (conjunction)
> 2. $|x| > c \Leftrightarrow x < -c \quad \text{or} \quad x > c$ (disjunction)

From these we have, more generally:

> If c is any positive number:
>
> 1. $|ax + b| < c \Leftrightarrow -c < ax + b < c$ (conjunction)
> 2. $|ax + b| > c \Leftrightarrow ax + b < -c$ or $ax + b > c$ (disjunction)

We will use these properties and the procedure for solving one-variable linear equations to solve inequalities involving absolute value symbols.

- To solve $|2x + 1| < 9$:

Apply the appropriate property:

$$|2x + 1| < 9 \Leftrightarrow -9 < 2x + 1 < 9$$

Solve the resulting double inequality; subtract 1 from each quantity:

$$-9 - 1 < 2x < 9 - 1$$

$$-10 < 2x < 8$$

Divide by 2:

$$-5 < x < 4$$

Result is a conjunction.

The solution set for $|2x + 1| < 9$ is the open interval $(-5, 4)$ illustrated in Figure 3.6.

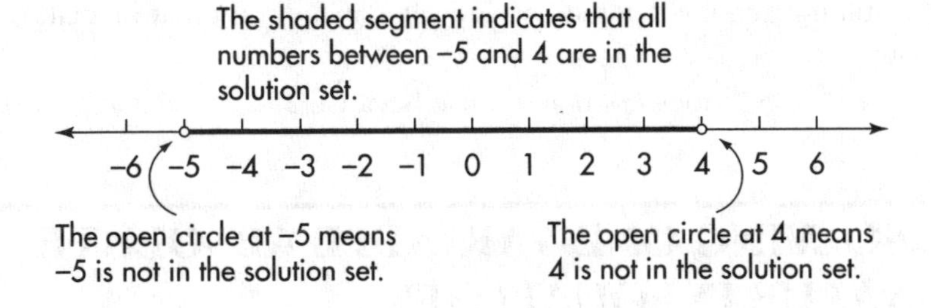

FIGURE 3.6

- To solve $|2x + 1| > 9$:

Apply the appropriate property:

$$|2x + 1| > 9 \Leftrightarrow 2x + 1 < -9 \quad \text{or} \quad 2x + 1 > 9$$

Solve each of the resulting inequalities:

$$\begin{aligned} 2x + 1 &< -9 &\quad \text{or} \quad 2x + 1 &> 9 \\ 2x &< -9 - 1 &\quad \text{or} \quad 2x &> 9 - 1 \\ 2x &< -10 &\quad \text{or} \quad 2x &> 8 \\ x &< -5 &\quad \text{or} \quad x &> 4 \end{aligned}$$

Result is a disjunction.

The solution set for $|2x + 1| > 9$ is $(-\infty, -5) \cup (4, \infty)$, illustrated in Figure 3.7.

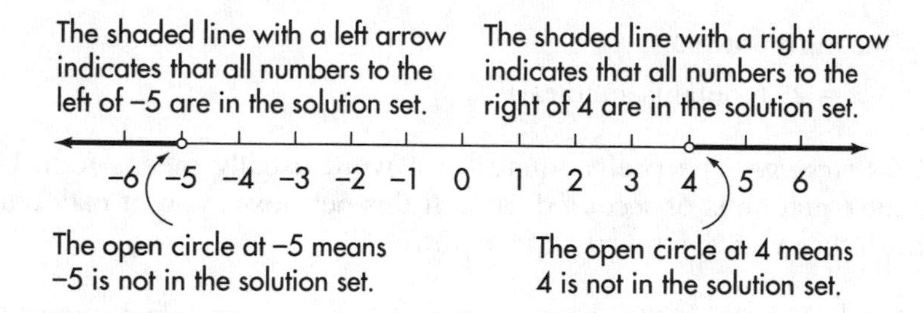

FIGURE 3.7

3.7 SOLVING ONE-VARIABLE QUADRATIC EQUATIONS

A **one-variable quadratic equation** is an equation that can be written in the standard form

$$ax^2 + bx + c = 0,$$

where x is the variable and a, b, and c are constants, with $a \neq 0$. The constants a, b, and c are called the **coefficients** of the quadratic equation. Some examples of one-variable quadratic equations follow:

- $x^2 + 8x - 20 = 0$
- $2x^2 - 7x + 5 = 0$
- $4x^2 - 9 = 0$
- $x^2 - 4x = -4$
- $-3x^2 + \frac{x}{3} + \frac{4}{5} = 0$
- $5.44x^2 + 4.08x + 2.72 = 0$

A one-variable quadratic equation has only *one* variable that occurs to the second power in *at least* one term, and (after terms have been simplified) there is *no* term in which the variable has an exponent greater than 2 or a fractional exponent (no roots or radicals). The following are *not* one-variable quadratic equations:

- $3x + 4 = 5x$ (is linear)
- $2x^3 + x^2 + 8x - 20 = 0$ (variable occurs to third power)
- $3x^2 + xy = 6$ (has two variables)
- $\frac{200}{t} + 25t^2 = 60$ (when simplified, the variable will be raised to the third power)
- $\sqrt{z} = 25$ (contains a radical)

A helpful message: The prefix "quad" in a word usually means four, but a *quad*ratic equation is of *second* degree. If this perplexes you, it may help to remember that x^2 gives the area of a s*qua*re.

To find the solution set for a quadratic equation, you must determine all values for the variable that make the quadratic equation a true statement. The solutions of a quadratic equation are also called its **roots.** A quadratic equation may have exactly *one* real root, *two* real unequal roots, or *no* real roots.

Three methods for solving quadratic equations are (1) by **factoring,** (2) by **completing the square,** and (3) by using the **quadratic formula.**

1. Solving by Factoring

To solve a quadratic equation by factoring:

1. Express the equation in standard form:

$$ax^2 + bx + c = 0$$

   - Remove grouping symbols, if any, using the distributive property; then simplify.
   - Eliminate fractions, if any, by multiplying both sides of the equation by the LCD of all the fractions in the equation; then simplify.
   - Undo indicated addition or subtraction to get all nonzero terms in descending powers of the variable on the left side, leaving only zero on the right; then simplify.
2. Factor the left side of the equation.
3. Set each factor containing the variable equal to zero.
4. Solve each of the resulting linear equations.
5. Check each root in the original equation.

Step 3 in the above procedure is based on the **property of zero products** for numbers:

> If the product of two quantities is zero, at least one of the quantities is zero. That is, if $ab = 0$, then $a = 0$ or $b = 0$, or both.

- To solve $x^2 + 8x - 20 = 0$ by factoring:

Express the equation in standard form:

$$x^2 + 8x - 20 = 0$$

Factor the left side:

$$(x + 10)(x - 2) = 0$$

Set each factor equal to zero and solve the resulting two equations:

$$x + 10 = 0 \quad \text{or} \quad x - 2 = 0$$

$$x = -10 \quad \text{or} \quad x = 2$$

Check each solution:

$$(-10)^2 + 8(-10) - 20 \stackrel{?}{=} 0 \qquad (2)^2 + 8(2) - 20 \stackrel{?}{=} 0$$

$$100 - 80 - 20 \stackrel{?}{=} 0 \qquad 4 + 16 - 20 \stackrel{?}{=} 0$$

$$0 \stackrel{\checkmark}{=} 0 \qquad 0 \stackrel{\checkmark}{=} 0$$

Thus, the solution set for $x^2 + 8x - 20 = 0$ is $\{-10, 2\}$.

- To solve $2x^2 - 7x + 5 = 0$ by factoring:

Express the equation in standard form:

$$2x^2 - 7x + 5 = 0$$

Factor the left side:

$$(2x - 5)(x - 1) = 0$$

Set each factor equal to zero and solve the resulting two equations:

$$2x - 5 = 0 \quad \text{or} \quad x - 1 = 0$$

$$2x = 5 \quad \text{or} \quad x = 1$$

$$x = 2.5$$

Check each solution:

$$2(2.5)^2 - 7(2.5) + 5 \stackrel{?}{=} 0 \qquad 2(1)^2 - 7(1) + 5 \stackrel{?}{=} 0$$

$$2(6.25) - 17.5 + 5 \stackrel{?}{=} 0 \qquad 2(1) - 7(1) + 5 \stackrel{?}{=} 0$$

$$12.5 - 17.5 + 5 \stackrel{?}{=} 0 \qquad 2 - 7 + 5 \stackrel{?}{=} 0$$

$$0 \stackrel{\checkmark}{=} 0 \qquad 0 \stackrel{\checkmark}{=} 0$$

Thus, the solution set for $2x^2 - 7x + 5 = 0$ is $\{1, 2.5\}$.

- To solve $(x - 5)(x + 2) = -12$ by factoring:

$$(x - 5)(x + 2) = -12$$

Express the equation in standard form:

$$x^2 - 3x - 10 = -12$$

$$x^2 - 3x + 2 = 0$$

Factor the left side:

$$(x - 2)(x - 1) = 0$$

Set each factor equal to zero and solve the resulting two equations:

$$x - 2 = 0 \quad \text{or} \quad x - 1 = 0$$

$$x = 2 \quad \text{or} \quad x = 1$$

Check each solution:

$$(2 - 5)(2 + 2) \stackrel{?}{=} -12 \qquad (1 - 5)(1 + 2) \stackrel{?}{=} -12$$

$$(-3)(4) \stackrel{?}{=} -12 \qquad (-4)(3) \stackrel{?}{=} -12$$

$$-12 \stackrel{\surd}{=} -12 \qquad -12 \stackrel{\surd}{=} -12$$

Thus, the solution set for $(x - 5)(x + 2) = -12$ is $\{1, 2\}$.

A helpful message: Some students begin solving $(x - 5)(x + 2) = -12$ by setting each factor on the left equal to -12. This is incorrect. The property of zero products can only be applied when the product is *zero,* not -12 or any other nonzero number!

2. Solving by Completing the Square

Before we look at the procedure for solving quadratic equations by completing the square, consider quadratic equations for which the coefficient of the first degree term is 0, such as $4x^2 - 9 = 0$. If we solve by factoring, we obtain

$$4x^2 - 9 = 0$$

$$(2x + 3)(2x - 3) = 0$$

$$2x + 3 = 0 \quad \text{or} \quad 2x - 3 = 0$$

$$x = -\frac{3}{2} \quad \text{or} \quad x = \frac{3}{2}$$

Check each solution:

$$4\left(\frac{3}{2}\right)^2 - 9 \stackrel{?}{=} 0 \qquad 4\left(-\frac{3}{2}\right)^2 - 9 \stackrel{?}{=} 0$$

$$4\left(\frac{9}{4}\right) - 9 \stackrel{?}{=} 0 \qquad 4\left(\frac{9}{4}\right) - 9 \stackrel{?}{=} 0$$

$$0 \stackrel{\surd}{=} 0 \qquad 0 \stackrel{\surd}{=} 0$$

Thus, $4x^2 - 9 = 0$ has two roots, $-\frac{3}{2}$ and $\frac{3}{2}$.

We will now look at two other methods for finding this solution. Isolating the term containing the variable, that is, "solving for x^2," we obtain:

$$4x^2 - 9 = 0$$
$$4x^2 = 9$$
$$x^2 = \frac{9}{4}$$

Method 1: Since the two squares, x^2 and $\frac{9}{4}$, are equal, their *principal* square roots are equal. Thus, we can "solve for x" by finding the *principal* square root of both sides of the equation (remember, $\sqrt{x^2} = |x|$):

$$\sqrt{x^2} = \sqrt{\frac{9}{4}}$$

$$|x| = \frac{3}{2}, \text{ which is true if and only if}$$

$$x = \frac{3}{2} \quad \text{or} \quad x = -\frac{3}{2}$$

Thus, the solution set for $4x^2 - 9 = 0$ is $\{-\frac{3}{2}, \frac{3}{2}\}$, the same as determined above.

Method 2: Since the two squares, x^2 and $\frac{9}{4}$, are equal, the definition of square root would lead us to conclude that x is one of the two square roots of $\frac{9}{4}$. Thus, we can "solve for x" by writing it as plus or minus the principal square root of $\frac{9}{4}$:

$$x = \pm\sqrt{\frac{9}{4}}$$

$$x = \pm\frac{3}{2}$$

Thus, the solution set for $4x^2 - 9 = 0$ is $\{-\frac{3}{2}, \frac{3}{2}\}$, again the same as already determined.

Which of these two methods you want to use is your personal choice. In the examples, we will use Method 1 because it reinforces the very important relationship that exists between square root and absolute value, namely, $\sqrt{x^2} = |x|$. However, you should not have any difficulty understanding these examples if you prefer Method 2.

A helpful message: As discussed in Chapter 1, the $\sqrt{}$ symbol is used to indicate the *principal* square root of a number, that is, the *nonnegative* (positive or zero) root. Since you do not know whether x is positive, negative, or zero, you must write $\sqrt{x^2} = |x|$ to be assured the right-hand member of this expression is *nonnegative.*

Regardless of which of the two methods above that you use, quadratic equations that can be written in the form:

$$x^2 = C,$$

where C is a number or algebraic expression *not containing* x, have the following solution:

$$x = \pm\sqrt{C}$$

If the quantity C is *zero,* there is *one* real root, which has the value zero; if *positive,* there are *two* unequal real roots; and if *negative,* there are *no* real roots. Some examples follow:

- Solve $25x^2 = 16$.

$$25x^2 = 16$$

Isolate the variable:

$$x^2 = \frac{16}{25}$$

Find the principle square root of both sides and solve for x:

$$\sqrt{x^2} = \sqrt{\frac{16}{25}}$$

$$|x| = \frac{4}{5}$$

$$x = \frac{4}{5} \quad \text{or} \quad x = -\frac{4}{5}$$

Check each solution:

$$25\left(\frac{4}{5}\right)^2 \stackrel{?}{=} 16 \qquad 25\left(-\frac{4}{5}\right)^2 \stackrel{?}{=} 16$$

$$25\left(\frac{16}{25}\right) \stackrel{?}{=} 16 \qquad 25\left(\frac{16}{25}\right) \stackrel{?}{=} 16$$

$$16 \stackrel{\checkmark}{=} 16 \qquad 16 \stackrel{\checkmark}{=} 16$$

Thus, the solution set for $25x^2 = 16$ is $\left\{-\frac{4}{5}, \frac{4}{5}\right\}$.

- Solve $36x^2 - 15 = 0$.

$$36x^2 - 15 = 0$$

Isolate the variable (solve for x^2):

$$36x^2 = 15$$

$$x^2 = \frac{15}{36}$$

Find the principal square root of both sides and solve for x:

$$\sqrt{x^2} = \sqrt{\frac{15}{36}}$$

$$\sqrt{x^2} = \frac{\sqrt{15}}{\sqrt{36}}$$

$$|x| = \frac{\sqrt{15}}{6}$$

$$x = \frac{\sqrt{15}}{6} \quad \text{or} \quad x = -\frac{\sqrt{15}}{6}$$

Check each solution:

$$36\left(\frac{\sqrt{15}}{6}\right)^2 \stackrel{?}{=} 15 \qquad 36\left(-\frac{\sqrt{15}}{6}\right)^2 \stackrel{?}{=} 15$$

$$36\left(\frac{15}{36}\right) \stackrel{?}{=} 15 \qquad 36\left(\frac{15}{36}\right) \stackrel{?}{=} 15$$

$$15 \stackrel{\surd}{=} 15 \qquad 15 \stackrel{\surd}{=} 15$$

Thus, the solution set for $36x^2 - 15 = 0$ is $\left\{-\frac{\sqrt{15}}{6}, \frac{\sqrt{15}}{6}\right\}$.

You can extend the procedure to include equations that have the form:

$$(x + k)^2 = C$$

where C is a number or algebraic expression *not containing* x.

For example:

- Solve $(x + 3)^2 = 16$:

Isolate the variable expression $(x + 3)^2$:

$$(x + 3)^2 = 16$$

Find the principal square root of both sides:

$$\sqrt{(x + 3)^2} = \sqrt{16}$$

$$|x + 3| = 4$$

$$x + 3 = 4 \quad \text{or} \quad x + 3 = -4$$

$$x = 1 \quad \text{or} \quad x = -7$$

Check each solution:

$$(1 + 3)^2 \stackrel{?}{=} 16 \qquad (-7 + 3)^2 \stackrel{?}{=} 16$$

$$(4)^2 \stackrel{?}{=} 16 \qquad (-4)^2 \stackrel{?}{=} 16$$

$$16 \stackrel{\surd}{=} 16 \qquad 16 \stackrel{\surd}{=} 16$$

Thus, the solution set for $(x + 3)^2 = 16$ is $\{-7, 1\}$.

We incorporate this notion in the following:

To solve a quadratic equation by completing the square:

1. Get all terms containing the variable on the left side, and all other terms on the right to obtain an equation of the form:

$$ax^2 + bx = C_0$$

2. If a is not 1, divide each term by a to obtain an equation of the form:

$$x^2 + \frac{b}{a}x = C_1,$$

where $C_1 = \frac{C_0}{a}$.

3. Add the square of half the coefficient of x to both sides to obtain an equation of the form:

$$\underbrace{x^2 + \frac{b}{a}x + \left(\frac{b}{2a}\right)^2}_{\text{perfect trinomial square}} = C,$$

where $C = C_1 + \left(\frac{b}{2a}\right)^2 = C_1 + \frac{b^2}{4a^2}$.

4. Factor the left side as the square of a binomial to obtain an equation of the form:

$$\left(x + \frac{b}{2a}\right)^2 = C$$

5. Find the principal square root of both sides:

$$\left|x + \frac{b}{2a}\right| = \sqrt{C}$$

6. Solve for the variable.
7. Check each root in the original equation.

A helpful message: The quantities C_0, C_1, and C, representing the right side of the equation in the above procedure, have subscripts to indicate the right side may (and likely will) change value as you proceed through the steps indicated.

- To solve $x^2 + 8x - 20 = 0$ by completing the square:

$$x^2 + 8x - 20 = 0$$

Get all terms containing the variable on the left side, and all other terms on the right:

$$x^2 + 8x = 20$$

Add the square of half of 8, or 4^2, to both sides:

$$x^2 + 8x + 4^2 = 20 + 4^2$$

$$x^2 + 8x + 4^2 = 20 + 16$$

$$x^2 + 8x + 4^2 = 36$$

Factor the left side:

$$(x + 4)^2 = 36$$

Find the principal square root of both sides:

$$|x + 4| = 6$$

Solve for x:

$$|x + 4| = 6$$

$$x + 4 = -6 \quad \text{or} \quad x + 4 = 6$$

$$x = -10 \quad \text{or} \quad x = 2$$

Check each solution:

$$(-10)^2 + 8(-10) - 20 \stackrel{?}{=} 0 \qquad (2)^2 + 8(2) - 20 \stackrel{?}{=} 0$$

$$100 - 80 - 20 \stackrel{?}{=} 0 \qquad 4 + 16 - 20 \stackrel{?}{=} 0$$

$$0 \stackrel{\surd}{=} 0 \qquad 0 \stackrel{\surd}{=} 0$$

Thus, the solution set for $x^2 + 8x - 20 = 0$ is $\{-10, 2\}$, the same as obtained before.

- To solve $2x^2 - 7x + 5 = 0$ by completing the square:

$$2x^2 - 7x + 5 = 0$$

Get all terms containing the variable on the left side, and all other terms on the right:

$$2x^2 - 7x = -5$$

Divide each term by 2, the coefficient of x^2:

$$x^2 - \frac{7}{2}x = -\frac{5}{2}$$

Add the square of half of $\frac{7}{2}$, or $\left(\frac{1}{2} \cdot \frac{7}{2}\right)^2 = \left(\frac{7}{4}\right)^2$, to both sides:

$$x^2 - \frac{7}{2}x + \left(\frac{7}{4}\right)^2 = -\frac{5}{2} + \left(\frac{7}{4}\right)^2$$

$$x^2 - \frac{7}{2}x + \left(\frac{7}{4}\right)^2 = -\frac{40}{16} + \frac{49}{16}$$

$$x^2 - \frac{7}{2}x + \left(\frac{7}{4}\right)^2 = \frac{9}{16}$$

Factor the left side:

$$\left(x - \frac{7}{4}\right)^2 = \frac{9}{16}$$

Find the principal square root of both sides:

$$\left|x - \frac{7}{4}\right| = \frac{3}{4}$$

Solve for x:

$$\left|x - \frac{7}{4}\right| = \frac{3}{4}$$

$$x - \frac{7}{4} = \frac{3}{4} \quad \text{or} \quad x - \frac{7}{4} = -\frac{3}{4}$$

$$x = \frac{3}{4} + \frac{7}{4} \quad \text{or} \quad x = -\frac{3}{4} + \frac{7}{4}$$

$$x = \frac{10}{4} = 2.5 \quad \text{or} \quad x = \frac{4}{4} = 1$$

Check each solution:

$$2(2.5)^2 - 7(2.5) + 5 \stackrel{?}{=} 0 \qquad 2(1)^2 - 7(1) + 5 \stackrel{?}{=} 0$$

$$2(6.25) - 17.5 + 5 \stackrel{?}{=} 0 \qquad 2(1) - 7(1) + 5 \stackrel{?}{=} 0$$

$$12.5 - 17.5 + 5 \stackrel{?}{=} 0 \qquad 2 - 7 + 5 \stackrel{?}{=} 0$$

$$0 \stackrel{\checkmark}{=} 0 \qquad 0 \stackrel{\checkmark}{=} 0$$

Thus, the solution set for $2x^2 - 7x + 5 = 0$ is $\{1, 2.5\}$, the same as obtained before.

3. Solving by Using the Quadratic Formula

You can solve the standard form for a quadratic equation by completing the square:

$$ax^2 + bx + c = 0$$

Get all terms containing the variable on the left side, and all other terms on the right:

$$ax^2 + bx = -c$$

Divide each term by a, the coefficient of x^2:

$$x^2 + \frac{b}{a}x = -\frac{c}{a}$$

Add the square of half of $\frac{b}{a}$, or $\left(\frac{b}{2a}\right)^2$, to both sides:

$$x^2 + \frac{b}{a}x + \left(\frac{b}{2a}\right)^2 = -\frac{c}{a} + \left(\frac{b}{2a}\right)^2$$

$$x^2 + \frac{b}{a}x + \left(\frac{b}{2a}\right)^2 = -\frac{c}{a} + \frac{b^2}{4a^2}$$

$$x^2 + \frac{b}{a}x + \left(\frac{b}{2a}\right)^2 = -\frac{4ac}{4a^2} + \frac{b^2}{4a^2}$$

$$x^2 + \frac{b}{a}x + \left(\frac{b}{2a}\right)^2 = \frac{b^2 - 4ac}{4a^2}$$

Factor the left side:

$$\left(x + \frac{b}{2a}\right)^2 = \frac{b^2 - 4ac}{4a^2}$$

Find the principal square root of both sides:

$$\left|x + \frac{b}{2a}\right| = \frac{\sqrt{b^2 - 4ac}}{2a}$$

Solve for x:

$$\left|x + \frac{b}{2a}\right| = \frac{\sqrt{b^2 - 4ac}}{2a}$$

$$x + \frac{b}{2a} = \frac{\sqrt{b^2 - 4ac}}{2a} \quad \text{or} \quad x + \frac{b}{2a} = -\frac{\sqrt{b^2 - 4ac}}{2a}$$

$$x = -\frac{b}{2a} + \frac{\sqrt{b^2 - 4ac}}{2a} \quad \text{or} \quad x = -\frac{b}{2a} - \frac{\sqrt{b^2 - 4ac}}{2a}$$

$$x = \frac{-b + \sqrt{b^2 - 4ac}}{2a} \quad \text{or} \quad x = \frac{-b - \sqrt{b^2 - 4ac}}{2a}$$

When $ax^2 + bx + c = 0$ is solved by completing the square, the result obtained is called the **quadratic formula:**

$$x = \frac{-b \pm \sqrt{b^2 - 4ac}}{2a}$$

Here a, b, and c are the coefficients of the quadratic equation $ax^2 + bx + c = 0$ in standard form. The $\pm$ (read "plus or minus") in the formula is used to combine what is actually two equations:

$$x = \frac{-b + \sqrt{b^2 - 4ac}}{2a} \quad \text{and} \quad x = \frac{-b - \sqrt{b^2 - 4ac}}{2a}$$

This means, of course, that you evaluate twice: once using a $+$ sign with $\sqrt{b^2 - 4ac}$ and once using a $-$ sign with $\sqrt{b^2 - 4ac}$.

The quantity $b^2 - 4ac$ under the $\sqrt{\ }$ sign is called the **discriminant** of the quadratic equation. If a, b, and c are real numbers, it can be used to characterize the solution set as follows:

The quadratic equation $ax^2 + bx + c = 0$ has

exactly *one* real root if $b^2 - 4ac = 0$;
two real unequal roots if $b^2 - 4ac > 0$;
no real roots if $b^2 - 4ac < 0$.

In real-world applications, the most efficient method for solving quadratic equations is usually to use the quadratic formula:

To solve a quadratic equation by using the quadratic formula:

1. Express the equation in standard form:

$$ax^2 + bx + c = 0$$

   - Remove grouping symbols, if any, using the distributive property; then simplify.
   - Eliminate fractions, if any, by multiplying both sides of the equation by the LCD of all the fractions in the equation; then simplify.
   - Undo indicated addition or subtraction to get all nonzero terms in descending powers of the variable on the left side, leaving only zero on the right; then simplify.
2. Determine the values of the coefficients a, b, and c.
3. Substitute into the quadratic formula.
4. Evaluate and simplify.
5. Check each root in the original equation.

A helpful message: In step 2, when determining the values of a, b, and c, do not forget to include the $-$ sign when it precedes a number. Also, when you substitute negative values for a, b, and c into the formula, it is a good idea to enclose the substitutions in parentheses.

- To solve $x^2 + 8x - 20 = 0$ by using the quadratic formula:

Express in standard form:

$$x^2 + 8x - 20 = 0$$

Determine the values of a, b, and c:

(Include the − sign.)

$$a = 1,\ b = 8,\ c = -20$$

Substitute into the quadratic formula:

$$x = \frac{-8 \pm \sqrt{8^2 - 4(1)(-20)}}{2(1)} = \frac{-8 \pm \sqrt{64 + 80}}{2} = \frac{-8 \pm \sqrt{144}}{2} = \frac{-8 \pm 12}{2}$$

Evaluate twice:

$$x = \frac{-8 + 12}{2} = \frac{4}{2} = 2 \quad \text{or} \quad x = \frac{-8 - 12}{2} = \frac{-20}{2} = -10$$

Check each solution:

$$(-10)^2 + 8(-10) - 20 \stackrel{?}{=} 0 \qquad (2)^2 + 8(2) - 20 \stackrel{?}{=} 0$$

$$100 - 80 - 20 \stackrel{?}{=} 0 \qquad 4 + 16 - 20 \stackrel{?}{=} 0$$

$$0 \stackrel{\surd}{=} 0 \qquad 0 \stackrel{\surd}{=} 0$$

Thus, the solution set for $x^2 + 8x - 20 = 0$ is $\{-10, 2\}$, the same as obtained before.

- To solve $2x^2 - 7x + 5 = 0$ by using the quadratic formula:

Express in standard form:

$$2x^2 - 7x + 5 = 0$$

Determine the values of a, b, and c:

(Include the − sign.)

$$a = 2,\ b = -7,\ c = 5$$

Substitute into the quadratic formula:

$$x = \frac{-(-7) \pm \sqrt{(-7)^2 - 4(2)(5)}}{2(2)} = \frac{7 \pm \sqrt{49 - 40}}{4} = \frac{7 \pm \sqrt{9}}{4} = \frac{7 \pm 3}{4}$$

Evaluate twice:

$$x = \frac{7 + 3}{4} = \frac{10}{4} = \frac{5}{2} = 2.5 \quad \text{or} \quad x = \frac{7 - 3}{4} = \frac{4}{4} = 1$$

Check each solution:

$$2(2.5)^2 - 7(2.5) + 5 \stackrel{?}{=} 0 \qquad 2(1)^2 - 7(1) + 5 \stackrel{?}{=} 0$$

$$2(6.25) - 17.5 + 5 \stackrel{?}{=} 0 \qquad 2(1) - 7(1) + 5 \stackrel{?}{=} 0$$

$$12.5 - 17.5 + 5 \stackrel{?}{=} 0 \qquad 2 - 7 + 5 \stackrel{?}{=} 0$$

$$0 \stackrel{\surd}{=} 0 \qquad 0 \stackrel{\surd}{=} 0$$

Thus, the solution set for $2x^2 - 7x + 5 = 0$ is $\{1, 2.5\}$, the same as obtained before.

A helpful message: Take care when writing the quadratic formula and substituting values into it. The fraction bar must be extended to include the entire numerator, $-b \pm \sqrt{b^2 - 4ac}$, and the $\sqrt{}$ symbol extended to include the quantity $b^2 - 4ac$:

- $x = \dfrac{-b \pm \sqrt{b^2 - 4ac}}{2a}$ is correct.
- $x = -b \pm \dfrac{\sqrt{b^2 - 4ac}}{2a}$ is incorrect.
- $x = \dfrac{-b}{2a} \pm \sqrt{b^2 - 4ac}$ is incorrect.
- $x = \dfrac{-b \pm \sqrt{b^2} - 4ac}{2a}$ is incorrect.

- To solve $4x^2 - 9 = 0$ by using the quadratic formula:

Express in standard form:

$$4x^2 - 9 = 0$$

Determine the values of a, b, and c:

(Include the − sign.)

$$a = 4,\ b = 0,\ c = -9$$

Substitute into the quadratic formula:

$$x = \frac{-(0) \pm \sqrt{(0)^2 - 4(4)(-9)}}{2(4)} = \frac{\pm\sqrt{144}}{8} = \frac{\pm 12}{8} = \frac{\pm 3}{2}$$

Evaluate twice:

$$x = \frac{3}{2} \quad \text{or} \quad x = \frac{-3}{2} = -\frac{3}{2}$$

Check each solution:

$$4\left(\frac{3}{2}\right)^2 - 9 \stackrel{?}{=} 0 \qquad 4\left(-\frac{3}{2}\right)^2 - 9 \stackrel{?}{=} 0$$

$$4\left(\frac{9}{4}\right)^2 - 9 \stackrel{?}{=} 0 \qquad 4\left(\frac{9}{4}\right) - 9 \stackrel{?}{=} 0$$

$$0 \stackrel{\surd}{=} 0 \qquad 0 \stackrel{\surd}{=} 0$$

Thus, the solution set for $4x^2 - 9 = 0$ is $\left\{\frac{3}{2}, -\frac{3}{2}\right\}$, the same as obtained before.

- To solve $-3x^2 + \frac{x}{3} + \frac{4}{5} = 0$ by using the quadratic formula:

Express in standard form:

$$-3x^2 + \frac{x}{3} + \frac{4}{5} = 0$$

Multiply each term by the LCD, 15:

$$15(-3x^2) + \frac{15}{1}\left(\frac{x}{3}\right) + \frac{15}{1}\left(\frac{4}{5}\right) + 15(0)$$

$$-45x^2 + 5x + 12 = 0$$

Determine the values of a, b, and c:

(Include the − sign.)

$$a = -45,\ b = 5,\ c = 12$$

Substitute into the quadratic formula:

$$x = \frac{-5 \pm \sqrt{(5)^2 - 4(-45)(12)}}{2(-45)} = \frac{-5 \pm \sqrt{25 + 2160}}{-90} = \frac{-5 \pm \sqrt{2185}}{-90}$$

Evaluate twice:

$$x = -\frac{-5 + \sqrt{2185}}{90} \cong -0.4638 \quad \text{or} \quad x = -\frac{-5 - \sqrt{2185}}{90}$$

exact root   approximate root

$$= \frac{5 + \sqrt{2185}}{90} \cong 0.5749$$

exact root   approximate root

A helpful message: When the $\sqrt{\ }$ quantity is irrational as in the above example, it is customary to write the exact roots, retaining the $\sqrt{\ }$ symbol; however, in practical problems, approximation of the roots using a scientific calculator is recommended.

Check each solution:

Checking the exact roots in this case can be rather tedious. It is more efficient just to go over your work several times or check using approximate roots as follows:

$$-3(-0.4638)^2 + \frac{-0.4638}{3} + \frac{4}{5} \stackrel{?}{=} 0 \qquad -3(0.5749)^2 + \frac{0.5749}{3} + \frac{4}{5} \stackrel{?}{=} 0$$

$$-0.6453 - 0.1546 + 0.8 \cong 0 \stackrel{\checkmark}{=} 0 \qquad -0.9915 + 0.1916 + 0.8 \cong 0 \stackrel{\checkmark}{=} 0$$

Thus, the solution set for $-3x^2 + \frac{x}{3} + \frac{4}{5} = 0$ is $\left\{\frac{-5+\sqrt{2185}}{90}, \frac{5+\sqrt{2185}}{90}\right\}$.

- To solve $2x^2 = 6x$ by using the quadratic formula:

$$2x^2 = 6x$$

Express in standard form:

$$2x^2 - 6x = 0$$

Determine the values of a, b, and c:

(Include the − sign.)

$$a = 2,\ b = -6,\ c = 0$$

Substitute into the quadratic formula:

$$x = \frac{-(-6) \pm \sqrt{(-6)^2 - 4(2)(0)}}{2(2)} = \frac{6 \pm \sqrt{36}}{4} = \frac{6 \pm 6}{4}$$

Evaluate twice:

$$x = \frac{6 + 6}{4} = \frac{12}{4} = 3 \quad \text{or} \quad x = \frac{6 - 6}{4} = \frac{0}{4} = 0$$

Check each solution:

$$2(3)^2 \stackrel{?}{=} 6(3) \qquad 2(0)^2 \stackrel{?}{=} 6(0)$$
$$2(9) \stackrel{?}{=} 18 \qquad 2(0) \stackrel{?}{=} 0$$
$$18 \stackrel{\checkmark}{=} 18 \qquad 0 \stackrel{\checkmark}{=} 0$$

Thus, the solution set for $2x^2 = 6x$ is $\{0, 3\}$.

A helpful message: You may be tempted to approach the previous problem by dividing both sides by x first:

$$2x^2 = 6x$$
$$2x = 6$$
$$x = 3$$

This is incorrect because it does not yield the full solution set. The best way to avoid this mistake is to abide by the following rule.

> *Never* divide both sides of an equation by the variable or by an expression containing the variable.

A helpful message: When you need to solve a quadratic equation for which no particular method is specified, solving by factoring or using the quadratic formula will usually be less time-consuming and tedious than solving by completing the square.

3.8 SOLVING OTHER EQUATIONS IN ONE-VARIABLE

Many equations that are not of the type discussed in the previous sections can be transformed into one of these forms. For example, consider equations involving fractions in which a denominator expression contains the variable, such as the following:

- $\frac{1}{2} + \frac{1}{3} = \frac{1}{t}, t \neq 0$
- $\frac{15}{x - 20} + \frac{40}{x} = 1.3, x \neq 0, x \neq 20$
- $\frac{x}{x - 3} = \frac{3}{x - 3}, x \neq 3$
- $\frac{42}{x} = \frac{2}{5}, x \neq 0$

These equations can be transformed by multiplying both sides of the equation by the LCD of all the fractions. The resulting equation (now free of fractions) can be solved using the methods previously discussed. Of course, you must keep in mind that an *excluded value* for the variable *cannot* be in the solution set.

- Solve $\frac{1}{2} + \frac{1}{3} = \frac{1}{t}$.

$$\frac{1}{2} + \frac{1}{3} = \frac{1}{t}$$

Multiply each term on both sides by the LCD, $6t$:

$$\frac{\overset{3}{\cancel{6}}t}{1}\left(\frac{1}{\cancel{2}}\right) + \frac{\overset{2}{\cancel{6}}t}{1}\left(\frac{1}{\cancel{3}}\right) = \frac{6\cancel{t}}{1}\left(\frac{1}{\cancel{t}}\right)$$

$$3t + 2t = 6 \quad \text{(linear equation)}$$

$$5t = 6$$

$$t = \frac{6}{5}$$

Check:

$$\frac{1}{2} + \frac{1}{3} \stackrel{?}{=} \frac{1}{\frac{6}{5}}$$

$$\frac{5}{6} \stackrel{\checkmark}{=} \frac{5}{6}$$

Thus, the solution set for $\frac{1}{2} + \frac{1}{3} = \frac{1}{t}$ is $\left\{\frac{6}{5}\right\}$.

- Solve $\frac{15}{x - 20} + \frac{40}{x} = 1.3$.

$$\frac{15}{x - 20} + \frac{40}{x} = 1.3$$

Multiply each term on both sides by the LCD, $x(x-20)$:

$$\frac{x\cancel{(x-20)}}{1}\cdot\frac{15}{\cancel{x-20}}+\frac{\cancel{x}(x-20)}{1}\cdot\frac{40}{\cancel{x}}=1.3x(x-20)$$

$$15x+40(x-20)=1.3x^2-26x$$

$$15x+40x-800=1.3x^2-26x$$

$$-1.3x^2+81x-800=0 \quad \text{(quadratic equation)}$$

$$x=\frac{-81\pm\sqrt{(81)^2-4(-1.3)(-800)}}{2(-1.3)}=\frac{-81\pm\sqrt{6561-4160}}{-2.6}$$

$$=\frac{-81\pm\sqrt{2401}}{-2.6}=\frac{-81\pm 49}{-2.6}$$

$$x=\frac{-81+49}{-2.6}=\frac{-32}{-2.6}=\frac{160}{13} \quad \text{or} \quad x=\frac{-81-49}{-2.6}=\frac{-130}{-2.6}=50$$

Check each solution:

$$\frac{15}{\frac{160}{13}-20}+\frac{40}{\frac{160}{13}}\stackrel{?}{=}1.3 \qquad \frac{15}{50-20}+\frac{40}{50}\stackrel{?}{=}1.3$$

$$\frac{13\cdot 15}{160-13\cdot 20}+\frac{13\cdot 40}{160}\stackrel{?}{=}1.3 \qquad \frac{15}{30}+\frac{40}{50}\stackrel{?}{=}1.3$$

$$-\frac{195}{100}+\frac{520}{160}\stackrel{?}{=}1.3 \qquad 1.3\stackrel{\surd}{=}1.3$$

$$1.3\stackrel{\surd}{=}1.3$$

Thus, the solution set for $\frac{15}{x-20}+\frac{40}{x}=1.3$ is $\left\{\frac{160}{13}, 50\right\}$.

- Solve $\dfrac{x}{x-3}=\dfrac{3}{x-3}$.

$$\frac{x}{x-3}=\frac{3}{x-3}$$

Multiply each term on both sides by the LCD, $x-3$:

$$\frac{\cancel{x-3}}{1}\cdot\frac{x}{\cancel{x-3}}=\frac{\cancel{x-3}}{1}\cdot\frac{3}{\cancel{x-3}}$$

The apparent solution $x=3$ cannot be a solution, since 3 is an excluded value for x.

Thus, the solution set for $\frac{x}{x-3}=\frac{3}{x-3}$ is $\varnothing$; that is, $\frac{x}{x-3}=\frac{3}{x-3}$ has no solution.

A helpful message: You should always check your solution to an equation. In the above example, the check would have revealed that $x=3$ leads to division by zero, which can never occur:

Check:

$$\frac{3}{3-3} \stackrel{?}{=} \frac{3}{3-3}$$

$$\frac{3}{0} \stackrel{?}{=} \frac{3}{0}$$

Stop the check, since division by 0 is indicated; thus, 3 cannot be a root.

- Solve $\frac{42}{x} = \frac{2}{5}$.

$$\frac{42}{x} = \frac{2}{5}$$

Multiply both sides by the LCD, $5x$:

$$\frac{5\cancel{x}}{1}\left(\frac{42}{\cancel{x}}\right) = \frac{\cancel{5}x}{1}\left(\frac{2}{\cancel{5}}\right)$$

$$210 = 2x$$

$$105 = x$$

Check:

$$\frac{42}{105} \stackrel{?}{=} \frac{2}{5}$$

$$\frac{2}{5} \stackrel{\checkmark}{=} \frac{2}{5}$$

Thus, the solution set for $\frac{42}{x} = \frac{2}{5}$ is $\{105\}$.

The equation $\frac{42}{x} = \frac{2}{5}$ is called a *proportion*. A **proportion** is a statement of equality between two ratios. A shortcut for multiplying both sides by the LCD in a proportion is to "cross-multiply":

$$\frac{42}{x} \times \frac{2}{5}$$

$$2x = 42(5)$$

$$2x = 210$$

$$x = 105$$

Some equations contain radicals or fractional exponents. Examples are the following:

- $\sqrt{x-2} = 11$
- $\sqrt{x+6} = x$
- $(x-5)^{\frac{1}{3}} = 4$

These equations can be transformed by raising both sides to an appropriate power. However, use caution when doing this because the solution set of the transformed equation may contain an **extraneous root,** a value which is *not* a solution of the original equation. Therefore, it is imperative that you check all answers obtained in the *original* equation.

- Solve $\sqrt{x-2} = 11;\ x \geq 2$.

$$\sqrt{x-2} = 11;\ x \geq 2$$

Square both sides:

$$(\sqrt{x-2})^2 = 11^2$$
$$x - 2 = 121$$
$$x = 123$$

Check in the *original* equation:

$$\sqrt{123-2} \stackrel{?}{=} 11$$
$$\sqrt{121} \stackrel{?}{=} 11$$
$$11 \stackrel{\surd}{=} 11$$

Thus, the solution set for $\sqrt{x-2} = 11$ is $\{123\}$.

- Solve $\sqrt{x+6} = x;\ x \geq -6$.

$$\sqrt{x+6} = x;\ x \geq -6$$

Square both sides:

$$(\sqrt{x+6})^2 = x^2$$
$$x + 6 = x^2$$
$$-x^2 + x + 6 = 0$$
$$x^2 - x - 6 = 0$$
$$(x-3)(x+2) = 0$$
$$x = 3 \quad \text{or} \quad x = -2$$

Check each answer in the *original* equation:

$\sqrt{3+6} \stackrel{?}{=} 3$	$\sqrt{-2+6} \stackrel{?}{=} -2$; This can never be true,
$\sqrt{9} \stackrel{?}{=} 3$	since $\sqrt{}$ always denotes a *nonnegative*
$3 \stackrel{\surd}{=} 3$	number; that is, $\sqrt{-2+6} = \sqrt{4} = 2$, not -2.
3 is a root.	-2 is an *extraneous* root.

Thus, the solution set for $\sqrt{x+6} = x$ is $\{3\}$.

- Solve $(x-5)^{\frac{1}{3}} = 4$.

$$(x-5)^{\frac{1}{3}} = 4$$

Raise both sides to the third power:

$$((x-5)^{\frac{1}{3}})^3 = 4^3$$
$$x - 5 = 64$$
$$x = 69$$

Check in the *original* equation.

$$(69-5)^{\frac{1}{3}} \stackrel{?}{=} 4$$
$$64^{\frac{1}{3}} \stackrel{?}{=} 4$$
$$4 \stackrel{\surd}{=} 4$$

Thus, the solution set for $(x-5)^{\frac{1}{3}} = 4$ is $\{69\}$.

Equations that are not quadratic equations, but can be written in the form of a quadratic, that is, in the form

$$au^2 + bu + c = 0$$

where u is a mathematical expression, can be solved using the methods for solving quadratic equations.

For example, $z^4 - 13z^2 + 36 = 0$ is quadratic in form with $u = z^2$. That is, $z^4 - 13z^2 + 36 = (z^2)^2 - 13(z^2) + 36 = 0$ can be written as $u^2 - 13u + 36 = 0$, where $u = z^2$.

This equation can be solved as follows:

Express the equation in standard form:

$$z^4 - 13z^2 + 36 = 0$$

Factor the left side:

$$(z^2 - 9)(z^2 - 4) = 0$$

Set each factor equal to zero and solve the resulting two *quadratic* equations:

$$z^2 - 9 = 0 \quad \text{or} \quad z^2 - 4 = 0$$
$$z^2 = 9 \quad \text{or} \quad z^2 = 4$$
$$|z| = 3 \quad \text{or} \quad |z| = 2$$
$$z = \pm 3 \quad \text{or} \quad z = \pm 2$$

Check each solution:

Check $z = 3$:

$$(3)^4 - 13(3)^2 + 36 \stackrel{?}{=} 0$$
$$81 - 117 + 36 \stackrel{?}{=} 0$$
$$0 \stackrel{\surd}{=} 0$$

Check $z = -3$:

$$(-3)^4 - 13(-3)^2 + 36 \stackrel{?}{=} 0$$
$$81 - 117 + 36 \stackrel{?}{=} 0$$
$$0 \stackrel{\surd}{=} 0$$

Check $z = 2$:

$$(2)^4 - 13(2)^2 + 36 \stackrel{?}{=} 0$$
$$16 - 52 + 36 \stackrel{?}{=} 0$$
$$0 \stackrel{\surd}{=} 0$$

Check $z = -2$:

$$(-2)^4 - 13(-2)^2 + 36 \stackrel{?}{=} 0$$
$$16 - 52 + 36 \stackrel{?}{=} 0$$
$$0 \stackrel{\surd}{=} 0$$

Thus, the solution set for $z^4 - 13z^2 + 36 = 0$ is $\{\pm 2, \pm 3\}$.

You may prefer to solve $z^4 - 13z^2 + 36 = 0$ by first substituting $u = z^2$ in the equation to obtain:

$$u^2 - 13u + 36 = 0$$

You can then solve for u:

Express the equation in standard form:

$$u^2 - 13u + 36 = 0$$

Factor the left side:

$$(u - 9)(u - 4) = 0$$

Set each factor equal to zero and solve the resulting equations:

$$u - 9 = 0 \quad \text{or} \quad u - 4 = 0$$

$$u = 9 \quad \text{or} \quad u = 4$$

Then you can substitute z^2 back in for u and proceed as above:

$$z^2 = 9 \quad \text{or} \quad z^2 = 4$$

$$|z| = 3 \quad \text{or} \quad |z| = 2$$

$$z = \pm 3 \quad \text{or} \quad z = \pm 2$$

Equations that can be written so that one side is a factorable polynomial can be solved by factoring completely, then, applying the zero property for products, setting each factor equal to zero. For example, $y^3 - y^2 - 6y = 0$ can be solved as follows:

$$y^3 - y^2 - 6y = 0$$

$$y(y^2 - y - 6) = 0$$

$$y(y - 3)(y + 2) = 0$$

$$y = 0,\ y = 3, \text{ or } y = -2$$

Check each solution:

Check $y = 0$:

$$(0)^3 - (0)^2 - 6(0) \stackrel{?}{=} 0$$

$$0 - 0 - 0 \stackrel{?}{=} 0$$

$$0 \stackrel{\checkmark}{=} 0$$

Check $y = 3$:

$$(3)^3 - (3)^2 - 6(3) \stackrel{?}{=} 0$$

$$27 - 9 - 18 \stackrel{?}{=} 0$$

$$0 \stackrel{\checkmark}{=} 0$$

Check $y = -2$:

$$(-2)^3 - (-2)^2 - 6(-2) \stackrel{?}{=} 0$$

$$-8 - 4 + 12 \stackrel{?}{=} 0$$

$$0 \stackrel{\checkmark}{=} 0$$

Thus, the solution set for $y^3 - y^2 - 6y = 0$ is $\{-2, 0, 3\}$.

3.9 SOLVING ONE-VARIABLE QUADRATIC INEQUALITIES

One-variable quadratic inequalities are inequalities that can be written in one of the following **standard forms:**

- $ax^2 + bx + c < 0$
- $ax^2 + bx + c \leq 0$
- $ax^2 + bx + c > 0$
- $ax^2 + bx + c \geq 0$

Before attempting to solve a quadratic inequality, put it in standard form and determine if the corresponding quadratic equation $ax^2 + bx + c = 0$ has *no* real roots; exactly *one* real root, call it r; or *two* real unequal roots, call them r_1 and r_2 (with $r_1 < r_2$). Table 3.1 summarizes the solution sets when $a > 0$:

The solution sets in the table are based on the rules for multiplying signed numbers, which can be summarized as follows:

> If two factors have the same sign, their product is positive; if they have opposite signs, their product is negative.

To see how this works, consider $x^2 + 8x - 20 < 0$. Since the left side can be factored, we have an equivalent inequality:

$$(x + 10)(x - 2) < 0$$

The product on the left is negative (< 0) when the two factors, $x + 10$ and $x - 2$, have opposite signs. From the work in previous sections, you know -10 and 2 are the roots of the quadratic equation $x^2 + 8x - 20 = 0$. As shown in Figure 3.8, these numbers determine three intervals on the number line, namely $(-\infty, -10)$, $(-10, 2)$, and $(2, \infty)$. You can see that $x + 10$ is negative when x is less than -10 and positive when x is greater than -10; and, similarly, $x - 2$ is negative when x is less than 2 and positive when x is greater than 2. Therefore, *the signs of the factors do not change throughout a particular interval.* This means the product of the factors $(x + 10)$ and $(x - 2)$ will change signs *only* at -10 and 2, the *zeros* of $x^2 + 8x - 20$. The results are shown in Figure 3.8. Therefore, the portion of the number line in which the two factors have opposite signs is between -10 and 2; thus, the interval $(-10, 2)$ is the solution to the inequality $x^2 + 8x - 20 < 0$.

The inequality $x^2 + 8x - 20 > 0$ would be solved in a similar manner, yielding $(-\infty, -10) \cup (2, \infty)$ as the solution set.

The solution set for $x^2 + 8x - 20 \leq 0$ would be $[-10, 2]$. The endpoints -10 and 2 are *included* because the quadratic is zero at its roots.

TABLE 3.1. SOLUTION SETS FOR QUADRATIC INEQUALITIES

Quadratic Inequality ($a > 0$)	Root(s) of $ax^2 + bx + c = 0$	Solution Set	Graph
$ax^2 + bx + c < 0$ $ax^2 + bx + c \leq 0$	no real roots	$\varnothing$	no solution
$ax^2 + bx + c > 0$ $ax^2 + bx + c \geq 0$	no real roots	$(-\infty, \infty)$	
$ax^2 + bx + c < 0$	one real root, r	$\varnothing$	no solution
$ax^2 + bx + c \leq 0$	one real root, r	$\{r\}$	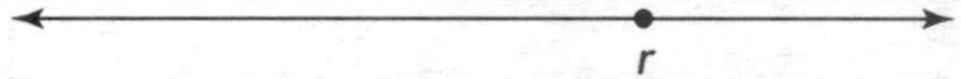
$ax^2 + bx + c > 0$	one real root, r	$(-\infty, r) \cup (r, \infty)$	
$ax^2 + bx + c \geq 0$	one real root, r	$(-\infty, \infty)$	
$ax^2 + bx + c < 0$	two real roots, r_1 and r_2	(r_1, r_2)	
$ax^2 + bx + c \leq 0$	two real roots, r_1 and r_2	(r_1, r_2)	
$ax^2 + bx + c > 0$	two real roots, r_1 and r_2	$(-\infty, r_1) \cup (r_2, \infty)$	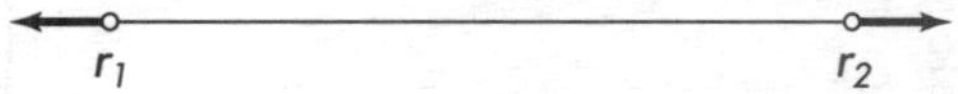
$ax^2 + bx + c \geq 0$	two real roots, r_1 and r_2	$(-\infty, r_1] \cup [r_2, \infty)$	

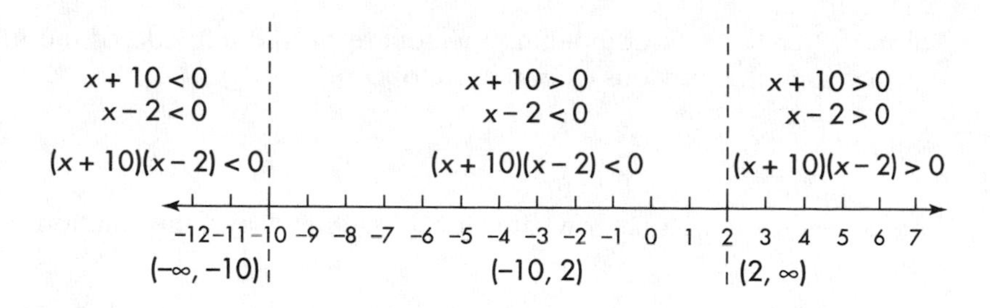

FIGURE 3.8

Similarly, the solution set for $x^2 + 8x - 20 \geq 0$ would be $(-\infty, -10] \cup [2, \infty)$. Here are some other examples of solving quadratic inequalities:

- Solve $x^2 - 6x + 9 < 0$.
  The equation $x^2 - 6x + 9 = 0$ has exactly one real root, 3; therefore, $x^2 - 6x + 9 = (x - 3)^2 < 0$ has no solution, since $(x - 3)^2$ can never be negative no matter what value x takes on.
- Solve $x^2 - 6x + 9 \leq 0$.
  The equation $x^2 - 6x + 9 = 0$ has exactly one real root, 3; therefore, $x^2 - 6x + 9 = (x - 3)^2 \leq 0$ has solution set $\{3\}$, since $(x - 3)^2$ is zero at 3 and is positive everywhere else. Only the number 3 makes the inequality true.
- Solve $x^2 - 6x + 9 \geq 0$.
  The equation $x^2 - 6x + 9 = 0$ has exactly one real root, 3; therefore, the solution set for $x^2 - 6x + 9 = (x - 3)^2 \geq 0$ is $(-\infty, \infty)$, since $(x - 3)^2$ is always nonnegative (positive or zero) no matter what value x assumes.
- Solve $x^2 - 6x + 9 > 0$.
  The equation $x^2 - 6x + 9 = 0$ has exactly one real root, 3; therefore, the solution set for $x^2 - 6x + 9 = (x - 3)^2 > 0$ is $(-\infty, 3) \cup (3, \infty)$, since $(x - 3)^2$ is positive at every real number except at the number 3, where it has the value zero. Therefore, 3 must be excluded from the solution set.
- Solve $x^2 + x + 1 > 0$. Since the discriminant of $x^2 + x + 1 = 0$ is $1 - 4 = -3$, which is less than zero, we know that $x^2 + x + 1 = 0$ has no real roots; however, we can complete the square on the left side of the inequality as follows:

$$x^2 + x + 1 > 0$$

$$x^2 + x > -1$$

$$x^2 + x + \frac{1}{4} > -1 + \frac{1}{4}$$

$$\left(x + \frac{1}{2}\right)^2 > -\frac{3}{4}$$

This is true for all replacements for x. Thus, the solution set is $(-\infty, \infty)$.

- Solve $x^2 + x + 1 \le 0$. Completing the square on the left side of the inequality as in the previous example, we obtain:

$$\left(x + \frac{1}{2}\right)^2 \le -\frac{3}{4}$$

  This is never true no matter what value x assumes. Thus, the solution set is $\varnothing$.

You shouldn't try to memorize the table given at the beginning of this section; rather, it is only necessary that you remember the following:

> If $ax^2 + bx + c$ $(a > 0)$ has no real roots, it is always positive; if it has exactly one real root, it is zero at that root; and if it has two real roots, it is negative between them, positive to the left of the leftmost root, positive to the right of the rightmost root, and zero only at its roots.

CHAPTER 3 SUMMARY

3.1 An **equation** is a statement that two mathematical expressions are equal. An equation may be true, or it may be false. A **one-variable linear equation** is an equation that can be written in the form $ax + b = 0$, $a \neq 0$. A **solution,** or **root,** of an equation is a number that when substituted for the variable makes the equation true. The set consisting of all the solutions of an equation is called its **solution set.** To **solve an equation** means to find its solution set. If the solution set consists of all real numbers, the equation is called an **identity.** If the solution set is empty, the equation has no solution. Equations that have the same solution set are called **equivalent.**

An equation is solved when the variable has a coefficient of 1 and is by itself on one and only one side of the equation. Two main tools used in solving equations are the following:

1. Addition or subtraction of the same quantity on both sides of the equation
2. Multiplication or division by the same *nonzero* quantity on both sides of the equation

Except for some special situations, if you *do the same thing to both sides* of an equation, the result will be an equivalent equation.

To solve a one-variable linear equation:

1. Remove grouping symbols, if any.
2. Eliminate fractions, if any.
3. Undo indicated addition or subtraction to isolate the variable on one side.

4. If necessary, factor the side containing the variable so that one of the factors is the variable.
5. Divide both sides by the coefficient of the variable.
6. Check the solution in the original equation.

3.2 An equation that expresses the relationship between two or more variables is called a **formula.** The procedure for solving one-variable linear equations can be used to solve a formula for a specific variable when the value(s) of the other variable(s) is (are) known. The procedure also can be used to solve a formula or **literal equation** (an equation with no numbers, only letters) for a specific variable in terms of the other variable(s). In general, isolate the specific variable and treat all other variable(s) as constants. This is called **changing the subject** of the formula or literal equation.

3.3 The procedure for solving one-variable linear equations can be used to solve a two-variable equation for one variable in terms of the other variable. Another common use for the procedure for solving one-variable linear equations is to transform equations into the form $y = mx + b$, where m and b are constants.

3.4 One-variable absolute value equations can be solved using the procedure for solving one-variable linear equations and the following:
$|ax + b| = 0 \Leftrightarrow ax + b = 0$; or
If c is any positive number:
$|ax + b| = c \Leftrightarrow$ either $ax + b = c$ or $ax + b = -c$

3.5 An **inequality** is a mathematical statement that two expressions are not equal. We use the symbols $>$ (greater than), $<$ (less than), $\geq$ (greater than or equal to), and $\leq$ (less than or equal to) to write one-variable linear inequalities. The **graph** of the solution set of the inequality can be illustrated on a number line. When solving inequalities, treat them just like equations *except* for one difference: If you multiply or divide both sides of the inequality by a *negative* number, *reverse* the direction of the inequality. Sometimes two statements of inequality apply to a variable expression simultaneously and, thus, can be combined into a **double inequality.**

3.6 One-variable absolute value inequalities can be solved using the procedure for solving one-variable linear equations and the following:
If c is any positive number:
$|ax + b| < c \Leftrightarrow -c < ax + b < c$ (**conjunction**); and
$|ax + b| > c \Leftrightarrow ax + b < -c$ or $ax + b > c$ (**disjunction**)

3.7 A **one-variable quadratic equation** is an equation that can be written in the **standard form** $ax^2 + bx + c = 0$. The constants a, b, and c are called the **coefficients** of the quadratic equation. The solutions of a quadratic equation are also called its **roots.** A quadratic equation may have exactly *one* real root, *two* real unequal roots, or *no* real roots. Three methods for solving quadratic equations are (1) by **factoring,** (2) by **completing the square,** and (3) by using the **quadratic formula.**

To solve a quadratic equation by factoring:

1. Express the equation in standard form: $ax^2 + bx + c = 0$
2. Factor the left side of the equation.
3. Set each factor containing the variable equal to zero.
4. Solve each of the resulting linear equations.
5. Check each root in the original equation.

Step 3 in this procedure is based on the **property of zero products** for numbers: If the product of two quantities is zero, at least one of the quantities is zero.

Quadratic equations that can be written in the form $x^2 = C$ have the solutions $x = \pm\sqrt{C}$. If the quantity C is *zero,* there is *one* real root, which has the value zero; if *positive,* there are *two* unequal real roots; and if *negative,* there are *no* real roots.

To solve a quadratic equation by completing the square:

1. Get all terms containing the variable on the left side, and all other terms on the right.
2. If the coefficient of the squared term is not 1, divide each term by it.
3. Add the square of half the coefficient of the middle term to both sides.
4. Factor the left side as the square of a binomial.
5. Find the principal square root of both sides.
6. Solve for the variable.
7. Check each root in the original equation.

To solve a quadratic equation by using the quadratic formula:

1. Express the equation in standard form: $ax^2 + bx + c = 0$
2. Determine the values of the coefficients a, b, and c.
3. Substitute into the quadratic formula: $x = \dfrac{-b \pm \sqrt{b^2 - 4ac}}{2a}$
4. Evaluate and simplify.
5. Check each root in the original equation.

The quantity $b^2 - 4ac$ is called the **discriminant** of the quadratic equation. The quadratic equation $ax^2 + bx + c = 0$ has exactly *one* real root if $b^2 - 4ac = 0$; *two* real unequal roots if $b^2 - 4ac > 0$; and *no* real roots if $b^2 - 4ac < 0$. *Never* divide both sides of an equation by the variable or by an expression containing the variable.

3.8 Many equations involving fractions in which a denominator expression contains the variable can be transformed into linear or quadratic equations by multiplying both sides of the equation by the LCD of all the fractions. An *excluded value* for the variable *cannot* be in the solution set. A **proportion** is a statement of equality between two ratios. A shortcut for multiplying both sides by the LCD in a proportion is to "cross-multiply." Many equations containing radicals or fractional exponents can be transformed into linear or quadratic equations by raising both sides to an appropriate power. Use caution when doing this because the solution set of the transformed equation may contain an **extraneous**

root, a value that is *not* a solution of the original equation. Therefore, check all answers obtained in the *original* equation.

Equations that are not quadratic equations but can be written in the form of a quadratic can be solved using the methods for solving quadratic equations. It naturally follows that equations that can be written so that the one side is a factorable higher degree polynomial can be solved by factoring completely, then setting each factor equal to zero.

3.9 The solution sets for quadratic inequalities are based on the rules for multiplying signed numbers: If two factors have the same sign, their product is positive; if they have opposite signs, their product is negative. If $ax^2 + bx + c$ $(a > 0)$ has no real roots, it is always positive; if it has exactly one real root, it is zero at that root; and if it has two real roots, it is negative between them, positive to the left of the leftmost root, positive to the right of the rightmost root, and zero only at its roots.

PRACTICE PROBLEMS FOR CHAPTER 3

Solve:

1. $3x + 50 = 35$
2. $25 - 3t = 3t + 13$
3. $7.5 = \frac{y}{7}$
4. $\frac{x}{2} - 1 = \frac{3x}{4} + 3$
5. $0.3x + 0.4(15 - x) = 5.2$
6. $\frac{1}{t} + \frac{1}{5} = 1$
7. $\frac{90}{x} = \frac{3}{4}$

Solve for the indicated variable:

8. $P = 2(l + w)$, for w
9. $A = \frac{h}{2}(b_1 + b_2)$, for b_1

Write in the form $y = mx + b$:

10. $3x - 2y + 4 = 0$
11. $3(x - 2) = -5(y - 8)$
12. $\frac{4}{x - 3} = \frac{2}{y + 3}$

Find the solution set and indicate it on a number line:

13. $2x + 1 > 5$
14. $3 + 4x \leq 15$

15. $3 - 4x \leq 15$
16. $-7 < 2x + 1 < 5$
17. $1 \leq 4x + 13 < 25$
18. $|5x + 2| = 12$
19. $|5x + 2| < 12$
20. $|5x + 2| \geq 12$

Solve using any convenient method:

21. $x^2 - 3x - 18 = 0$
22. $6x^2 = 17x + 5$
23. $(x + 1)(x + 3) = 15$
24. $4x^2 = 49$
25. $3x^2 = 15x$
26. $x^2 + 3x = 20$
27. $\dfrac{4}{2x - 3} = \dfrac{2(x - 1)}{5}$
28. $\sqrt{x - 3} = 5$
29. $(x + 2)^{\frac{1}{3}} = 5$
30. $\sqrt{3x + 10} = x$

Find the solution set and indicate it on a number line:

31. $x^2 - 10x + 25 \leq 0$
32. $x^2 - 10x + 25 > 0$
33. $x^2 - 10x + 25 \geq 0$
34. $x^2 - 3x - 18 \leq 0$
35. $x^2 - 3x - 18 \geq 0$

SOLUTIONS TO PRACTICE PROBLEMS

(Checks will not be shown.)

Solve:

1. $$\begin{aligned} 3x + 50 &= 35 \\ 3x &= 35 - 50 \\ 3x &= -15 \\ x &= -5 \end{aligned}$$

2. $$\begin{aligned} 25 - 3t &= 3t + 13 \\ -3t - 3t &= 13 - 25 \\ -6t &= -12 \\ t &= 2 \end{aligned}$$

3. $$\begin{aligned} 7.5 &= \frac{y}{7} \\ 7(7.5) &= \frac{\cancel{7}}{1}\left(\frac{y}{\cancel{7}}\right) \\ 52.5 &= y \end{aligned}$$

4. $$\frac{x}{2} - 1 = \frac{3x}{4} + 3$$

$$\frac{\cancel{4}^{2}}{1}\left(\frac{x}{\cancel{2}}\right) - 4(1) = \frac{\cancel{4}}{1}\left(\frac{3x}{\cancel{4}}\right) + 4 \cdot 3$$

$$2x - 4 = 3x + 12$$

$$-x = 16$$

$$x = -16$$

5. $$0.3x + 0.4(15 - x) = 5.2$$

$$0.3x + 6 - 0.4x = 5.2$$

$$-0.1x = -0.8$$

$$x = 8$$

6. $$\frac{1}{t} + \frac{1}{5} = 1$$

$$\frac{5\cancel{t}}{1}\left(\frac{1}{\cancel{t}}\right) + \frac{\cancel{5}t}{1}\left(\frac{1}{\cancel{5}}\right) = 5t(1)$$

$$5 + t = 5t$$

$$-4t = -5$$

$$t = 1.25$$

7. $$\frac{90}{x} = \frac{3}{4}$$

$$3x = 360$$

$$x = 120$$

Solve for the indicated variable:

8. $$P = 2(l + w), \text{ for } w$$

$$P = 2l + 2w$$

$$P - 2l = 2w$$

$$\frac{P - 2l}{2} = w$$

$$w = \frac{P - 2l}{2}$$

9. $$A = \frac{h}{2}(b_1 + b_2), \text{ for } b_1$$

$$A = \frac{hb_1}{2} + \frac{hb_2}{2}$$

$$2A = hb_1 + hb_2$$

$$2A - hb_2 = hb_1$$

$$\frac{2A - hb_2}{h} = b_1$$

$$b_1 = \frac{2A - hb_2}{h}$$

Write in the form $y = mx + b$:

10. $3x - 2y + 4 = 0$
$$-2y = -3x - 4$$
$$y = \frac{3}{2}x + 2$$

11. $3(x - 2) = -5(y - 8)$
$$3x - 6 = -5y + 40$$
$$5y = -3x + 46$$
$$y = -\frac{3}{5}x + 9.2$$

12. $\dfrac{4}{x - 3} = \dfrac{2}{y + 3}$
$$4(y + 3) = 2(x - 3)$$
$$4y + 12 = 2x - 6$$
$$4y = 2x - 18$$
$$y = \frac{1}{2}x - 4.5$$

Find and illustrate the solution set on a number line:

13. $2x + 1 > 5$
$$2x > 4$$
$$x > 2$$

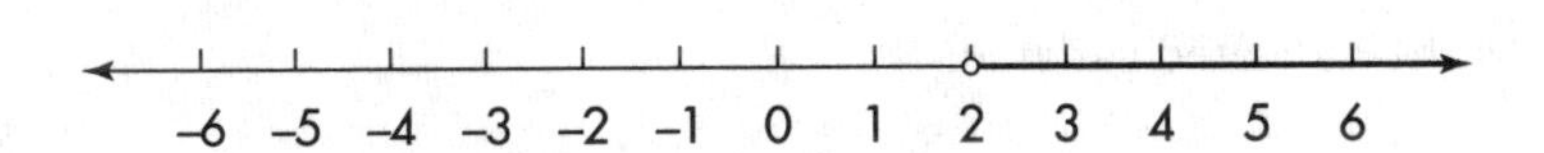

14. $3 + 4x \leq 15$
$$4x \leq 12$$
$$x \leq 3$$

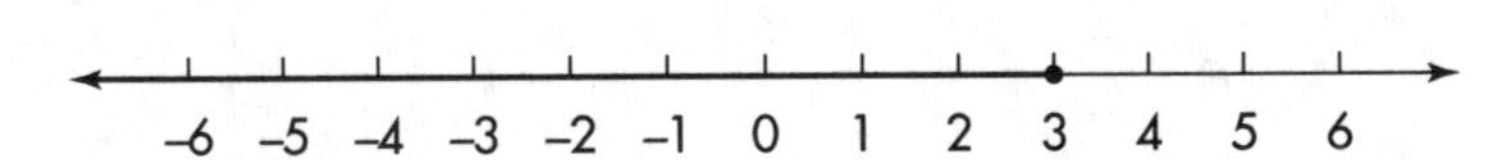

15. $3 - 4x \leq 15$
$$-4x \leq 12$$
$$x \geq -3$$

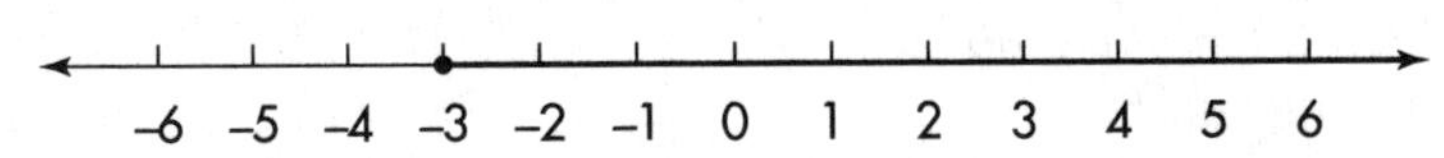

16. $-7 < 2x + 1 < 5$
$$-8 < 2x < 4$$
$$-4 < x < 2$$

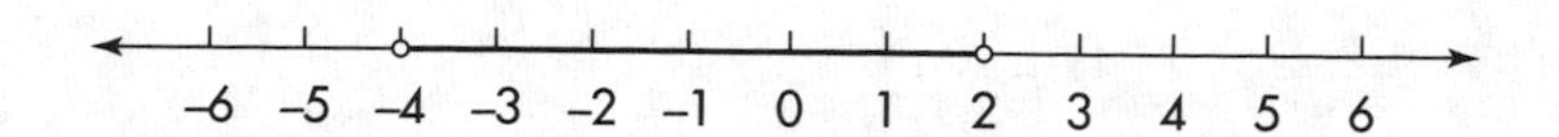

17. $$\begin{aligned} 1 &\le 4x + 13 < 25 \\ -12 &\le \quad 4x \quad < 12 \\ -3 &\le \quad x \quad < 3 \end{aligned}$$

−6 −5 −4 −3 −2 −1 0 1 2 3 4 5 6

18. $|5x + 2| = 12$

$$|5x + 2| = 12 \Leftrightarrow 5x + 2 = 12 \quad \text{or} \quad 5x + 2 = -12$$
$$5x = 10 \quad \text{or} \quad 5x = -14$$
$$x = 2 \quad \text{or} \quad x = -2.8$$

−6 −5 −4 −3 −2 −1 0 1 2 3 4 5 6

19. $|5x + 2| < 12$

$$|5x + 2| < 12 \Leftrightarrow -12 < 5x + 2 < 12$$
$$-14 < \quad 5x \quad < 10$$
$$-2.8 < \quad 5x \quad < 2$$

−6 −5 −4 −3 −2 −1 0 1 2 3 4 5 6

20. $|5x + 2| \ge 12$

$$|5x + 2| \ge 12 \Leftrightarrow 5x + 2 \le -12 \quad \text{or} \quad 5x + 2 \ge 12$$
$$5x \le -14 \quad \text{or} \quad 5x \ge 10$$
$$x \le -2.8 \quad \text{or} \quad x \ge 2$$

−6 −5 −4 −3 −2 −1 0 1 2 3 4 5 6

Solve using any convenient method:

21. $x^2 - 3x - 18 = 0$

$$(x - 6)(x + 3) = 0$$
$$x - 6 = 0 \quad \text{or} \quad x + 3 = 0$$
$$x = 6 \quad \text{or} \quad x = -3$$

22. $6x^2 = 17x + 5$

$$6x^2 - 17x - 5 = 0$$
$$x = \frac{-(-17) \pm \sqrt{(-17)^2 - 4(6)(-5)}}{2(6)}$$
$$= \frac{17 \pm \sqrt{289 + 120}}{12}$$

$$= \frac{17 \pm \sqrt{409}}{12}$$

$$x = \frac{17 + \sqrt{409}}{12} \text{ or } x = \frac{17 - \sqrt{409}}{12}$$

23. $(x + 1)(x + 3) = 15$

$$x^2 + 4x + 3 = 15$$
$$x^2 + 4x - 12 = 0$$
$$(x + 6)(x - 2) = 0$$
$$x + 6 = 0 \quad \text{or} \quad x - 2 = 0$$
$$x = -6 \quad \text{or} \quad x = 2$$

24. $4x^2 = 49$

$$x^2 = \frac{49}{4}$$

$$|x| = \sqrt{\frac{49}{4}}$$

$$|x| = \frac{\sqrt{49}}{\sqrt{4}}$$

$$|x| = \frac{7}{2}$$

$$x = \pm\frac{7}{2}$$

25. $3x^2 = 15x$

$$3x^2 - 15x = 0$$
$$3x(x - 5) = 0$$
$$3x = 0 \quad \text{or} \quad x - 5 = 0$$
$$x = 0 \quad \text{or} \quad x = 5$$

26. $x^2 + 3x = 20$

$$x^2 + 3x - 20 = 0$$

$$x = \frac{-3 \pm \sqrt{3^2 - 4(1)(-20)}}{2(1)}$$

$$= \frac{-3 \pm \sqrt{9 + 80}}{2}$$

$$= \frac{-3 \pm \sqrt{89}}{2}$$

$$x = \frac{-3 + \sqrt{89}}{2} \text{ or } x = \frac{-3 - \sqrt{89}}{2}$$

27. $$\frac{4}{2x-3} = \frac{2(x-1)}{5}$$
$$2(x-1)(2x-3) = 20$$
$$(x-1)(2x-3) = 10$$
$$2x^2 - 5x + 3 = 10$$
$$2x^2 - 5x - 7 = 0$$
$$x = \frac{-(-5) \pm \sqrt{(-5)^2 - 4(2)(-7)}}{2(2)}$$
$$= \frac{5 \pm \sqrt{25 + 56}}{4} = \frac{5 \pm \sqrt{81}}{4} = \frac{5 \pm 9}{4}$$
$$x = \frac{5+9}{4} = \frac{14}{4} = 3.5 \text{ or } x = \frac{5-9}{4} = \frac{-4}{4} = -1$$

28. $$\sqrt{x-3} = 5$$
$$(\sqrt{x-3})^2 = 5^2$$
$$x - 3 = 25$$
$$x = 28$$

You *must* check in the original equation:

$$\sqrt{28-3} \stackrel{?}{=} 5$$
$$\sqrt{25} \stackrel{?}{=} 5$$
$$5 \stackrel{\surd}{=} 5$$

29. $$(x+2)^{\frac{1}{3}} = 5$$
$$((x+2)^{\frac{1}{3}})^3 - 5^3$$
$$x + 2 = 125$$
$$x = 123$$

You *must* check in the original equation:

$$(123 + 2)^{\frac{1}{3}} \stackrel{?}{=} 5$$
$$(125)^{\frac{1}{3}} \stackrel{?}{=} 5$$
$$5 \stackrel{\surd}{=} 5$$

30. $$\sqrt{3x+10} = x$$
$$(\sqrt{3x+10})^2 = x^2$$
$$3x + 10 = x^2$$
$$-x^2 + 3x + 10 = 0$$
$$x^2 - 3x - 10 = 0$$
$$(x-5)(x+2) = 0$$
$$x - 5 = 0 \quad \text{or} \quad x + 2 = 0$$
$$x = 5 \quad \text{or} \quad x = -2$$

You *must* check in the original equation:

$\sqrt{3 \cdot 5 + 10} \stackrel{?}{=} 5$ $\quad$ $\sqrt{3 \cdot (-2) + 10} \stackrel{?}{=} -2$; This can never be true;

$\sqrt{25} \stackrel{?}{=} 5$ $\quad$ thus, -2 is an extraneous root.

$5 \stackrel{\checkmark}{=} 5$

Thus, 5 is a root.

Find the solution set and indicate it on a number line:

31. $x^2 - 10x + 25 = 0$ has one real root, 5. Therefore, $x^2 - 10x + 25 = (x - 5)^2 \leq 0$ has solution set $\{5\}$; that is, 5 is the *only* number that makes the inequality $x^2 - 10x + 25 = (x - 5)^2 \leq 0$ true.

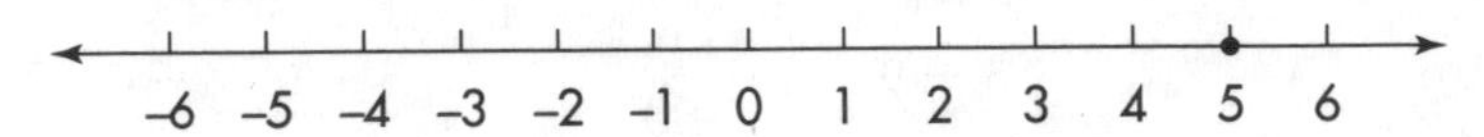

32. $x^2 - 10x + 25 = 0$ has one real root, 5. Therefore, $x^2 - 10x + 25 = (x - 5)^2 > 0$ has solution set $(-\infty, 5) \cup (5, \infty)$; that is, every real number *except* 5 makes the inequality $x^2 - 10x + 25 = (x - 5)^2 > 0$ true.

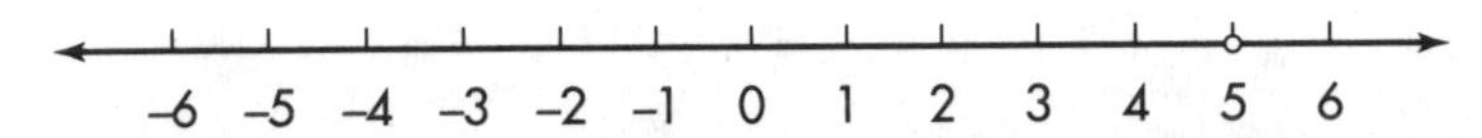

33. $x^2 - 10x + 25 = 0$ has one real root, 5. Therefore, $x^2 - 10x + 25 = (x - 5)^2 \geq 0$ has solution set $(-\infty, \infty)$; that is, every real number makes the inequality $x^2 - 10x + 25 = (x - 5)^2 \geq 0$ true.

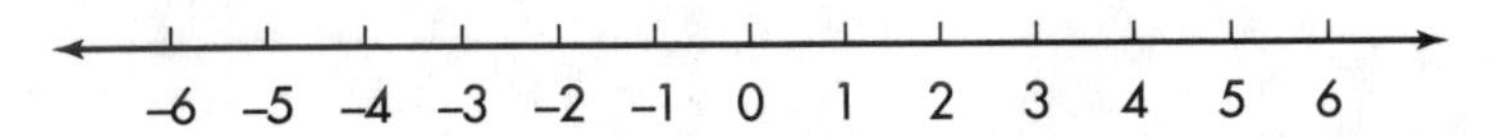

34. $x^2 - 3x - 18 = 0$ has two roots, -3 and 6. Therefore, the solution set for $x^2 - 3x - 18 = (x + 3)(x - 6) \leq 0$ is $[-3, 6]$.

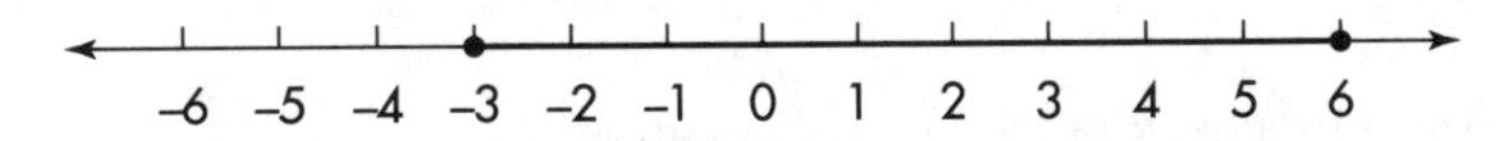

35. $x^2 - 3x - 18 = 0$ has two roots, -3 and 6. Therefore, the solution set for $x^2 - 3x - 18 = (x + 3)(x - 6) \geq 0$ is $(-\infty, -3] \cup [6, \infty)$.

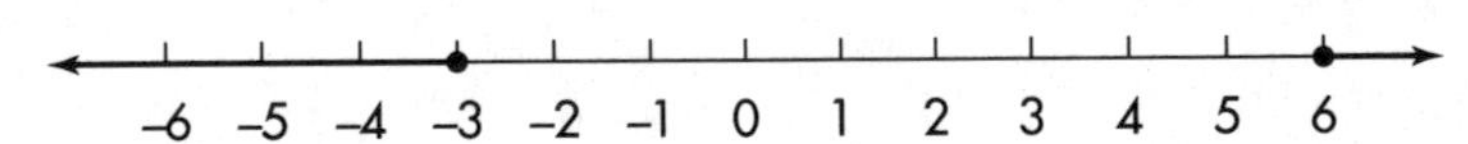

4
PROBLEM SOLVING

4.1 TRANSLATING INTO MATHEMATICAL LANGUAGE

To be successful in problem solving, you must be able to write word statements and phrases as *mathematical* statements and phrases. A table of common word phrases and their corresponding mathematical phrases follows. You can use a variety of letters to represent an unknown quantity. Frequently x is used, but it does not have to be.

TABLE 4.1 MATHEMATICAL PHRASES

Word Phrase	Mathematical Phrase
Addition	
the *sum* of x and y	$x + y$
a number *plus* 10	$t + 10$
a number *added* to 6	$6 + n$
5 *more than* a number	$x + 5$
a number *exceeded by* 50	$n + 50$
a number *increased by* 2	$y + 2$
an amount *increased by* 5%	$x + 5\%(x) = x + 0.05x$
Subtraction	
the *difference* between x and y	$x - y$
the *difference* between y and x	$y - x$
a number *subtracted from* 6	$6 - n$
6 *subtracted from* a number	$n - 6$
a number *minus* 10	$t - 10$
10 *minus* a number	$10 - t$
5 *less than* a number	$x - 5$
a number *decreased by* 2	$y - 2$
a number *diminished by* 7	$z - 7$
an amount *decreased by* 5%	$x - 5\%(x) = x - 0.05x$
Multiplication	
a number *multiplied by* 3	$3x$
7 *times* a number	$7a$
twice a number	$2y$
the *product* of 4 and a number	$4x$

TABLE 4.1 Continued

$\frac{3}{4}$ *of* a number	$\frac{3}{4}x$ or $\frac{3x}{4}$
half of a number	$\frac{1}{2}x$ or $\frac{x}{2}$ or $0.5x$
5 percent *of* a number	$5\%(x) = 0.05x$
Division	
a number *divided by* 5	$\frac{x}{5}$
5 *divided by* a number	$\frac{5}{x}$
the *quotient* of a number and 4	$\frac{y}{4}$
the *quotient* of 4 and a number	$\frac{4}{y}$
the *ratio* of a number and 3	$\frac{n}{3}$
the *ratio* of 3 and a number	$\frac{3}{n}$
a number *over* 10	$\frac{z}{10}$

A helpful message: Subtraction and division are not commutative; that is, in general, $a - b \neq b - a$ and $\frac{a}{b} \neq \frac{b}{a}$. Consequently, you must read problems carefully so that you write the numbers in a subtraction or division mathematical phrase in the correct order.

4.2 WORKING WITH UNITS OF MEASUREMENT

Application problems usually involve units of measurement, such as inches, pounds, years, dollars, and so forth. Every measurement includes both a number and a unit. In mathematical computations involving units of measurement, the units term must be included as part of the completely defined measure and must undergo the same mathematical operations.

You can add or subtract units of measurement only if they are both measures of weight, both measures of distance, and so forth. Multiplication or division of units of measurement should occur only when the resulting product or quotient has meaning.

4.3 STEPS IN PROBLEM SOLVING

The focus in this section is not on any particular type of problem; rather, the emphasis is on learning a strategy for attacking application problems. A general procedure for solving problems using algebra consists of the following steps:

PROBLEM SOLVING STRATEGY

1. Read the problem carefully.
2. Read the question and decide what you are to find.
3. Represent all unknowns with variable expressions.
4. Write an equation(s) representing the facts.
5. Solve and check the equation(s).
6. Verbalize the solution and check it in the context of the problem.

Bear in mind that problem solving seldom occurs in a step-by-step fashion. You may find yourself going back to a previous step or skipping steps outlined above. Nevertheless, the steps in the procedure can assist you in understanding and solving a multitude of problems.

1. *Read the problem carefully* to understand it and find out what it is about. Don't try to solve it at this point. Look for and underline key words to help you break the problem down. List all the information you can find in the problem. Try restating the facts to make them more specific. Are the facts consistent with your knowledge of the real world?

   Try to visualize in your mind the situation described in the problem. To help with this, you can do one or more of the following:
   - Draw a picture, diagram, or graph.
   - Make a table or chart.
   - Make a physical model or act out the problem, if possible.
2. *Read the question and decide what you are to find.* Look for a sentence that contains words or phrases like "what is," "find," "how many," "determine," and the like. This is usually (but not always) the last sentence in the problem. Draw a double line under this part of the problem. If the problem will involve more than one step, list the steps in the order in which they should be done.

   Identify the information that you will need to use in the problem. Ask yourself, "Are any facts missing? Are there definitions or facts needed that I should know? Is there information given that I don't need? Is there a formula for this problem that I should know? Is there a concept needed I should recall?"

   If the problem involves units of measurement, determine what should be the units for the answer. Often, this can be the key to deciding how to work the problem.

3. *Represent all unknowns with variable expressions.* When representing an unknown by a variable, make an explicit statement of the form, "Let $x = \ldots$," so that you will be sure to know what your variable(s) represent(s). If you are solving using one-variable equations, but the problem has two or more variables, you should be judicious in selecting the unknown to be represented by the variable.

   In problems involving one or more unknowns, sometimes one (or more) of the unknowns is described in terms of another unknown. For such situations, designate the variable as the unknown used in the description. For example, you can represent "The length is 3 meters more than the width" by the following:

   Let w = the width. (since length is described in terms of width)

   Then express the "described" unknown in terms of the designated variable by translating the description:

$$\underbrace{\text{The length}}_{\text{length}} \quad \underbrace{\text{is}}_{=} \quad \underbrace{\text{3 meters more than the width.}}_{3 + w}$$

   Let $w + 3$ = the length.

   Sometimes, when you have multiple unknowns, it is not obvious which should be designated "x." Don't let that circumstance unnerve you. Just pick an unknown and call it "x." Then reread the problem to determine how to represent any other unknowns in terms of x. To do this, look for relationships that exist between the unknown numbers. Sometimes, it is convenient to let the variable represent the *smaller* (or *smallest*) unknown.

   Another helpful device is to temporarily use two variables in an equation that expresses a given relationship, such as:

   "There are 33 coins in all."

$$x + y = 33$$

   Then solve for one of the variables in terms of the other variable:

$$y = 33 - x$$

   Use what you get as the representation of the desired variable.

4. *Write an equation(s) representing the facts.* Read the problem again, this time translating each word phrase into a mathematical phrase and using the variable representations determined in Step 3. The word *is* or some other form of the verb *to be* usually indicates equality (=), as do *equals, results in, yields,* and similar verbs. Place the mathematical phrases into an equation that represents the facts in the problem. Look for a formula or relationship (that you have not already used in Step 3) to help you arrange the mathematical phrases in the equation. Check that the equation makes sense, that it shows the relationship accurately. If units are

involved, check that the indicated calculations will result in the proper units for the answer.

If you have trouble getting started on Step 4:

- Use "a whole equals the sum of its parts."
- Look for a pattern.
- Use a similar, but simpler, related problem.
- Try guessing and checking an answer against the facts given in the problem.
- Try using what you know about the proper units needed for the answer.

5. *Solve and check the equation(s).* Use appropriate procedures to solve and check the equation(s). You may find it convenient to omit units while you are solving the equation, since you have already checked that the solution will result in an answer that has the proper units. Do keep in mind, however, that the units term is part of a completely defined measure and is subject to the same mathematical operations as the number to which it is attached.
6. *Verbalize the solution and check it in the context of the problem.* Check the solution against the facts in the problem statement. You may also want to check the solution in any pictures, diagrams, tables, charts, and so forth, that you used for organizing information. Have you answered all the questions asked in the problem? Does the solution satisfy the facts of the problem? Is it reasonable? This step should not be neglected because by verbalizing the solution in the context of the problem, you can better decide whether it makes sense.

The following examples of problem solving will demonstrate the use of the six-step problem-solving strategy. When looking at these examples, realize there are usually multiple ways to solve a problem. The examples show one way, but you may think of others.

Ratio-Proportion-Percent Problems

Solve: The specific gravity of a substance can be computed as the ratio of the weight of a given volume of that substance to the weight of an equal volume of water. If zinc has specific gravity 7.29, find the weight in pounds of 1 ft^3 of zinc. Water weighs 62.4 lb/ft^3.

Solution:

Underline key words:

The specific gravity of a substance can be computed as the ratio of the weight of a given volume of that substance to the weight of an equal volume of water. If zinc has specific gravity 7.29, and water weighs 62.4 lb/ft^3, find the weight in pounds of 1 ft^3 of zinc.

List all the information. Try restating the facts to make them more specific:

Specific gravity is the ratio of the weight of a substance to the weight of an equal volume of water.

Zinc has specific gravity 7.29.
Water weighs 62.4 lb/ft^3.

Definition needed:

A **ratio** is another name for the quotient of two quantities.

Pound (lb) is a unit of weight.

Cubic foot (ft^3) is a unit of volume.

Try restating the facts to make them more specific:

$$\text{Specific gravity of zinc} = \frac{\text{weight of 1 ft}^3\text{ of zinc}}{\text{weight of 1 ft}^3\text{ of water}}$$

What are you to find?

. . . find the weight in pounds of 1 ft^3 of zinc. *Find the weight (in pounds) of 1 ft^3 of zinc.*

Represent all unknowns with variable expressions:

Let x = weight (in pounds) of 1 ft^3 of zinc.

Write an equation using the facts and definitions:

$$\text{Specific gravity of zinc} = \frac{\text{what a given volume of zinc weighs}}{\text{what the same volume of water weighs}}$$

$$= \frac{\text{weight of 1 ft}^3\text{ of zinc}}{\text{weight of 1 ft}^3\text{ of water}}$$

$$7.29 = \frac{x}{62.4\text{ lb}}$$

Solve the equation:

$$7.29 = \frac{x}{62.4\text{ lb}}$$

$$x = (7.29)(62.4\text{ lb})$$

$$x \cong 454.9\text{ lb}$$

Check the solution in the equation:

$$7.29 \stackrel{?}{=} \frac{454.9\text{ lb}}{62.4\text{ lb}}$$

$$7.29 \stackrel{\checkmark}{=} \frac{454.9}{62.4} \cong 7.29$$

Verbalize the solution:

Solution: The weight of 1 ft^3 of zinc is 454.9 lb (approximately).

Check it in the context of the problem:

The specific gravity of zinc can be computed as the ratio of 454.9 lb (what 1 ft^3 of zinc weighs) to 62.4 lb (what 1 ft^3 of water weighs). This is about 450 divided by 60, which is about 7.5. The answer seems reasonable in the context of the problem. ✓

Solve: Susie, a sales clerk in a computer store, makes a 6% commission on all items she sells. Last week, she earned $570 in commission. What was the total of her sales for the week?

Solution:

Underline key words:

Susie, a sales clerk in a computer store, makes a 6% commission on all items she sells. Last week, she earned $570 in commission. What was the total of her sales for the week?

List all the information. Try restating the facts to make them more specific:

Susie's commission rate is 6%.
Susie's commission is based on her total sales.
Susie's commission was $570.

Definition needed:

Percent means "per hundred." When used in computations, percents are changed to fractions or decimals. Thus, $6\% = \frac{6}{100} = 0.06$.

Formula needed:

$P = RB$, where

P is the **percentage,** the "part of the whole."

R is the **rate,** the number with "%" or the word "percent" attached.

B is the **base,** the "whole amount."

The **key idea** in percent problems is that *a specified percent of an amount is that percent multiplied times the amount.*

An effective strategy is first to identify P, R, and B; then write and solve an equation for the unknown quantity.

A helpful message: In application problems, a percent without a base is usually meaningless. Make sure you identify the base associated with each percent mentioned in a problem.

Try restating the facts to identify P, R, and B:

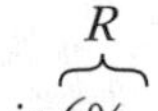

Susie's commission rate is $\overbrace{6\%}^{R}$.

$$\text{Susie's commission is based on her } \overbrace{\text{total sales}}^{B}.$$

$$\overbrace{\text{Susie's commission}}^{P} \text{ was \$570.}$$

What are you to find?

What was the total of her sales for the week? *Find Susie's total sales for the week.*

Represent all unknowns with variable expressions:

Let B = amount of Susie's total sales (in dollars).

Write an equation using the formula:

$$\overbrace{\text{Susie's commission}}^{P} = \overbrace{6\%}^{R} \cdot \overbrace{\text{(her total sales)}}^{B}$$

$$\$570 = 6\% \cdot B$$

$$\$570 = 0.06B$$

(\$ = \$; The units are correct.)

Solve the equation:

$$0.06B = \$570$$

$$B = \$9{,}500$$

Check the solution in the equation:

$$0.06(\$9{,}500) \stackrel{?}{=} \$570$$

$$\$570 \stackrel{\checkmark}{=} \$570$$

Verbalize the solution:

Solution: Susie's sales for the week totaled \$9,500.

Check it in the context of the problem:

Susie has total sales of \$9,500. She makes a 6% commission on this amount. This is 6% of \$9,500, or \$570. ✓

Integer and Age Problems

Solve: Twice the sum of three consecutive integers is 42 more than twice the sum of the first and third integers. Find the integers.

Solution:

Underline key words:

Twice the sum of three consecutive integers is 42 more than twice the sum of the first and third integers. Find the integers.

Diagram the facts, restating to make them more specific:

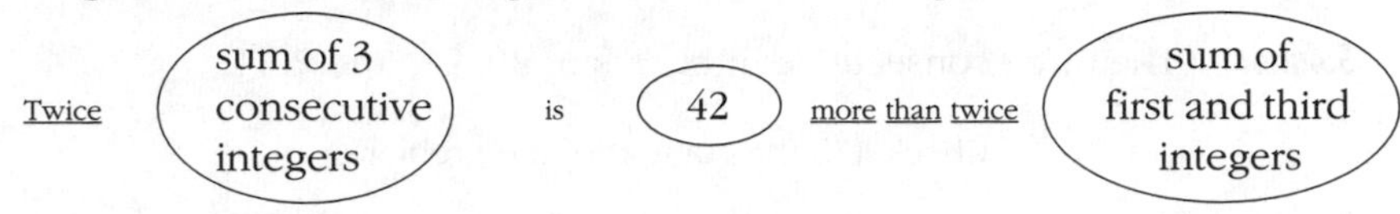

Definition needed:

Consecutive integers are integers that differ by 1.

What are you to find?

Find the integers. *Find the three consecutive integers.*

Represent all unknowns with variable expressions:

Let n = first integer (smallest).
Let $n + 1$ = second integer.
Let $(n + 1) + 1 = n + 2$ = third integer.

Put this information into your diagram:

Twice $n + (n + 1) + (n + 2)$ is 42 more than twice $n + (n + 2)$

Write an equation using your diagram as a guide:

Twice $n + (n + 1) + (n + 2)$ is 42 more than twice $n + (n + 2)$

$$2[n + (n + 1) + (n + 2)] \quad = \quad 42 \quad + \quad 2[n + (n + 2)]$$

Solve the equation for all unknowns:

$$\begin{aligned}
2[n + (n + 1) + (n + 2)] &= 42 + 2[n + (n + 2)] \\
2(3n + 3) &= 42 + 2(2n + 2) \\
6n + 6 &= 42 + 4n + 4 \\
2n &= 40 \\
n &= 20 = \text{first integer} \\
n + 1 &= 20 + 1 = 21 = \text{second integer} \\
n + 2 &= 20 + 2 = 22 = \text{third integer}
\end{aligned}$$

The solution is 20, 21, and 22.

Check the solution in the equation:

$$\begin{aligned}
2[20 + (20 + 1) + (20 + 2)] &\stackrel{?}{=} 42 + 2[20 + (20 + 2)] \\
2[20 + 21 + 22] &\stackrel{?}{=} 42 + 2[20 + 22] \\
2[63] &\stackrel{?}{=} 42 + 2[42] \\
126 &\stackrel{\checkmark}{=} 126
\end{aligned}$$

Verbalize the solution:

Solution: The three consecutive integers are 20, 21, and 22.

Check it in the context of the problem:

20, 21, and 22 are consecutive integers. ✓ Their sum is 63. Twice this is 126.

The sum of the first integer, 20, and the third integer, 22, is 42. Twice this is 84.

42 more than 84 is 126; this is the same as twice the sum of the integers. ✓

Solve: Wayne is twice as old as Elvia. Five years from now the sum of their ages will be 64. How old are Wayne and Elvia now?

Solution:

Underline key words:

Wayne is twice as old as Elvia. Five years from now the sum of their ages will be 64. How old are Wayne and Elvia now?

List all the information. Try restating the facts to make them more specific:

Wayne's age now is twice Elvia's age now.
Five years from now, the sum of their ages will be 64.
The key idea in age problems is that comparison of ages is usually done within a specified time period—either in the present, future, or past. It is helpful to make a table listing the ages at the different time periods given in the problem.

Make a table to organize the facts given in the problem.

When?	**Elvia's age (in years)**	**Wayne's age (in years)**	**Sum (in years)**
Now	Elvia's age now	Wayne's age now = twice Elvia's age	Elvia's age now + Wayne's age now
5 years from now	Elvia will be 5 years older.	Wayne will be 5 years older.	64

What are you to find?

How old are Wayne and Elvia now? *Find Wayne's and Elvia's ages (in years) now.*

Represent all unknowns with variable expressions:

Since Wayne's age now is described in terms of Elvia's age now:

Let x = Elvia's age now (in years).

Express Wayne's age in terms of x:

Wayne's age now is twice Elvia's age now.

Wayne's age now = $2x$

Let $2x$ = Wayne's age now (in years).

Update the table.

When?	Elvia's age (in years)	Wayne's age (in years)	Sum (in years)
Now	x	$2x$	$x + 2x$
5 years from now	Elvia will be 5 years older: $x + 5$	Wayne will be 5 years older: $2x + 5$	64

Write an equation using the information in the table:

$$\underbrace{\text{Elvia's age 5 yr from now}}_{x+5} + \underbrace{\text{Wayne's age 5 yr from now}}_{2x+5} = \underbrace{\text{Sum of their ages 5 yr from now}}_{64}$$

(years + years = years; The units are correct.)

Solve the equation, omitting the units for convenience:

$$x + 5 + 2x + 5 = 64$$
$$3x = 54$$
$$x = 18 \text{ (Elvia's age now in years)}$$
$$2x = 36 \text{ (Wayne's age now in years)}$$

Check the solution in the equation:

$$18 + 5 + 2(18) + 5 \stackrel{?}{=} 64$$
$$64 \stackrel{\checkmark}{=} 64$$

Verbalize the solution:

Solution: Wayne is 36 years old now and Elvia is 18 years old now.

Check it in the context of the problem:

Wayne is 36, which is twice 18, Elvia's age. ✓

Five years from now, Wayne will be 41 (36 plus 5) and Elvia will be 23 (18 plus 5); the sum of their ages will be 41 plus 23, which is 64. ✓

Check in the table:

When?	Elvia's age (in years)	Wayne's age (in years)	Sum (in years)
Now	18	$36 = 2(18) =$ twice Elvia's age ✓	not needed
5 years from now	$18 + 5 = 23$	$36 + 5 = 41$	$23 + 41 = 64$ ✓

Coin and Mixture Problems

Solve: A collection of quarters and dimes amounts to $4.35. If there are 33 coins in all, find the number of each coin.

Solution:

Underline key words:

A collection of quarters and dimes amounts to $4.35. If there are 33 coins in all, find the number of each coin.

List all the information. Try restating the facts to make them more specific:

The total value of a collection of quarters and dimes is $4.35.

There are 33 coins (quarters and dimes) altogether.

Facts needed:

The value of one quarter is $0.25 and the value of one dime is $0.10.

Concept needed:

The key idea in coin problems is that *the value for each type of coin is determined by multiplying the value of each coin times the number of coins you have of that denomination*. You must organize the data carefully to attend to this.

It is helpful to make a table that lists the number and value of the coins.

Make a table to organize the information:

Denomination	**Quarters**	**Dimes**	**Total**
Value per coin	\$0.25	\$0.10	not applicable
Number of coins	number of quarters	number of dimes	33
Value	\$0.25(no. of quarters)	\$0.10(no. of dimes)	\$4.35

What are you to find?

. . . find the number of each coin. *Find the number of quarters and the numbers of dimes in the collection.*

Represent all unknowns with variable expressions:

Neither unknown is expressed in terms of the other unknown. Try using two variables (temporarily):

Let x = number of quarters in the collection.

Let y = number of dimes in the collection.

From the facts, we can write:

$x + y = 33$, which we can solve for y to obtain:

$y = 33 - x$ = number of dimes in the collection

Update the table:

Denomination	**Quarters**	**Dimes**	**Total**
Value per coin	\$0.25	\$0.10	not applicable
Number of coins	x	$33 - x$	33
Value (in dollars)	\$0.25(no. of quarters) = $0.25x$	\$0.10(no. of dimes) = $0.10(33 - x)$	\$4.35

Write an equation from the information in the chart, being sure to check the units:

$$\underbrace{\text{Total value of quarters}} + \underbrace{\text{Total value of dimes}} = \underbrace{\text{Total value of all the coins}}$$

$$\$0.25x + \$0.10(33 - x) = 4.35$$

(\$ + \$ = \$; The units are correct.)

Solve the equation for all unknowns, omitting the units for convenience:

$$0.25x + 0.10(33 - x) = 4.35$$
$$0.25x + 3.30 - 0.10x = 4.35$$
$$0.15x = 1.05$$
$$x = 7 \text{ (the number of quarters)}$$
$$33 - x = 33 - 7 = 26 \text{ (the number of dimes)}$$

Check the solution in the equation:

$$\$0.25(7) + \$0.10(33 - 7) \stackrel{?}{=} \$4.35$$
$$\$0.25(7) + \$0.10(26) \stackrel{?}{=} \$4.35$$
$$\$1.75 + \$2.60 \stackrel{?}{=} \$4.35$$
$$\$4.35 \stackrel{\surd}{=} \$4.35$$

Verbalize the solution:

Solution: There are 7 quarters and 26 dimes in the collection.

Check it in the context of the problem:

The value of 7 quarters is \$0.25 times 7, which is \$1.75.
The value of 26 dimes is \$0.10 times 26, which is \$2.60.
The collection of quarters and dimes has a total value of \$1.75 plus \$2.60, which is \$4.35. ✓
7 quarters and 26 dimes equals 33 coins in all. ✓

Check in the table:

Denomination	**Quarters**	**Dimes**	**Total**
Value per coin	\$0.25	\$0.10	not applicable
Number of coins	7	26	7 + 26 = 33 ✓
Value	\$0.25(7) = \$1.75	\$0.10(26) = \$2.60	\$1.75 + \$2.60 = \$4.35 ✓

Solve: How many ounces of distilled water must be added to 30 ounces of an 80% alcohol solution to yield a 50% solution?

Solution:

Underline key words:

How many ounces of <u>distilled water</u> must be added to 30 ounces of an <u>80% alcohol solution</u> to yield a <u>50% solution</u>?

List all the information:

The 30 ounce solution is 80% alcohol.

We need to make a solution that is 50% alcohol by adding distilled water to the 80% solution.

Facts needed:

Distilled water contains no alcohol. It is a 0% alcohol solution.

Are the facts consistent with your knowledge of the real world?

The concentration of a given substance in a solution is the amount of the substance per volume of solution. Adding distilled water to a 80% solution will make it weaker (it will dilute the solution.). A 50% solution is weaker than a 80% solution, so the facts are consistent with common knowledge.

Concept needed:

The amount of a given substance in a solution (or mixture) equals *the percent of substance contained in the solution* (or mixture) *times the total amount of the solution* (or mixture).

It is helpful to draw a picture when working mixture problems. It should show three containers: the two original containers (*before mixing*) and a container for the new solution (*after mixing*).

Usually, the **key idea** in a mixture problem is that *the amount of a given substance you have before mixing should equal the amount of that substance you will have after mixing.*

Alcohol is the substance that should remain constant in this problem.

Draw a picture to help you visualize the situation:

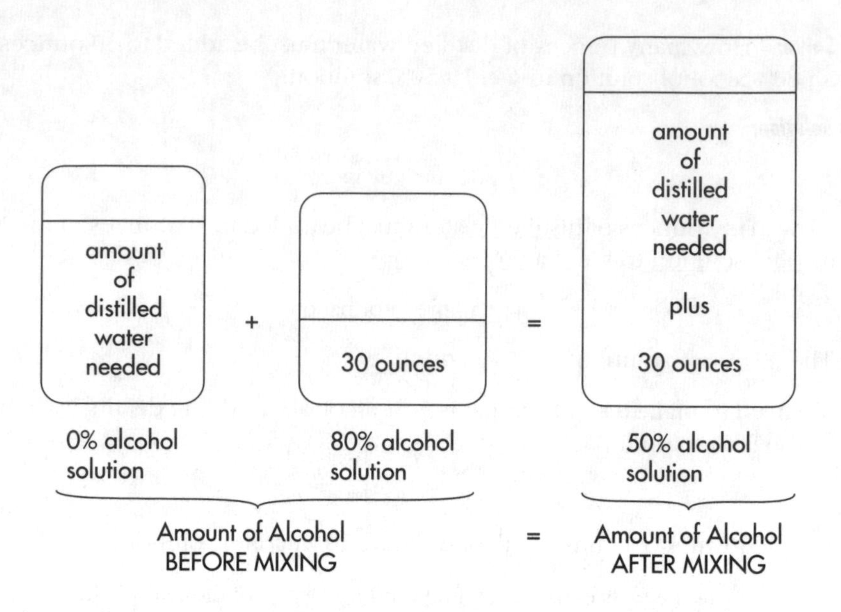

What are you to find?

How many ounces of distilled water must be added to 30 ounces of an 80% alcohol solution to yield a 50% solution? *Find how many ounces of distilled water must be added.*

Represent all unknowns with variable expressions:

From the picture, you can see that you do not know how much distilled water is to be added *and* you do not know how much total solution you will have after you have added it. Therefore, there are two unknowns, but the second can be found from the first:

Let x = number of ounces of distilled water needed.

Let $x + 30$ oz = total amount of final solution (in ounces).

Update your picture:

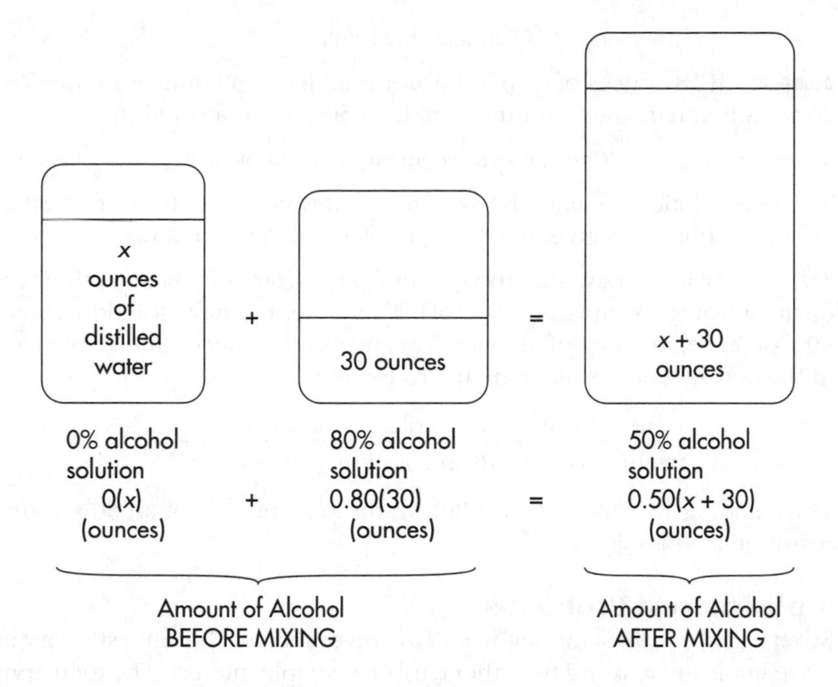

Write an equation from the facts given, being sure to check the units:

$$\underbrace{\begin{array}{c}\text{Amount of alcohol}\\ \text{before mixing}\end{array}} = \underbrace{\begin{array}{c}\text{Amount of alcohol}\\ \text{after mixing}\end{array}}$$

$$0x + 0.80(30 \text{ oz}) = 0.50(x + 30 \text{ oz})$$

$$(\text{ounces} + \text{ounces} = \text{ounces; The units are correct.})$$

Solve the equation for all unknowns, omitting the units for convenience:

$$\begin{aligned} 0.80(30) &= 0.50(x + 30) \\ 24 &= 0.5(30 + x) \\ 24 &= 15 + 0.5x \\ -0.5x &= 15 - 24 \\ -0.5x &= -9 \\ x &= 18 \text{ ounces of distilled water} \\ x + 30 &= 48 \text{ ounces of 50\% alcohol solution} \end{aligned}$$

Check the solution in the equation:

$$\begin{aligned} 0.80(30 \text{ oz}) &\stackrel{?}{=} 0.50(18 \text{ oz} + 30 \text{ oz}) \\ 24 \text{ oz} &\stackrel{?}{=} 0.5(48 \text{ oz}) \\ 24 \text{ oz} &\stackrel{\checkmark}{=} 24 \text{ oz} \end{aligned}$$

Verbalize the solution:

Solution: If 18 ounces of distilled water is added to 30 ounces of an 80% alcohol solution the new solution will be a 50% alcohol solution.

Check it in the context of the problem:

The new solution contains 18 oz of distilled water plus 30 oz of an 80% alcohol solution. This gives a total of 48 oz in the new solution.

The alcohol in the new solution comes from the 80% alcohol solution (since distilled water contains no alcohol). Therefore, the new solution contains 80% of 30 oz, or 24 oz of alcohol. This means the concentration of alcohol in the new solution is given by the following:

$$\frac{\text{no. of ounces of alcohol}}{\text{total no. of ounces in solution}} = \frac{24 \text{ oz}}{48 \text{ oz}} = 50\% \checkmark$$

Thus, adding 18 ounces of distilled water to 30 ounces of an 80% solution results in a 50% solution. ✓

Simple Interest Problems

Solve: Paul receives money from two investments. One investment earns 12% simple interest and the other earns 6% simple interest. The total annual return on the two investments if $660. If the amount invested at 6% is $2,000 more than the amount invested at 12%, what is the total amount invested?

Solution:

Underline key words:

Paul receives money from two investments. One investment earns 12% simple interest and the other earns 6% simple interest. The total annual return on the two investments is $660. If the amount invested at 6% is $2,000 more than the amount invested at 12%, what is the total amount invested?

List all the information:

There are two investments, one at 12% simple interest and the other at 6% simple interest.

The total annual return on the two investments is $660.

The amount invested at 6% is $2,000 more than the amount invested at 12%.

Definition needed:

The **annual return** on an investment is how much money the investment earned in a period of 1 year.

Try restating the facts to make them more specific:

The total annual interest on the two investments is $660.

Formula needed:

Simple interest is $I = Prt$, where I is the interest earned on an amount P, called the **principal,** invested at a given rate r for t years. For a *one year investment,* the simple interest formula is $I = Pr$.

The **key idea** in a simple interest problem is that *the interest earned by an investment is determined by the rate and length of time invested at that rate.* You must organize the data carefully to attend to this. It will be helpful to make a table or chart.

Make a table to organize the facts:

	12% Investment	**6% Investment**	**Total**
Amount of investment	amt. invested at 12%	amt. invested at 6% = $2,000 more than amt. invested at 12%	sum of two investments
Simple interest rate	12% = 0.12	6% = 0.06	not needed
Time invested	1 year	1 year	1 year
Annual interest earned	0.12(amt. invested at 12%)	0.06(amt. invested at 6%)	$660

What are you to find?

. . . <u>what is the total amount invested</u>? *Find the total amount of the two investments.*

If the problem will involve more than one step, list the steps in the order in which they should be done:

To find the total amount invested:

Step 1. *Find the amount invested at 12% and the amount invested at 6%.*
Step 2. *Add these two amounts.*

Step 1: Represent all unknowns with variable expressions:

Since the amount invested at 6% is described in terms of the amount invested at 12%:

Let x = amount (in dollars) invested at 12%

$\underbrace{\text{The amt. invested at 6\%}}$ is $\underbrace{\text{\$2,000 more than the amt. invested at 12\%}}$

The amt. invested at 6% = $2,000 + x

Let x + $2,000 = amount (in dollars) invested at 6%.

Update the table:

	12% Investment	6% Investment	Total
Amount of investment	x (dollars)	x + \$2,000	x + (x + \$2,000) (Step 2)
Simple interest rate	12% = 0.12	6% = 0.06	Do not add rates unless they have the *same* base.
Time invested	1 year	1 year	1 year
Annual interest earned	0.12x (dollars)	0.06(x + \$2,000)	\$660

Write an equation using the information in the table, being sure to check the units:

$$\underbrace{\text{Annual Interest earned at 12\%}}_{0.12x} + \underbrace{\text{Annual Interest earned at 6\%}}_{0.06(x + \$2{,}000)} = \underbrace{\text{Total Annual Interest earned}}_{\$660}$$

(\$ + \$ = \$; The units are correct.)

Solve the equation, omitting the units for convenience:

$$0.12x + 0.06(x + 2{,}000) = 660$$
$$0.12x + 0.06x + 120 = 660$$
$$0.18x = 540$$
$$x = \$3{,}000 \text{ invested at } 12\%$$
$$x + \$2{,}000 = \$5{,}000 \text{ invested at } 6\%$$

Check the equation:

$$0.12(\$3{,}000) + 0.06(\$3{,}000 + \$2{,}000) \stackrel{?}{=} \$660$$
$$0.12(\$3{,}000) + 0.06(\$5{,}000) \stackrel{?}{=} \$660$$
$$\$360 + \$300 \stackrel{?}{=} \$660$$
$$\$660 \stackrel{\checkmark}{=} \$660$$

Step 2: Add the amounts of the two investments:

Total amount invested = \$3,000 + \$5,000 = \$8,000

Verbalize the solution:

Solution: The amount invested at 12% is \$3,000 and the amount invested at 6% is \$5,000. The total amount invested is \$8,000.

Check it in the context of the problem:

The $3,000 invested for 1 year earns 12% of $3,000. This is $360. The $5,000 invested for 1 year earns 6% of $5,000. This is $300. The total annual return on the two investments is $360 plus $300, which is $660. ✓

$5,000, the amount invested at 6%, is $2,000 more than $3,000, the amount invested at 12%. ✓

Therefore, the total amount invested is $3,000 plus $5,000, which is $8,000. ✓

Check the solution in the table:

	12% Investment	**6% Investment**	**Total**
Amount of investment	$3,000	$3,000 + $2,000 = $5,000 ✓	$3,000 + $5,000 = $8,000 ✓
Simple interest rate	12% = 0.12	6% = 0.06	Do not add rates unless they have the *same* base.
Time invested	1 year	1 year	1 year
Annual interest earned	0.12($3,000) = $360	0.06($5,000) = $300	$360 + $300 = $660 ✓

After you feel adept in problem solving, you may skip some of the steps or do them mentally.

Distance-Rate-Time Problems

Solve: Two cars leave a restaurant, one traveling due north at 65 mph and the other due south at 55 mph. In how many hours will the cars be 624 miles apart?

Solution:

Formula needed:

$d = rt$, where

d is the distance traveled at a uniform rate of speed r for a given time t.

The **key idea** in a distance-rate-time problem is that *a given distance traveled is determined by the rate and the time traveled at that rate.* You must organize the data so that you keep track of this. It will be helpful to make a chart or table.

Represent all unknowns with variable expressions:

Let t = number of hours the cars will travel to be 624 miles apart.

Draw a diagram:

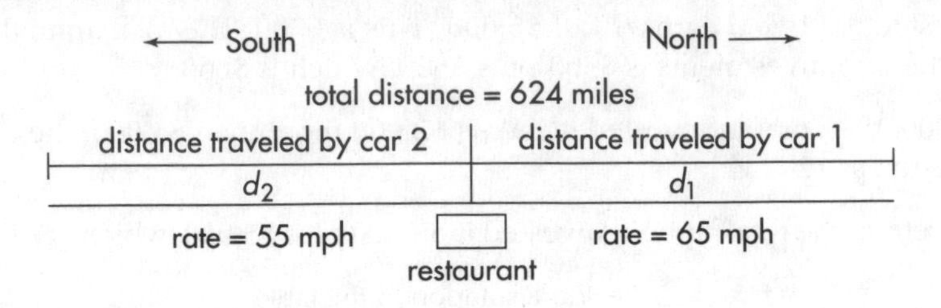

Make a chart:

Vehicle	**Distance**	**Rate**	**Time**
Car 1	$d_1 = (65 \text{ mph})t$	65 mph	t (hours)
Car 2	$d_2 = (55 \text{ mph})t$	55 mph	t (hours)
Total	624 miles	Do not add rates for different vehicles.	t (hours)

Write an equation using the information in the chart:

$$\underbrace{\text{Distance traveled by car 1}} + \underbrace{\text{Distance traveled by car 2}} = \underbrace{\text{Total distance traveled}}$$

$$65\ (\text{mph})t + (55\ \text{mph})t = 624 \text{ miles}$$

$$\left(\left(\frac{\text{miles}}{\cancel{\text{hr}}}\right)\cancel{\text{hr}} + \left(\frac{\text{miles}}{\cancel{\text{hr}}}\right)\cancel{\text{hr}} = \text{miles; The units are correct.}\right)$$

Solve the equation, omitting the units for convenience:

$$65t + 55t = 624$$
$$120t = 624$$
$$t = 5.2 \text{ hours}$$

Check the solution in the equation:

$$65(\text{mph})(5.2\text{ hr}) + (55\text{ mph})(5.2\text{ hr}) \stackrel{?}{=} 624 \text{ miles}$$
$$338 \text{ miles} + 286 \text{ miles} \stackrel{?}{=} 624 \text{ miles}$$
$$624 \text{ miles} \stackrel{\checkmark}{=} 624 \text{ miles}$$

Solution: The two cars will be 624 miles apart in 5.2 hours.

Check the solution in the context of the problem:

The first car travels 5.2 hours at 65 mph. This is 338 miles. The second car travels 5.2 hours at 55 mph. This is 286 miles. The total distance between them after 5.2 hours is 338 miles plus 286 miles, which is 624 miles. ✓

Check the solution in the chart:

Vehicle	Distance	Rate	Time
Car 1	$d_1 = (65 \text{ mph})(5.2 \text{ hr})$ $= 338 \text{ miles}$	65 mph	5.2 hours
Car 2	$d_2 = (55 \text{ mph})(5.2 \text{ hr})$ $= 286 \text{ miles}$	55 mph	5.2 hours
Total	338 mi + 286 mi = 624 miles ✓	Do not add rates for different vehicles.	5.2 hours

Work Problems

Solve: Ennis can paint a room in 6 hours working alone. Tim can do the same job working alone in 4 hours. How many hours will it take them working together to paint the room?

Solution:

Concept needed:

The **key idea** in work problems is that the rate at which work is done equals the amount of work accomplished divided by the amount of time worked; that is:

$$\text{rate} = \frac{\text{amount of work done}}{\text{time worked}}$$

If more than one worker is involved, it is helpful to make a chart showing the information about each worker. Then, usually, you will be able to determine the part of the task that can be done by each worker.

A helpful message: If a task is completed, the amount of work done is the *whole* task, so use "1 task" to represent the amount of work done.

Represent all unknowns with variable expressions:

Let $t =$ number of hours it will take for Ennis and Tim to paint the room working together.

Make a chart:

Worker	Rate	Amount of work	Time
Ennis	$\frac{1 \text{ room}}{6 \text{ hours}} = \frac{1}{6}$ room/hr	$\left(\frac{1}{6} \text{ room/hr}\right)t$	t (hours)
Tim	$\frac{1 \text{ room}}{4 \text{ hours}} = \frac{1}{4}$ room/hr	$\left(\frac{1}{4} \text{ room/hr}\right)t$	t (hours)
Total	Do not add rates for different workers.	1 room	t (hours)

Write an equation using the information in the chart:

$$\underbrace{\text{Part of room painted by Ennis}} + \underbrace{\text{Part of room painted by Tim}} = \text{Whole room painted}$$

$$\left(\frac{1}{6}\frac{\text{room}}{\text{hr}}\right)t + \left(\frac{1}{4}\frac{\text{room}}{\text{hr}}\right)t = 1 \text{ room}$$

$$\left(\left(\frac{\text{room}}{\cancel{\text{hr}}}\right)\cancel{\text{hr}} + \left(\frac{\text{room}}{\cancel{\text{hr}}}\right)\cancel{\text{hr}} = \text{room; The units are correct.}\right)$$

Solve the equation, omitting the units for convenience:

$$\left(\frac{1}{6}\right)t + \left(\frac{1}{4}\right)t = 1$$

$$4t + 6t = 24$$

$$10t = 24$$

$$t = 2.4 \text{ hours}$$

Check the solution in the equation:

$$\left(\frac{1}{6}\frac{\text{room}}{\cancel{\text{hr}}}\right)(2.4\ \cancel{\text{hr}}) + \left(\frac{1}{4}\frac{\text{room}}{\cancel{\text{hr}}}\right)(2.4\ \cancel{\text{hr}}) \stackrel{?}{=} 1 \text{ room}$$

$$0.4 \text{ room} + 0.6 \text{ room} \stackrel{?}{=} 1 \text{ room}$$

$$1.0 \text{ room} \stackrel{\checkmark}{=} 1 \text{ room}$$

Solution: It will take 2.4 hours for Ennis and Tim to paint the room working together.

Check the solution in the context of the problem:

Since Ennis can do the whole job in 6 hours, his rate is $\frac{1}{6}$ room/hr. If he works 2.4 hours, the amount of the room he paints is $\frac{1}{6}$ room/hr times 2.4

hr, or 0.4 of the room. Since Tim can do the whole job in 4 hours, his rate is $\frac{1}{4}$ room/hr. If he works 2.4 hours, the amount of the room he paints is $\frac{1}{4}$ room/hr times 2.4 hr, or 0.6 of the room. Together, they complete 0.4 of the room plus 0.6 of the room, which is 1.0 room painted. ✓

Check the solution in the chart:

Worker	Rate	Amount of work	Time
Ennis	$\frac{1 \text{ room}}{6 \text{ hours}} = \left(\frac{1}{6} \text{ room/hr}\right)$	$\left(\frac{1}{6} \text{ room/hr}\right)(2.5 \text{ hr}) =$ 0.4 room	2.4 hours
Tim	$\frac{1 \text{ room}}{4 \text{ hours}} = \left(\frac{1}{4} \text{ room/hr}\right)$	$\left(\frac{1}{4} \text{ room/hr}\right)(2.4 \text{ hr}) =$ 0.6 room	2.4 hours
Total	Do not add rates for different workers.	0.4 room + 0.6 room = 1.0 room = 1 room ✓	2.4 hours

A helpful message: Notice that the worker who takes *less* time working alone works at a *faster* rate and, therefore, completes more of the job than the other worker.

Geometry Problems

Solve: The length of a rectangle is 3 meters more than its width. If the area is 10 square meters, find the dimensions of the rectangle.

(Note: See Chapter 6 for a solution using two variables.)

Solution:

Formula needed:

(See Appendix C for a list of formulas used in geometry.)

In geometry problems, using the correct formula is very important.

The area of a rectangle with length l and width w is $A = lw$.

Represent all unknowns with variable expressions:

Let $w =$ the width of the rectangle (in meters).

Let 3 meters + $w =$ the length of the rectangle (in meters).

Making a diagram is usually essential for proper understanding of a geometry problem.

Make a diagram:

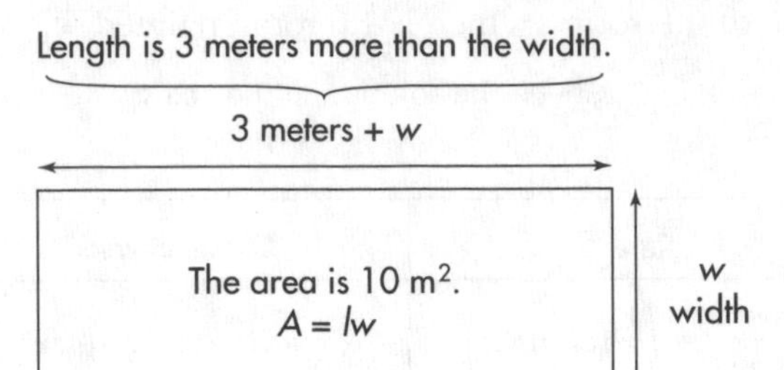

Write an equation using the information in the diagram:

$$(3 \text{ meters} + w)w = 10 \text{ m}^2$$

((meters + meters)meters = (meters)meters = m^2; The units are correct.)

Solve the equation, omitting the units for convenience:

$$(3 + w)w = 10$$
$$3w + w^2 = 10$$
$$w^2 + 3w - 10 = 0$$
$$(w - 2)(w + 5) = 10$$

$w = 2$ meters = width or $w = -5$ meters (reject because width should not be negative)

3 meters $+ w = 3$ meters $+ 2$ meters $= 5$ meters = length

Check the equation:

$$(3 \text{ meters} + w)w \stackrel{?}{=} 10 \text{ m}^2$$
$$(3 \text{ meters} + 2 \text{ meters})(2 \text{ meters}) \stackrel{?}{=} 10 \text{ m}^2$$
$$(5 \text{ meters})(2 \text{ meters}) \stackrel{?}{=} 10 \text{ m}^2$$
$$10 \text{ m}^2 \stackrel{\checkmark}{=} 10 \text{ m}^2$$

Solution: The rectangle has dimensions 5 meters by 2 meters.

Check the solution in the context of the problem:

Three meters more than the width is 3 meters plus 2 meters, which is 5 meters, the same as the length of the rectangle. ✓

The area of the rectangle is 5 meters times 2 meters, which is 10 m^2. ✓

Check the diagram:

5 meters is 3 meters more than 2 meters. ✓

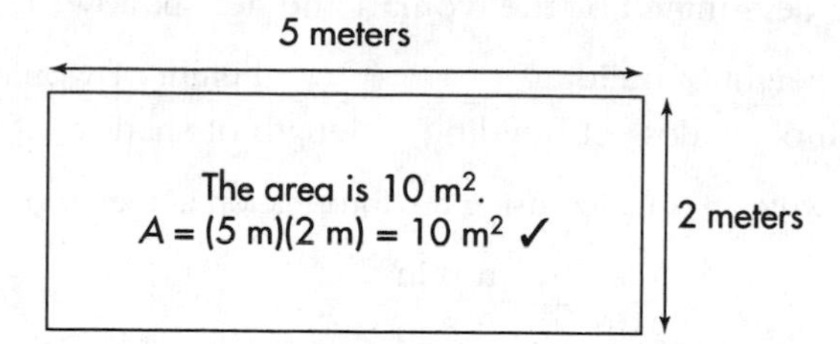

Solve: When a building casts a shadow 25 feet long, a woman who is 5 feet 6 inches tall casts a shadow 4 feet 8 inches long. What is the approximate height of the building?

Solution:

Represent all unknowns with variable expressions:

Let h = the height of the building (in feet).

Make a diagram: (It does not have to be drawn to scale):

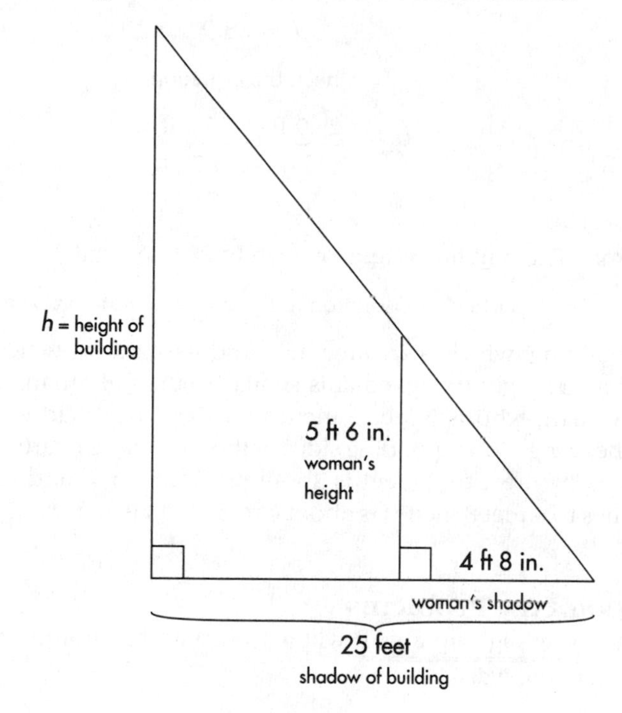

Concept needed:

(See Appendix C for a list of geometric concepts.)

If two triangles are similar, their corresponding angles are equal and corresponding sides are proportional.

The building and its shadow determine a right triangle that is similar to the right triangle determined by the woman and her shadow. Therefore,

$$\frac{\text{height of building}}{\text{length of shadow of building}} = \frac{\text{height of woman}}{\text{length of shadow of woman}}$$

Write an equation using the information in the diagram:

$$\frac{h}{25 \text{ ft}} = \frac{5 \text{ ft } 6 \text{ in}}{4 \text{ ft } 8 \text{ in}}$$

$$\frac{h}{25 \text{ ft}} = \frac{5.5 \text{ ft}}{4.75 \text{ ft}}$$

$$\left(\frac{\text{ft}}{\text{ft}} = \frac{\text{ft}}{\text{ft}}; \text{ The units are correct.}\right)$$

Solve the equation, omitting the units for convenience:

$$4.75h = (25)(5.5)$$

$$4.75h = 137.5$$

$$h \cong 28.9 \text{ feet}$$

Check the equation:

$$\frac{28.9 \text{ ft}}{25 \text{ ft}} \overset{\checkmark}{\cong} \frac{5.5 \text{ ft}}{4.75 \text{ ft}}$$

$$1.16 \overset{\checkmark}{\cong} 1.16$$

Solution: The building is approximately 28.9 feet tall.

Check the solution in the context of the problem:

The building, which is 28.9 feet tall, and its shadow, which is 25 feet long, determine a right triangle that is similar to the right triangle determined by the woman, who is 5 feet 6 inches tall, and her shadow, which is 4 feet 8 inches long. Corresponding sides of the two triangles are proportional. The ratio of 28.9 feet to 25 feet is about 30 to 25, or $\frac{6}{5}$, and the ratio of 5 feet 6 inches to 4 feet 8 inches is about 6 to 5, or $\frac{6}{5}$ also.

Trigonometry Problems

Solve: How high up a wall will a 13 ft wire reach if the wire is anchored 12 ft from the wall?

Solution: Represent all unknowns with variable expressions:

Let a = the height (in feet) the wire reaches up the wall.

Making a diagram is usually essential for proper understanding of a trigonometry problem.

Make a diagram:

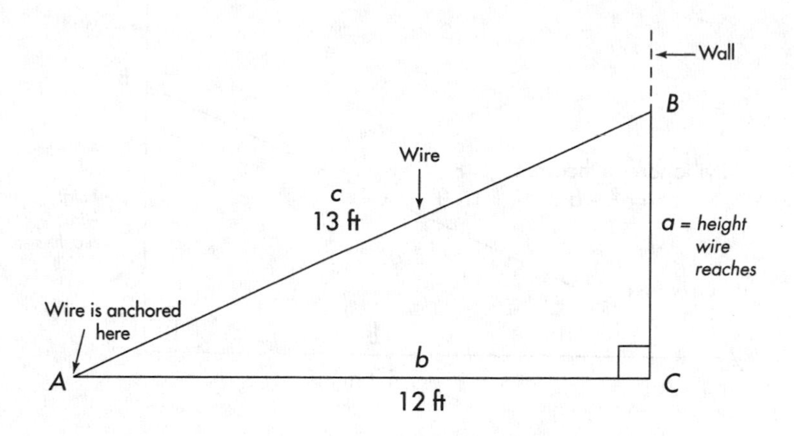

Formula needed:

(See Appendix C for a list of formulas used in trigonometry.)

In trigonometry problems, using the correct formula is very important.

The wire and the wall form a right triangle.

Given right triangle *ABC*:

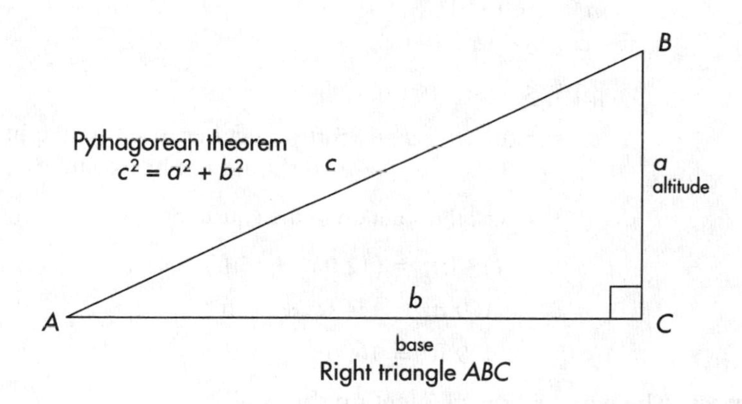

Right triangle *ABC*

> The Pythagorean theorem states that in a right triangle the square of the hypotenuse equals the sum of the squares of the other two sides, or $c^2 = a^2 + b^2$.

Update the diagram:

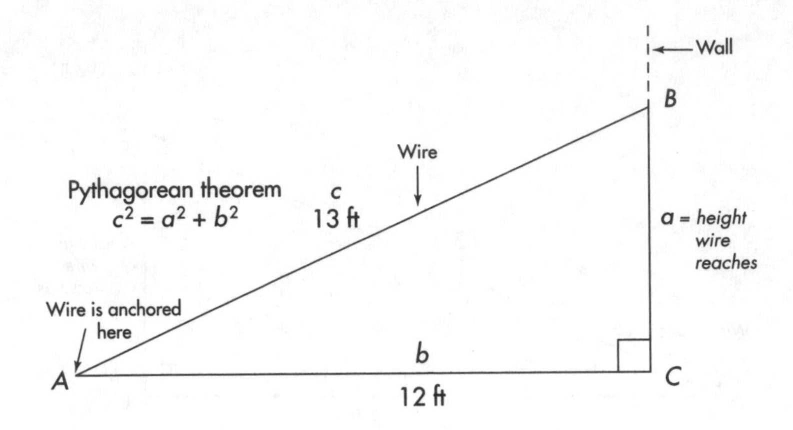

Write an equation using the information in the diagram:

$$(13 \text{ ft})^2 = (12 \text{ ft})^2 + a^2$$

$$((\text{ft})^2 = \quad (\text{ft})^2 + (\text{ft})^2\text{; The units are correct.})$$

Solve the equation, omitting the units for convenience:

$$(13)^2 = (12)^2 + a^2$$

$$(12)^2 + a^2 = (13)^2$$

$$144 + a^2 = 169$$

$$a^2 = 169 - 144$$

$$a^2 = 25$$

$$|a| = 5$$

$a = 5$ ft or $a = -5$ ft (reject because the height should not be negative)

Check the solution in the equation:

$$(13 \text{ ft})^2 \stackrel{?}{=} (12 \text{ ft})^2 + (5 \text{ ft})^2$$

$$169 \text{ ft}^2 \stackrel{?}{=} 144 \text{ ft}^2 + 25 \text{ ft}^2$$

$$169 \text{ ft}^2 \stackrel{\checkmark}{=} 169 \text{ ft}^2$$

Solution: The wire will reach 5 feet up the wall.

Check the solution in the context of the problem:

The wall and the wire determine a right triangle. The wire is the hypotenuse of the right triangle. The base is 12 feet, the distance to where the wire is anchored. The altitude is 5 feet, the height up the wall. By the Pythagorean theorem, the square of the hypotenuse, $(13 \text{ ft})^2$, which is 169 ft^2, equals the sum of the squares of the other two sides, 144 ft^2 plus 25 ft^2, which is also 169 ft^2. ✓

Check the diagram:

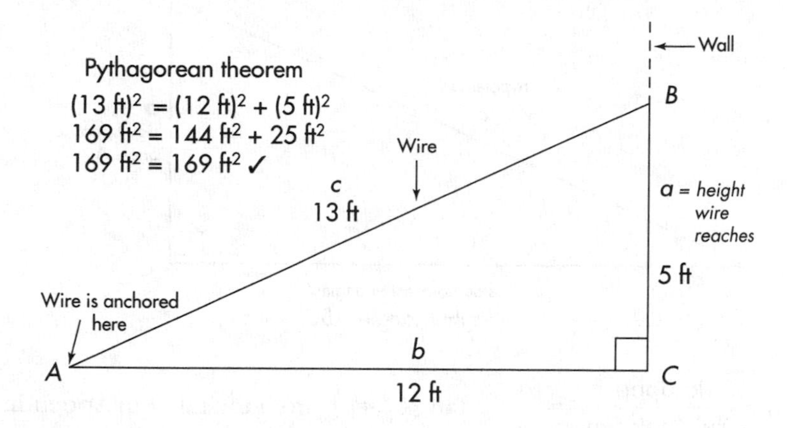

Solve: If a building casts a shadow 100 feet long when the sun is 30 degrees above the horizon, what is the height of the building?

Solution:

Represent all unknowns with variable expressions:

Let h = the height of the building (in feet).

Make a diagram:

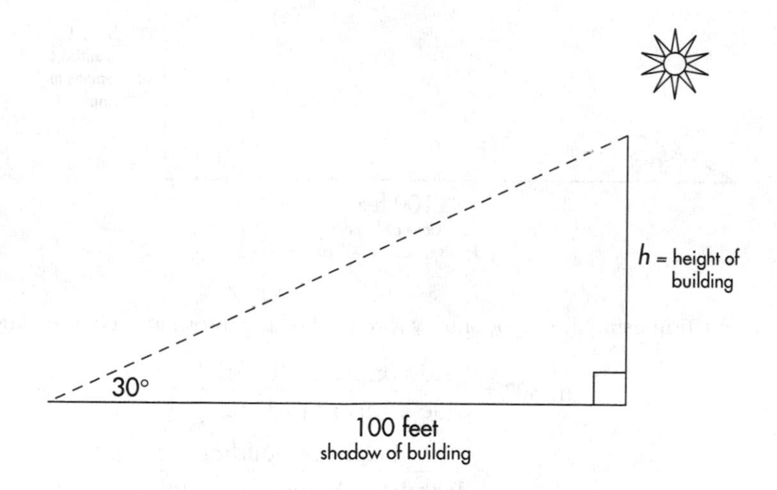

Write an equation using the information in the diagram:

The building and its shadow determine a right triangle.

Formula needed:

(See Appendix C for a list of formulas used in trigonometry.)

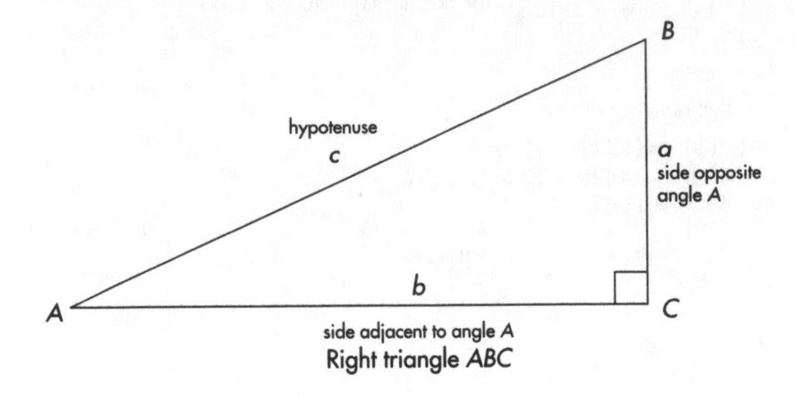

Right triangle ABC

$\tan A = \frac{\text{side opposite}}{\text{side adjacent}} = \frac{a}{b}$ $\tan 30° = \frac{1}{2}$ (from the table in Appendix C)

Update the diagram:

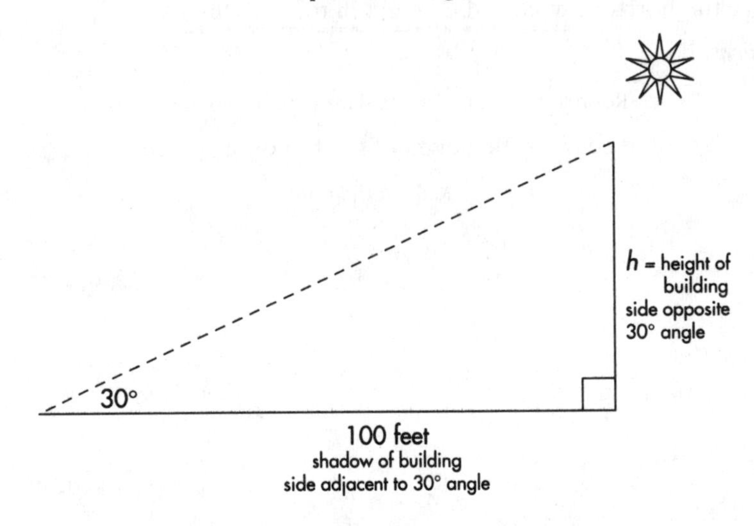

Write an equation using the trigonometry formula and the information in the diagram:

$$\tan 30° = \frac{\text{side opposite } 30° \text{ angle}}{\text{side adjacent to } 30° \text{ angle}}$$

$$= \frac{\text{height of building}}{\text{length of shadow of building}}$$

$$= \frac{h}{100 \text{ ft}}$$

Since $\tan 30° = \frac{1}{2}$, we have

$$\frac{1}{2} = \frac{h}{100 \text{ ft}}$$

Solve the equation:

$$2h = 100 \text{ ft}$$

$$h = 50 \text{ feet}$$

Check the equation:

$$\frac{1}{2} \stackrel{?}{=} \frac{50 \text{ ft}}{100 \text{ ft}}$$

$$\frac{1}{2} \stackrel{\checkmark}{=} \frac{1}{2}$$

Solution: The building is 50 feet tall.

Check the solution in the context of the problem:

When the sun is 30 degrees above the horizon, the building, which is 50 feet tall, and its shadow, which is 100 feet long, determine a right triangle. In a right triangle the tangent of an angle equals the ratio of the side opposite the angle to the side adjacent. The ratio of 50 feet to 100 feet is $\frac{1}{2}$; the tangent of 30° is also $\frac{1}{2}$. ✓

Statistical Problems

Solve: In order to earn a grade of A in her biology class, Yolanda must have an average of no less than 92 on four exams. She has made grades of 93, 87, and 89 on the first three exams. What is the lowest score she can make on the fourth exam and still receive a grade of A?

Solution:

Formula needed:

(See Appendix C for a list of formulas used in statistics.)

$$\text{mean (or average)} = \frac{\text{sum of the numbers}}{\text{how many numbers you have}}$$

Represent all unknowns with variable expressions:

Let x = Yolanda's score on the fourth exam.

Write an equation using the information in the problem:

$$\text{mean} = \frac{\text{sum of the 4 exam scores}}{4} = 92$$

$$\frac{93 + 87 + 89 + x}{4} = 92$$

Solve the equation:

$$\frac{\not{4}}{1}\left(\frac{93 + 87 + 89 + x}{\not{4}}\right) = 4(92)$$

$$93 + 87 + 89 + x = 368$$

$$269 + x = 368$$

$$x = 99$$

Check the solution in the equation:

$$\frac{93 + 87 + 89 + 99}{4} \stackrel{?}{=} 92$$

$$\frac{368}{4} \stackrel{?}{=} 92$$

$$92 \stackrel{\checkmark}{=} 92$$

Solution: Yolanda needs to make at least 99 on the fourth exam to make an A in the class.

CHAPTER 4 SUMMARY

4.1 To be successful in problem solving, you must be able to write word statements and phrases as *mathematical* statements and phrases. You can use a variety of letters to represent an unknown quantity. Frequently x is used, but it does not have to be.

4.2 Application problems usually involve **units of measurement,** such as inches, pounds, years, dollars, and so forth. In mathematical computations involving units of measurement, the units term must be included as part of the completely defined measure and must undergo the same mathematical operations.

4.3 A general procedure for **problem solving** is as follows:

1. Read the problem carefully.
2. Read the question and decide what you are to find.
3. Represent all unknowns with variable expressions.
4. Write an equation(s) representing the facts.
5. Solve and check the equation(s).
6. Verbalize the solution and check it in the context of the problem.

Problem solving seldom occurs in a step-by-step fashion. You may find yourself going back to a previous step or skipping steps.

The key idea in **percent** problems is that a specified percent of an amount is that percent multiplied times the amount. An effective strategy is first to identify P, R, and B; then write and solve an equation for the unknown quantity.

The key idea in **age** problems is that comparison of ages is usually done within a specified time period—either in the present, future, or past. It is helpful to make a table listing the ages at the different time periods given in the problem.
The key idea in **coin** problems is the total value for each type of coin is determined by multiplying the value of each coin times the number of coins you have of that denomination.
When solving **mixture** problems, the amount of a given substance in a solution (or mixture) equals the percent of substance contained in the solution (or mixture) times the total amount of the solution (or mixture). It is helpful to draw a picture showing three containers: the two original containers (*before mixing*) and a container for the new solution (*after mixing*). Usually, the key idea in a mixture problem is that the amount of a given substance you have *before mixing* should equal the amount of that substance you will have *after mixing.*
The key idea in a **simple interest** problem is that the interest earned by a given investment is determined by the interest rate and the length of time invested at that rate.
The key idea in a **distance-rate-time** problem is that a given distance traveled is determined by the rate and length of time traveled at that rate.
The key idea in **work** problems is that the rate at which work is done equals the amount of work accomplished divided by the amount of time worked; that is, rate $= \frac{\text{amount of work done}}{\text{time worked}}$.
In geometry and trigonometry problems, using the correct formula is very important. Making a diagram is usually essential for proper understanding of a geometry or trigonometry problem.

PRACTICE PROBLEMS FOR CHAPTER 4

1. One number is four times a second number. The sum of the numbers is 30. Find the two numbers.
2. Joe Don is 11 years older than Pablo. The sum of their ages is 39. What are the ages of each?
3. Annette has $10.50 in dimes and quarters. She has 14 more dimes than quarters. How many of each coin does she have?
4. A grocer mixes candy worth $2.50 per pound with candy worth $3.75 per pound to make a mixture of 90 pounds to sell at $3.00 per pound. How many pounds of each should he use?
5. Two cars leave a restaurant, one traveling due north at 70 mph and the other due south at 60 mph. In how many hours will the cars be 390 miles apart?
6. The length of a rectangular pasture is 14 yards more than its width. The perimeter is 180 yards. Find the dimensions of the pasture.

7. Michael invests $5,800, part at 6% annual interest and part at 9% annual interest. If this annual yield on the investment is $456, how much did he invest at each rate?
8. The price of a computer is increased by 50% to $1,245. Find the original price.
9. The length of a rectangle is 5 feet more than its width. Its area is 24 square feet. Find its dimensions.
10. The sum of Donna's and Brandon's ages is 36. The product of their ages is 323. Find their ages.
11. Tara can rake the yard in 3 hours working alone. Candi can do the same job working alone in 2 hours. How many hours will it take them working together to rake the yard?
12. How high up a cliff will a 40 ft beam reach if the end of the beam is touching the ground 24 ft from the base of the cliff?
13. A mixture contains (by weight) cornmeal and wheat bran in the ratio 2:3, respectively. Compute the total number of ounces in a mixture that contains 15 oz of cornmeal.
14. The area of a trapezoid-shaped field is 8,575 m^2. If the longer base of the field is 140 m and the altitude is 70 m, what is the length of the shorter base?

SOLUTIONS TO PRACTICE PROBLEMS

1. One number is four times a second number. The sum of the numbers is 30. Find the two numbers.

Solution: Let $n =$ 2nd number.
Let $4n =$ 1st number.

$$4n + n = 30$$
$$5n = 30$$
$$n = 6$$
$$4n = 24$$

Solution: The numbers are 24 and 6.

2. Joe Don is 11 years older than Pablo. The sum of their ages is 39. What are the ages of each?

Solution: Let $n =$ Pablo's age in years.
Let 11 years $+ n =$ Joe Don's age in years.

$$n + (11 \text{ years} + n) = 39 \text{ years}$$
$$n + (11 + n) = 39$$
$$2n + 11 = 39$$
$$2n = 28$$
$$n = 14 \text{ years}$$
$$11 \text{ years} + n = 11 \text{ years} + 14 \text{ years} = 25 \text{ years}$$

Solution: Joe Don is 25 years old and Pablo is 14 years old.

3. Annette has \$10.50 in dimes and quarters. She has 14 more dimes than quarters. <u>How many of each coin does she have?</u>

Solution: Let x = number of quarters.
Let $14 + x$ = number of dimes.

Denomination	Quarters	Dimes	Total
Value per coin	\$0.25	\$0.10	not applicable
Number of coins	x	$14 + x$	$x + (14 + x)$
Value (in dollars)	\$0.25(no. of quarters) = \$0.25x	\$0.10(no. of dimes) = \$0.10$(14 + x)$	\$10.50

$$\underbrace{\text{Total value of quarters}}_{\$0.25x} + \underbrace{\text{Total value of dimes}}_{\$0.10(14 + x)} = \underbrace{\text{Total value of all the coins}}_{\$10.50}$$

$$0.25x + 0.10(14 + x) = 10.50$$

$$0.25x + 1.4 + 0.1x = 10.50$$

$$0.35x = 9.10$$

$$x = 26 \text{ quarters}$$

$$14 + x = 14 + 26 = 40 \text{ dimes}$$

Solution: Annette has 40 dimes and 26 quarters.

4. A grocer mixes candy worth \$2.50 per pound with candy worth \$3.75 per pound to make a mixture of 90 pounds to sell at \$3.00 per pound. <u>How many pounds of each should he use?</u>

Solution: Neither unknown is expressed in terms of the other unknown. Try using two variables (temporarily):

Let x = number of pounds of \$2.50 per pound candy

Let y = number of pounds of \$3.75 per pound candy

From the facts, we can write:

$$x + y = 90 \text{ lb}$$

We can solve for y to obtain:

$$y = 90 \text{ lb} - x = \text{number of pounds of \$3.75 per pound candy}$$

Thus,

Let x = number of pounds of \$2.50 per pound candy.

Let 90 lb − x = number of pounds of \$3.75 per pound candy.

Kind	\$2.50/lb candy	\$3.75/lb candy	\$3.00/lb candy
Cost	\$2.50/lb	\$3.75/lb	\$3.00/lb
Number of pounds	x	90 lb − x	90 lb
Value	(\$2.50/lb)x	(\$3.75/lb)(90 lb − x)	(\$3.00/lb)(90 lb)

$$\underbrace{(\$2.50/\text{lb})x}_{\text{Value of 2.50/lb candy}} + \underbrace{(\$3.75/\text{lb})(90\text{ lb} - x)}_{\text{Value of 3.75/lb candy}} = \underbrace{(\$3.00/\text{lb})(90\text{ lb})}_{\text{Value of 3.00/lb candy (mixture)}}$$

$$2.50x + 3.75(90 - x) = (3.00)(90)$$

$$2.50x + 337.5 - 3.75x = 270$$

$$-1.25x = -67.5$$

$$x = 54 \text{ lb of \$2.50 per pound candy}$$

$$90 \text{ lb} - x = 90 \text{ lb} - 54 \text{ lb} = 36 \text{ lb of \$3.75 per pound candy}$$

Solution: The grocer should use 54 pounds of \$2.50/lb candy and 36 pounds of \$3.75/lb candy to make a 90 lb mixture to sell at \$3.00 per pound.

5. Two cars leave a restaurant, one traveling north at 70 mph and the other south at 60 mph. In how many hours will the cars be 390 miles apart?

Solution:

Let t = number of hours the cars will travel to be 390 miles apart.

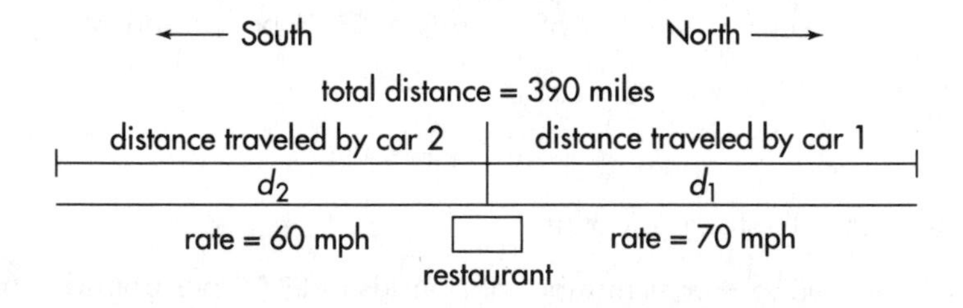

Vehicle	Distance	Rate	Time
Car 1	$d_1 = (70 \text{ mph})t$	70 mph	t (hours)
Car 2	$d_2 = (60 \text{ mph})t$	60 mph	t (hours)
Total	390 miles	Do not add rates for different vehicles.	t (hours)

$$\underbrace{\text{Distance traveled by car 1}}_{70(\text{mph})t} + \underbrace{\text{Distance traveled by car 2}}_{(60 \text{ mph})t} = \underbrace{\text{Total distance traveled}}_{390 \text{ miles}}$$

$$70t + 60t = 390$$

$$130t = 390$$

$$t = 3 \text{ hours}$$

Solution: The two cars will be 390 miles apart in 3 hours.

6. The length of a rectangular pasture is 14 yards more than its width. The perimeter is 180 yards. Find the dimensions of the pasture.
Solution:

Let w = the width of the pasture (in yards).

Let 14 yd + w = the length of the pasture (in yards).

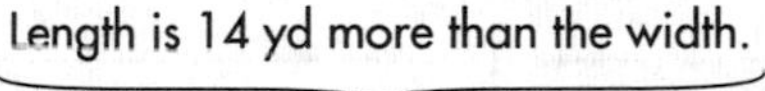

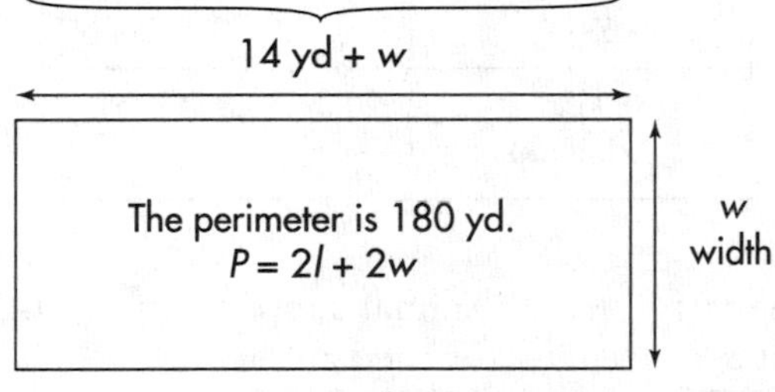

$$2(14 \text{ yd} + w) + 2w = 180 \text{ yd}$$

$$2(14 + w) + 2w = 180$$

$$28 + 2w + 2w = 180$$

$$4w = 152$$

$$w = 38 \text{ yards}$$

$$14 \text{ yd} + w = 52 \text{ yards}$$

Solution: The pasture has length 52 yards and width 38 yards.

7. Michael invests \$5,800, part at 6% annual interest and part at 9% annual interest. If his annual yield on the investment is \$456, how much did he invest at each rate?

Solution: Neither unknown is expressed in terms of the other unknown. Try using two variables (temporarily):

Let x = amount (in dollars) invested at 6%.

Let y = amount (in dollars) invested at 9%.

From the facts, we can write:

$$x + y = \$5{,}800$$

We can solve for y to obtain:

$$y = \$5{,}800 - x = \text{amount (in dollars) invested at 9\%}$$

Thus,

Let x = amount (in dollars) invested at 6%.

Let $\$5{,}800 - x$ = amount (in dollars) invested at 9%.

	6% Investment	**9% Investment**	**Total**
Amount of investment	x (dollars)	$\$5{,}800 - x$	\$5,800
Simple interest rate	6% = 0.06	9% = 0.09	Do not add rates unless they have the *same* base
Time invested	1 year	1 year	1 year
Annual Interest earned	$0.06x$ (dollars)	$0.09(\$5{,}800 - x)$	\$456

$$\underbrace{\text{Annual interest earned at 6\%}}_{0.06x} + \underbrace{\text{Annual interest earned at 9\%}}_{0.09(\$5{,}800 - x)} = \underbrace{\text{Total annual interest earned}}_{\$456}$$

$$0.06x + 0.09(5{,}800 - x) = 456$$
$$0.06x + 522 - 0.09x = 456$$
$$-0.03x = -66$$
$$x = \$2{,}200$$
$$\$5{,}800 - x = \$3{,}600$$

Solution: Michael invested \$2,200 at 6% and \$3,600 at 9%.

8. The price of a computer is increased by 50% to $1,245. Find the original price.

Solution: Let x = the original price of the computer.

$$x + 0.50x = \$1{,}245$$
$$1.5x = \$1{,}245$$
$$x = \$830$$

Solution: The original price was $830.

9. The length of a rectangle is 5 feet more than its width. Its area is 24 square feet. Find its dimensions.

Solution:

Let w = the width of the rectangle (in feet).

Let 5 ft + w = the length of the rectangle (in feet).

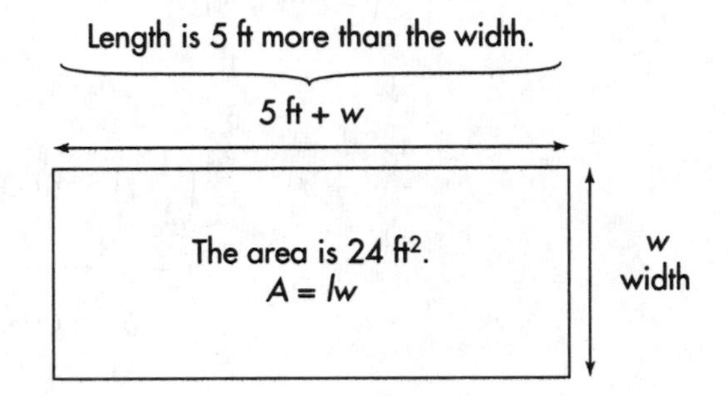

$$(5 \text{ ft} + w)w = 24 \text{ ft}^2$$
$$(5 + w)w = 24$$
$$5w + w^2 = 24$$
$$w^2 + 5w - 24 = 0$$
$$(w - 3)(w + 8) = 0$$
$$w = 3 \text{ feet} = \text{width} \quad \text{or} \quad w = -8 \text{ feet (reject)}$$
$$5 \text{ ft} + w = 5 \text{ ft} + 3 \text{ feet} = 8 \text{ feet} = \text{length}$$

Solution: The rectangle has length 8 feet and width 3 feet.

10. The sum of Donna's and Brandon's ages is 36. The product of their ages is 323. Find their ages.

Solution: Let x = Donna's age.
Let $36 - x$ = Brandon's age.

$$x(36 - x) = 323$$
$$36x - x^2 = 323$$
$$-x^2 + 36x - 323 = 0$$

$$x^2 - 36x + 323 = 0$$
$$(x - 19)(x - 17) = 0$$
$$x = 19 \text{ yr} \quad \text{or} \quad x = 17 \text{ yr}$$
$$36 - x = 17 \text{ yr} \quad \text{or} \quad 36 - x = 19 \text{ yr}$$

Two Solutions: Donna is 17 years old and Brandon is 19 years old or Donna is 19 years old and Brandon is 17 years old.

11. Tara can rake the yard in 3 hours working alone. Candi can do the same job working alone in 2 hours. How many hours will it take them working together to rake the yard?

Solution: Let t = the time it will take them working together to rake the yard.

$$\underbrace{\text{Part of yard raked by Tara}} + \underbrace{\text{Part of yard raked by Candi}} = \underbrace{\text{Whole yard raked}}$$

$$\left(\frac{1}{3}\text{ yard/hr}\right)t + \left(\frac{1}{2}\text{ yard/hr}\right)t = 1 \text{ yard}$$

$$\frac{1}{3}t + \frac{1}{2}t = 1$$
$$\frac{\overset{2}{\cancel{6}}}{1}\left(\frac{1}{\cancel{3}_1}t\right) + \frac{\overset{3}{\cancel{6}}}{1}\left(\frac{1}{\cancel{2}_1}t\right) = 6(1)$$
$$2t + 3t = 6$$
$$5t = 6$$
$$t = 1.2 \text{ hours}$$

Solution: It will take 1.2 hours for Tara and Candi to rake the yard working together.

12. How high up a cliff will a 40 ft beam reach if the end of the beam is touching the ground 24 ft from the base of the cliff?

Solution:

Let a = the height (in feet) the beam reaches up the cliff.

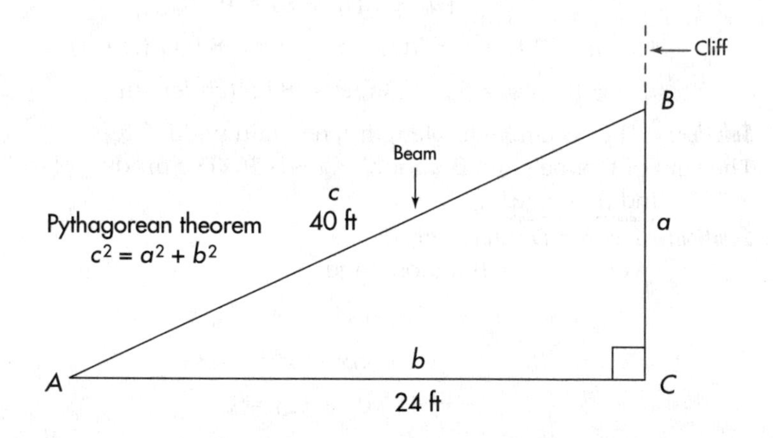

$$(40\text{ ft})^2 = (24\text{ ft})^2 + a^2$$
$$(40)^2 = (24)^2 + a^2$$
$$1{,}600 = 576 + a^2$$
$$1{,}600 - 576 = a^2$$
$$1{,}024 = a^2$$
$$a^2 = 1{,}024$$
$$a = 32\text{ feet} \quad \text{or} \quad a = -32\text{ feet (reject)}$$

Solution: The beam will reach 32 feet up the cliff.

13. A mixture contains (by weight) cornmeal and wheat bran in the ratio 2 : 3, respectively. Compute the total number of ounces in a mixture that contains 15 oz of cornmeal.

Solution:

Step 1:

Let x = number of ounces of wheat bran in the mixture.

$$\frac{15\text{ oz}}{x} = \frac{2}{3}$$
$$2x = 45\text{ oz}$$
$$x = 22.5\text{ ounces}$$

The mixture contains 22.5 ounces of wheat bran.

Step 2:

$$15\text{ oz} + 22.5\text{ oz} = 37.5\text{ oz}$$

Solution: The total number of ounces in the mixture is 37.5 ounces.

14. The area of a trapezoidal-shaped field is 8,575 m^2. If the longer base of the field is 140 m and the altitude is 70 m, what is the length of the shorter base?

Solution:

Let x = length of the shorter base in meters.

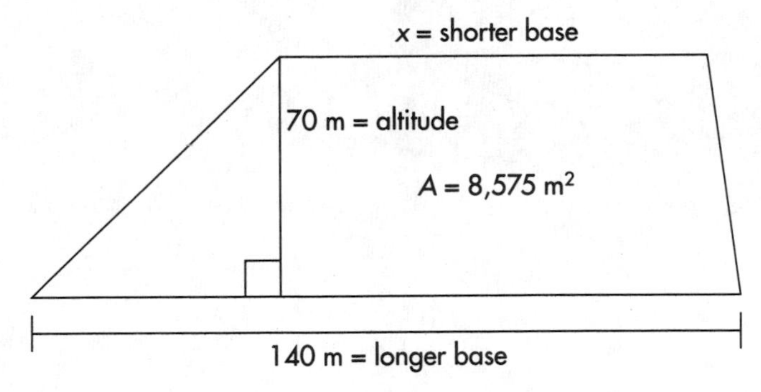

Formula for the area of a trapezoid: $A = \frac{1}{2}(a + b)h$

$$\frac{1}{2}(140 \text{ meters} + x)(70 \text{ meters}) = 8{,}575 \text{ m}^2$$

$$\frac{1}{2}(140 + x)(70) = 8{,}575$$

$$(140 + x)(35) = 8{,}575$$

$$4{,}900 + 35x = 8{,}575$$

$$35x = 3{,}675$$

$$x = 105 \text{ meters}$$

Solution: The length of the shorter base of the field is 105 meters.

5
FUNCTIONS, RELATIONS, AND THEIR GRAPHS

5.1 THE CARTESIAN COORDINATE PLANE

If two copies of the real number line are placed perpendicular to each other, that is, one horizontal and one vertical, so that they intersect at the zero point on each line, these two lines form the **axes** of a rectangular coordinate system called the **Cartesian coordinate plane.** The horizontal real line with positive direction to the right is called the **horizontal axis,** or the ***x*-axis,** and the vertical real line with positive direction upward is called the **vertical axis,** or the ***y*-axis.** The vertical and horizontal axes determine the plane of the rectangular coordinate system. Their point of intersection is called the **origin.**

With this coordinate system you can associate each point in the plane with an **ordered pair** (x, y) of real numbers x and y, called its **coordinates.** The order in an ordered pair is important. Therefore,

$$\text{If } x \neq y, \text{ then } (x, y) \neq (y, x)$$

In the ordered pair (x, y), the **first element** x, called the **abscissa** or ***x*-coordinate,** tells the *directed* (right or left) distance of the point from the vertical axis. The **second element** y, called the **ordinate** or ***y*-coordinate,** tells the *directed* (up or down) distance of the point from the horizontal axis. For example, (3, 2) is an ordered pair, with first element 3 and second element 2. To locate the point in the plane corresponding to (3, 2), start at the origin and move 3 units to the right, and then from that point move 2 units up (see Figure 5.1).

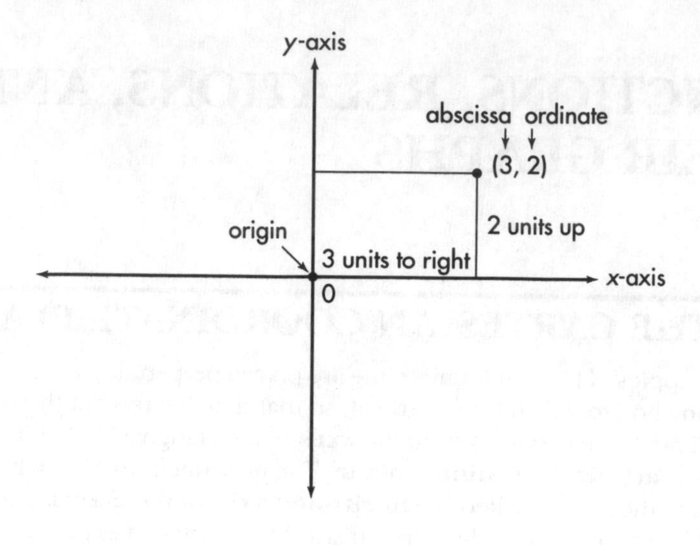

FIGURE 5.1

The Cartesian coordinate plane in Figure 5.2 shows the location of the points (4, −3), (−3, 3), (−5, −4), (2, 3), and (−3, 4).

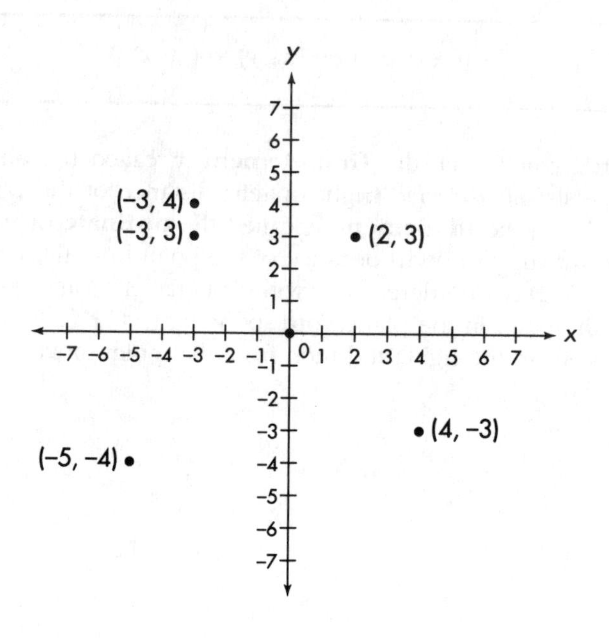

FIGURE 5.2

Two ordered pairs are **equal** if and only if they have *exactly* the same coordinates; that is,

EQUALITY OF ORDERED PAIRS

$(a, b) = (u, v)$ if and only if $a = u$ and $b = v$

The plane in which the coordinate system lies is divided into four sections called **quadrants.** They are named with the Roman numerals **I, II, III,** and **IV.** The numbering process begins in the upper right section and proceeds counterclockwise, as shown in Figure 5.3.

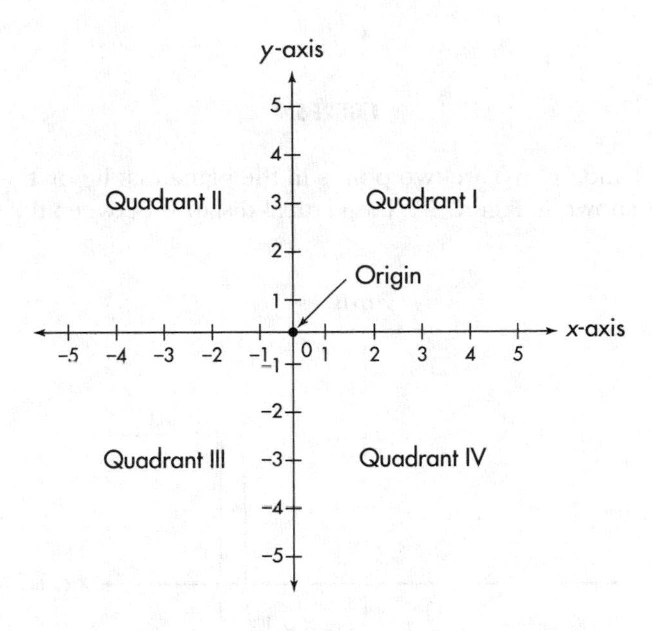

FIGURE 5.3

The Distance Formula

If (x_1, y_1) and (x_2, y_1) are two points in the plane that lie on the same horizontal line as shown in Figure 5.4, the horizontal distance between the two points is $|x_2 - x_1|$.

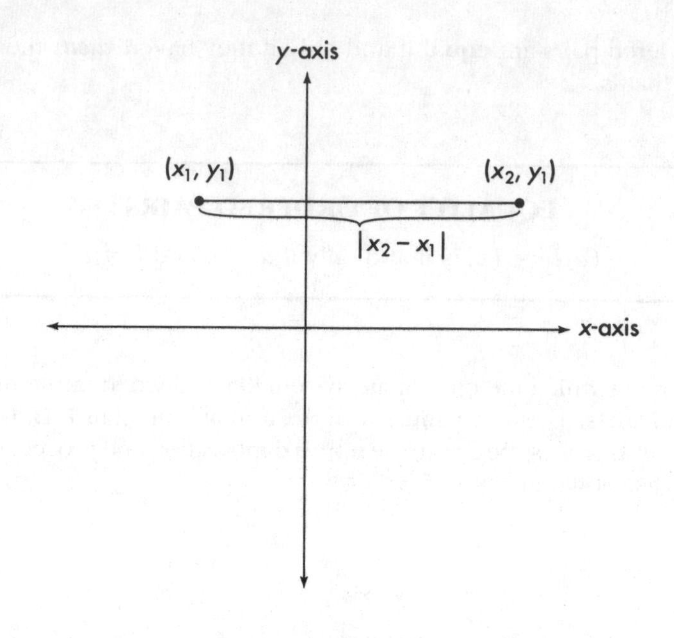

FIGURE 5.4

If (x_1, y_1) and (x_1, y_2) are two points in the plane that lie on the same vertical line as shown in Figure 5.5, the vertical distance between the two points is $|y_2 - y_1|$.

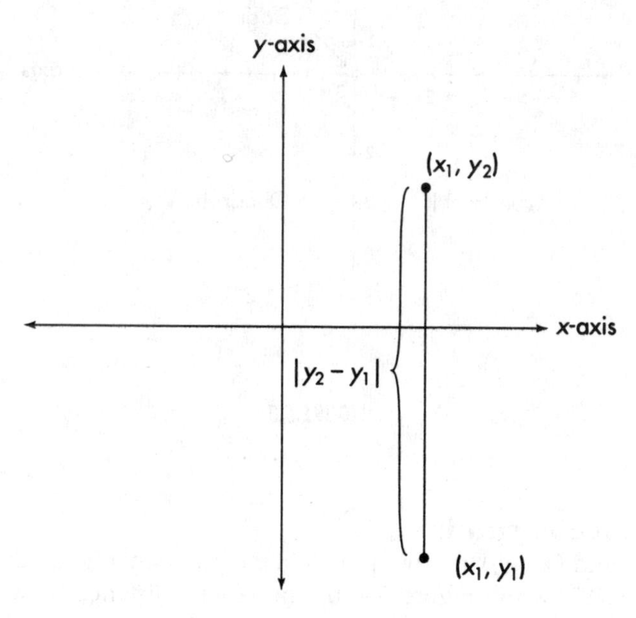

FIGURE 5.5

Now suppose (x_1, y_1) and (x_2, y_2) are two points in the plane, not lying on the same horizontal or vertical line, that are separated by a distance d. A right triangle can be formed that has d as the hypotenuse and the points (x_1, y_1), (x_2, y_2), and (x_2, y_1) as vertices (as shown in Figure 5.6).

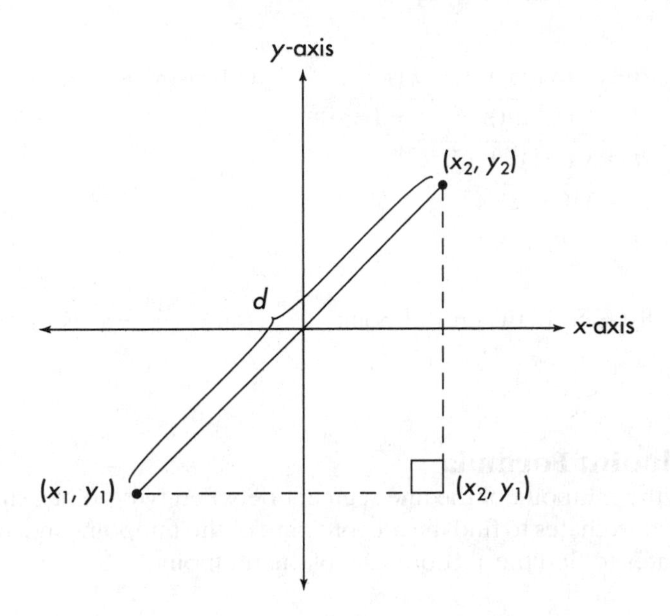

FIGURE 5.6

The points (x_2, y_1) and (x_1, y_1) lie on the same horizontal line; thus, the length of the horizontal side of the triangle is $|x_2 - x_1|$. Similarly, the points (x_2, y_1) and (x_2, y_2) lie on the same vertical line; thus, the length of the vertical side of the triangle is $|y_2 - y_1|$. By the Pythagorean Theorem,

$$\begin{aligned} d^2 &= |x_2 - x_1|^2 + |y_2 - y_1|^2 \\ &= (x_2 - x_1)^2 + (y_2 - y_1)^2 \quad \text{(because } |x_2 - x_1|^2 = (x_2 - x_1)^2 \text{ and } |y_2 - y_1|^2 = (y_2 - y_1)^2\text{)} \end{aligned}$$

This yields the following formula:

THE DISTANCE FORMULA

If (x_1, y_1) and (x_2, y_2) are two points in the plane, the distance d between them is given by:

$$d = \sqrt{(x_2 - x_1)^2 + (y_2 - y_1)^2}$$

A helpful message: The distance formula is very important in mathematics, so you should take time to memorize it.

For example, to find the distance between the points (4, −3) and (3, 2), substitute into the formula and simplify:

$$d = \sqrt{(x_2 - x_1)^2 + (y_2 - y_1)^2} \quad \text{(Either point can be } (x_1, y_1).\text{)}$$
$$d = \sqrt{(3 - 4)^2 + (2 - (-3))^2}$$
$$d = \sqrt{(-1)^2 + (5)^2}$$
$$d = \sqrt{1 + 25}$$
$$d = \sqrt{26} \cong 5.1$$

Thus, the distance between the points (3, 2) and (4, −3) is approximately 5.1 units.

The Midpoint Formula

To find the midpoint of the line segment between two points, simply average the x-coordinates to find the x-coordinate of the midpoint and average the y-coordinates to find the y-coordinate of the midpoint:

> If (x_1, y_1) and (x_2, y_2) are two points in the plane, the midpoint of the line segment between (x_1, y_1) and (x_2, y_2) has coordinates given by:
>
> $$\left(\frac{x_1 + x_2}{2}, \frac{y_1 + y_2}{2}\right)$$

For example, to find the midpoint of the line segment between the points (−5, −6) and (3, 2), substitute into the formula and simplify (Figure 5.7):

$$\text{midpoint} = \left(\frac{x_1 + x_2}{2}, \frac{y_1 + y_2}{2}\right) \quad \text{(Either point can be } (x_1, y_1).\text{)}$$
$$\text{midpoint} = \left(\frac{-5 + 3}{2}, \frac{-6 + 2}{2}\right) = \left(\frac{-2}{2}, \frac{-4}{2}\right) = (-1, -2)$$

A helpful message: Notice that in the midpoint formula, corresponding coordinates are *added,* not *subtracted* as in the distance formula.

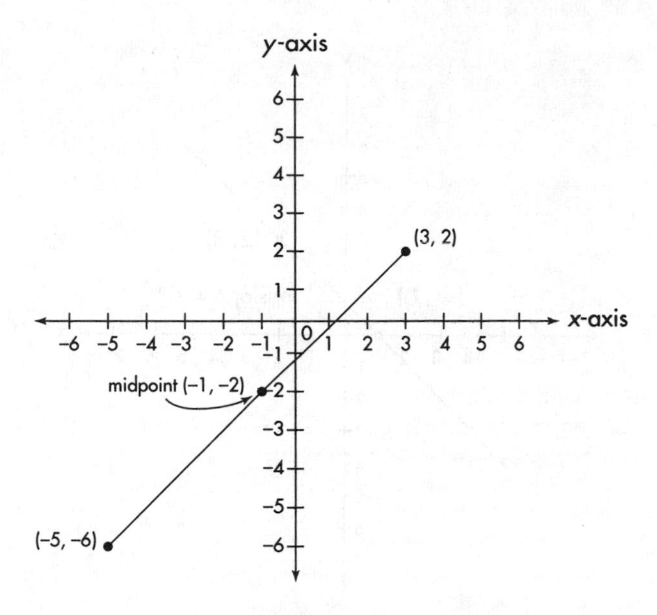

FIGURE 5.7

The Slope of a Line

When you move from one point on a line to another point on the line and compare the vertical change to the horizontal change, you are measuring the **slope** of the line. If (x_1, y_1) and (x_2, y_2) are the coordinates of two points on the line, the formula for the slope of the line is the following:

$$m = \frac{y_2 - y_1}{x_2 - x_1} \qquad (x_1 \neq x_2)$$

A helpful message: The slope is often called the "rise over the run." Use this to help you remember to put the vertical distance (rise) over the horizontal distance (run), so slope $= \frac{\text{rise}}{\text{run}}$.

For example, in Figure 5.8, the slope of the line that passes through the points $(-1, 0)$ and $(2, 3)$ is given by:

$$m = \frac{y_2 - y_1}{x_2 - x_1} \qquad \text{(Either point can be } (x_1, y_1).\text{)}$$

$$m = \frac{0 - 3}{-1 - 2} = \frac{-3}{-3} = 1$$

When a line slopes *upward* to the right, its slope is *positive* (Figure 5.9a), and when a line slopes *downward* to the right, its slope is *negative* (Figure 5.9b).

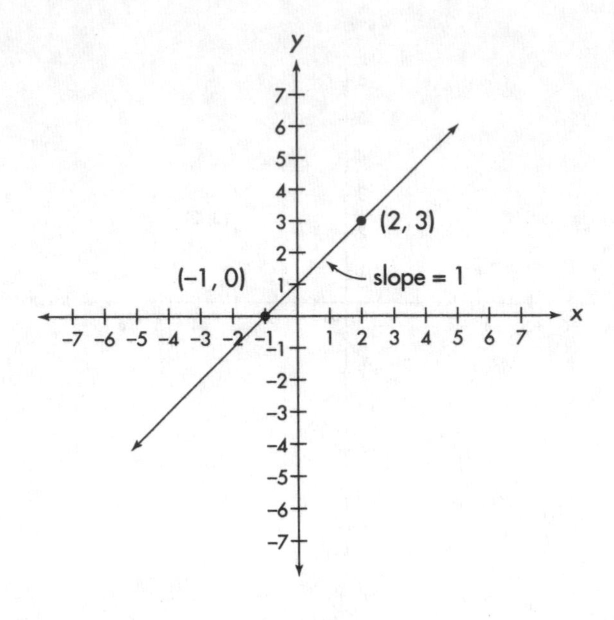

FIGURE 5.8

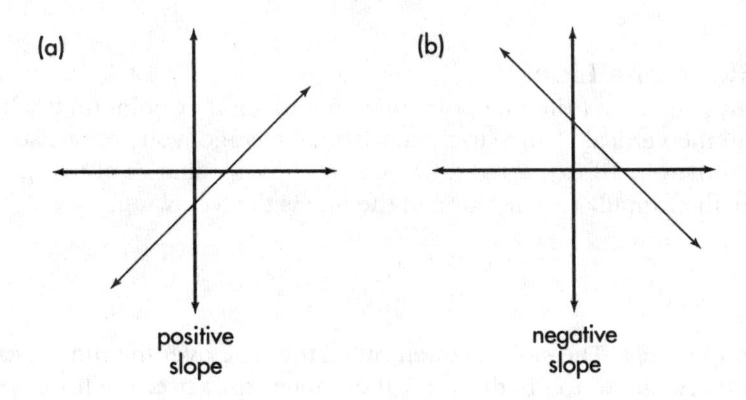

FIGURE 5.9

When the y-coordinates are equal, as would be the case for horizontal lines, the slope is *zero* because the numerator is zero.

When $y_1 = y_2$, slope is zero.

$$m = \frac{y_2 - y_1}{x_2 - x_1} = \frac{0}{x_2 - x_1} = 0$$

Therefore, *all horizontal lines have slope 0* (Figure 5.10a). When the x-coordinates are equal, as would be the case for vertical lines, the slope is *undefined* because the denominator is zero.

When $x_1 = x_2$, slope is undefined.

$$m = \frac{y_2 - y_1}{x_2 - x_1} = \frac{y_2 - y_1}{0} = \text{undefined}$$

Therefore, we say *all vertical lines have no slope* (Figure 5.10b).

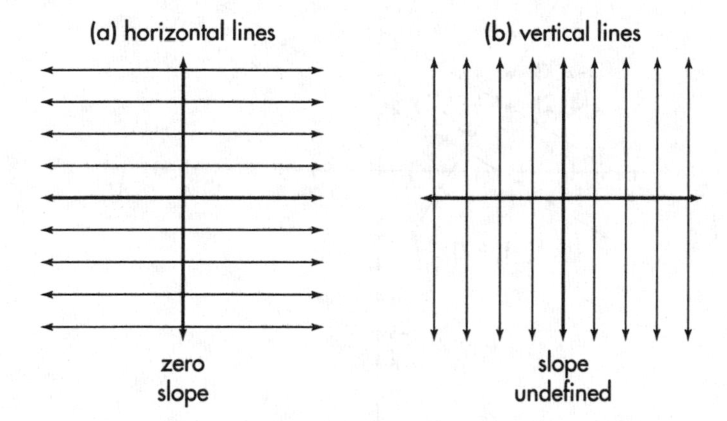

FIGURE 5.10

It can be shown that:

> If two lines are parallel, their slopes are equal; and if two lines are perpendicular, their slopes are negative reciprocals of each other.

For example, in Figure 5.11, the line that passes through the points (3, 4) and (−2, 1) has slope:

$$m = \frac{4 - 1}{3 + 2} = \frac{3}{5}$$

This line is parallel to the line that passes through the points (6, 8) and (−4, 2), which has slope:

$$m_1 = \frac{8 - 2}{6 + 4} = \frac{6}{10} = \frac{3}{5} = m$$

However, in Figure 5.11, the line through (3, 4) and (−2, 1) is perpendicular to the line through the points (−4, 3) and (−1, −2), which has slope:

$$m_2 = \frac{3 + 2}{-4 + 1} = -\frac{5}{3} = -\frac{1}{m}$$

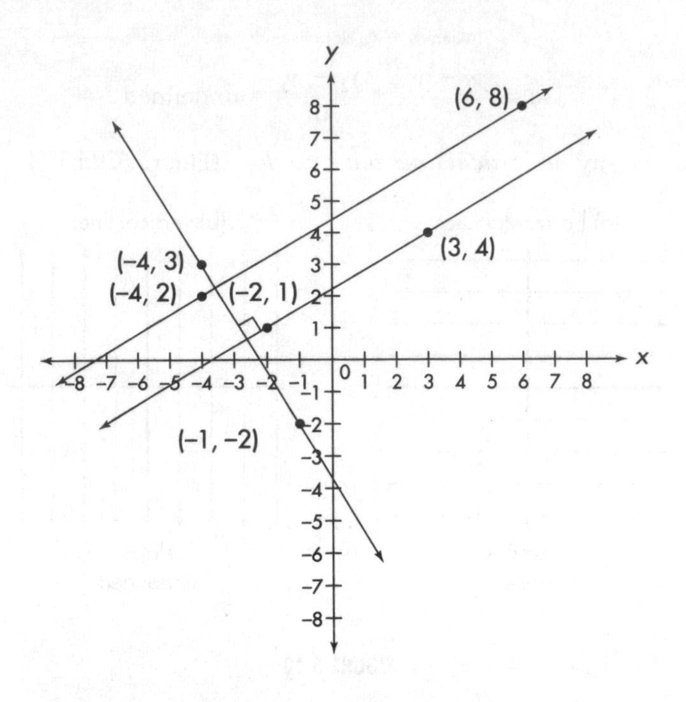

FIGURE 5.11

5.2 RELATIONS AND FUNCTIONS

The **Cartesian product** of two sets A and B, denoted $A \times B$, is the set of all ordered pairs (x, y) such that $x \in A$ and $y \in B$. In algebra, the most important Cartesian product is the set of all possible ordered pairs of real numbers, denoted $R \times R$, or simply R^2. The set R^2 is represented by the Cartesian coordinate plane.

Any subset of R^2 is a **relation** in R^2. The set of all first components in the ordered pairs in a relation is called the **domain** of the relation, and the set of all second components is called the **range** of the relation.

For example,

- $S = \{(-3, 9), (-2, 4), (-1, 1), (0, 0), (1, 1), (2, 4), (3, 9)\}$ is a relation with domain $\{-3, -2, -1, 0, 1, 2, 3\}$ and range $\{0, 1, 4, 9\}$.
- $T = \{(0, 0), (1, -1), (1, 1), (4, -2), (4, 2), (9, -3), (9, 3)\}$ is a relation with domain $\{0, 1, 4, 9\}$ and range $\{-3, -2, -1, 0, 1, 2, 3\}$.

There is a special kind of relation that is fundamental to all of mathematics. This special kind of relation is called a **function:**

DEFINITION OF FUNCTION

A function is a relation in which each first component is paired with *one and only one* second component.

All functions are relations, but not all relations are functions. In the examples above, both S and T are relations, but only S is a function. How do they differ?

In $S = \{(-3, 9), (-2, 4), (-1, 1), (0, 0), (1, 1), (2, 4), (3, 9)\}$, every element in the domain is paired with its square in the range. Since every number has one and only one square, the relation S is a function.

In $T = \{(0, 0), (1, -1), (1, 1), (4, -2), (4, 2), (9, -3), (9, 3)\}$, every element in the domain is paired with its square root in the range. Since every number (except 0) has two square roots, this relation is not a function.

When you look at the difference in terms of ordered pairs, you can see that in S, no two different ordered pairs have the same first component; however, this is not the case in T:

$S = \{(-3, 9), (-2, 4), (-1, 1), (0, 0), (1, 1), (2, 4), (3, 9)\}$ is a function.

$T = \{(0, 0), (1, -1), (1, 1), (4, -2), (4, 2), (9, -3), (9, 3)\}$ is *not* a function.

same first component (for $(1, -1), (1, 1)$); same first component (for $(4, -2), (4, 2)$); same first component (for $(9, -3), (9, 3)$)

This leads to the following alternate definition for function:

A function is a relation in which no two different ordered pairs have the same first component; that is, if (a, b) and (a, c) are elements in the function then $b = c$.

A helpful message: You likely noticed that $(-1, 1)$ and $(1, 1)$, $(-2, 4)$ and $(2, 4)$, and $(-3, 9)$ and $(3, 9)$ are pairs of elements in S that have the same second components. These pairs do not violate the definition of a function because the definition does not say that the second components must be different.

5.3 FUNCTIONAL NOTATION

Functions are usually denoted by lowercase letters such as f, g, h, and so on. A function that contains a **finite** number of ordered pairs may be defined in several common ways.

1. By listing its **ordered pairs:**

$$f = \{(0, 5), (1, 7), (2, 9), (3, 11), (4, 13), (5, 15), (6, 17)\}$$

2. By a **table:**

x	0	1	2	3	4	5	6
y	5	7	9	11	13	15	17

3. By an **arrow diagram:**

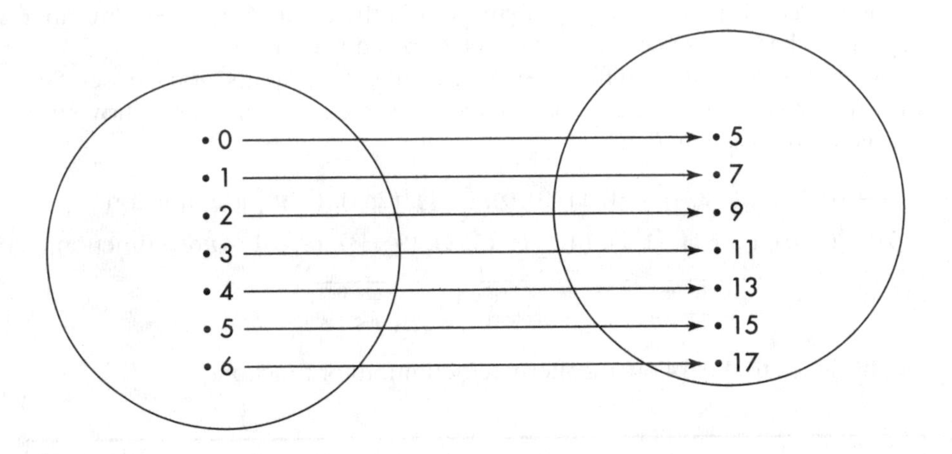

4. By using two **number lines:**

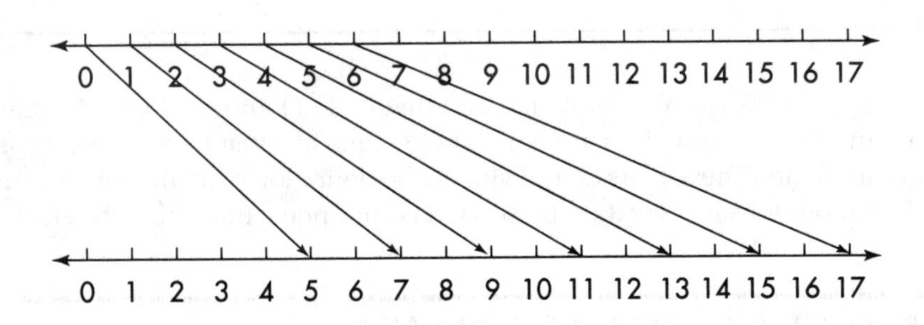

5. By a **rule:**
The function f is the set of ordered pairs (x, y) such that $x \in \{0, 1, 2, 3, 4, 5, 6\}$ and y is obtained by multiplying x by 2 and then adding 5 to the product.

6. By an **equation:**
   The function f is the set of ordered pairs (x, y) such that $x \in \{0, 1, 2, 3, 4, 5, 6\}$ and $y = 2x + 5$.
7. By a **graph:**

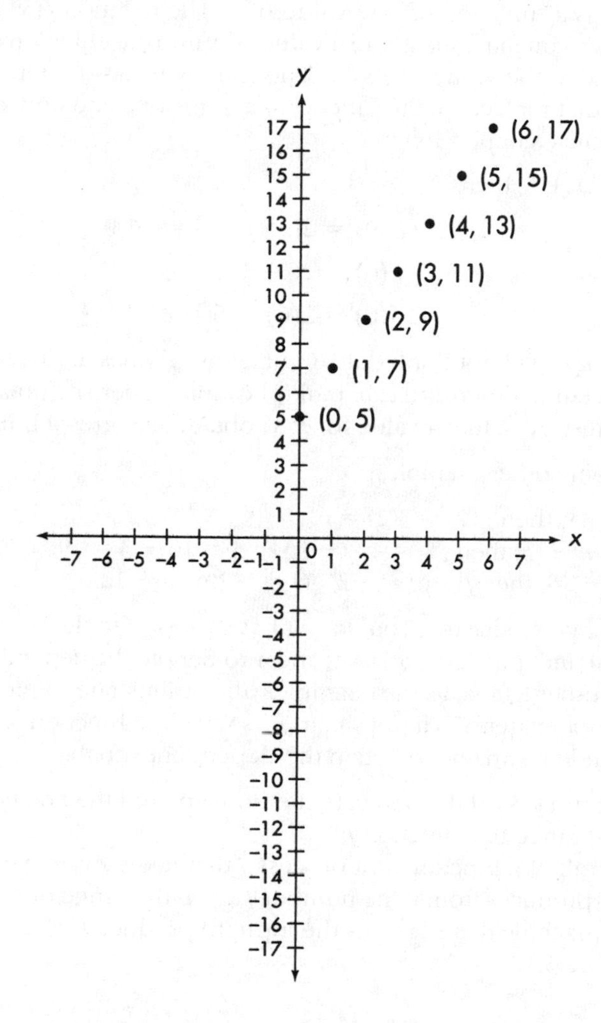

When the number of ordered pairs in a function is **infinite,** the function is usually defined by either an **equation** or a **graph.**

In the function defined by the equation $y = 2x + 5$, where $x \in R$,

when $x = -9$, $y = 2(-9) + 5 = -13$;
when $x = 0$, $y = 2(0) + 5 = 5$;
when $x = 6$, $y = 2(6) + 5 = 17$;
and so on.

Because the value of y depends on which value of x is substituted into the equation, y is called the **dependent variable** and x is called the **independent variable.**

To emphasize the dependency of y on x, we write $y = f(x)$, which means the value of y is a "function of" the value of x. The notation $f(x)$ is read "f of x" and indicates you must substitute a value of x into the equation defining the function to find y, the **value** of f at x. The statement $y = f(x)$ means that for each element that replaces x, the function f assigns one and only one replacement for y. Some examples follow:

- If $f(x) = 2x + 5$, then

$$f(-9) = 2(-9) + 5 = -13$$
$$f(0) = 2(0) + 5 = 5$$
$$f(6) = 2(6) + 5 = 17$$

A helpful message: Do not think that $f(x)$ means "f times x." The parentheses used in this notation do *not* indicate multiplication; rather, the notation $f(x)$ indicates you must substitute a value for x to obtain its corresponding y-value.

Here are some other examples:

- If $g(x) = 3x$, then $g(2) = 3(2) = 6$.
- If $f(u) = u^2 + 3$, then $f(-4) = (-4)^2 + 3 = 16 + 3 = 19$.
- If $f(t) = -2|t|$, then $f(-6) = -2|-6| = -2(6) = -12$.

A helpful message: Notice that you do not have to use f to denote the function, x to denote the independent variable, and y to denote the dependent variable. You may use other letters, as the examples above illustrate. Which letters you choose does not matter. What does matter is what the function tells you *to do* to the independent variable to obtain the dependent variable.

The notation D_f is used to indicate the domain, and the notation R_f is used to indicate the range of a function f.

You can think of a function as a process f that takes a number x in the domain of f and produces from it the number $f(x)$ in the range of f. Visualize the function as a machine that uses x as the input to produce $f(x)$ as the output as shown in Figure 5.12.

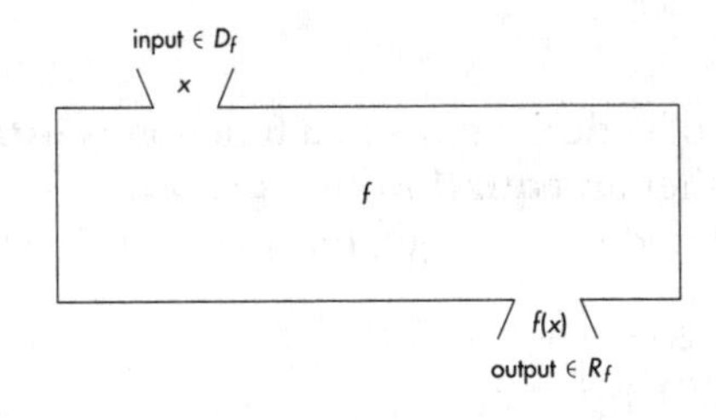

FIGURE 5.12

The equation $y = f(x)$ defines a set of ordered pairs for which $y = f(x)$ is true. The ordered pairs of the function are written in the form (x, y) or $(x, f(x))$. Thus, for $y = f(x) = 2x + 5$:

$$(-9, -13) \in f, \text{ since } f(-9) = 2(-9) + 5 = -13.$$
$$(0, 5) \in f, \text{ since } f(0) = 2(0) + 5 = 5.$$
$$(6, 17) \in f, \text{ since } 2(6) + 5 = 17.$$

For simplicity, we often refer to a function by the equation that describes it. For example, the function f that is the set of ordered pairs defined by the equation $y = 2x + 5$ may be described simply as "the function $y = 2x + 5$" or "the function $f(x) = 2x + 5$."

Other notations for expressing f follow:

- $f = \{(x, y) \mid y = 2x + 5\}$ is read as "the function f is the set of ordered pairs (x, y) such that $y = 2x + 5$."
- $f\colon x \rightarrow 2x + 5$ or $x \xrightarrow{f} 2x + 5$ is read as:
  "the function f maps x into $2x + 5$";
  "the function takes the argument x to $2x + 5$";
  "under function f, x is assigned to $2x + 5$"; or
  "under function f, the image of x is $2x + 5$."

Since a function is a set of ordered pairs, its graph can be determined in a coordinate plane. Each ordered pair is represented by a point in the plane. The elements of the domain are shown on the x-axis, while elements of the range are shown on the y-axis. By definition, each element in the domain of the function is paired with exactly one element in the range; thus, any vertical line in the plane *will intersect the graph of the function no more than once.* This is known as the **vertical line test.** For example, in Figure 5.13, graphs (a) and (b) are functions, but (c) and (d) are not.

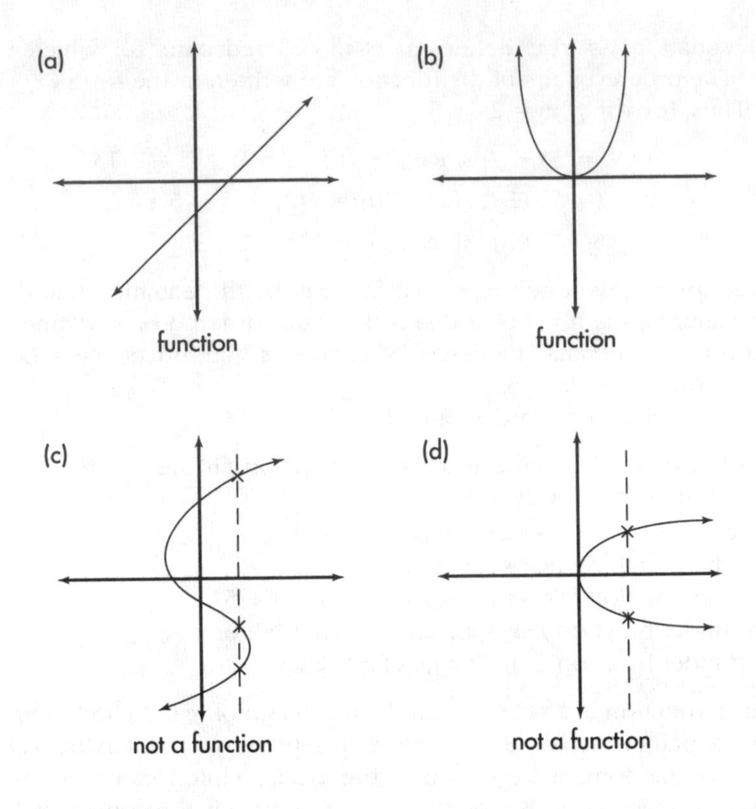

FIGURE 5.13

Notice that a vertical line would cut the curve in more than one place in (c) and (d). Thus, we can state the vertical line test as follows:

VERTICAL LINE TEST FOR FUNCTIONS

If any vertical line can be drawn so that it cuts the graph of a relation in more than one point, the relation is *not* a function.

If a function f is defined by an equation and no domain is specified, the largest possible subset of the real numbers for which each first component causes the corresponding function value to be a *real* number is its **domain of definition.** Hereafter, in any place where a function is discussed, we shall consider its domain of definition to be the domain of the function. For example,

- The function $f(x) = 2x + 5$ is defined for all real numbers R; thus, $D_f = \{x \in R\}$.

- The function $g(x) = \frac{x-2}{x+3}$ is undefined when $x + 3 = 0$; thus, $D_g = \{x$ such that $x \in R, x \neq -3\}$.
- The function $h(x) = \sqrt{x-5}$ is not a real number when $x - 5 < 0$; thus, $D_h = \{x$ such that $x \in R, x \geq 5\}$.

> In determining the domain of a function defined by an equation, start with the set of real numbers and eliminate any numbers that would make the equation undefined. If a rational expression is involved, exclude any numbers that would make a denominator zero. If a radical with an *even* index is involved, omit all values for which the expression under the radical is negative.

Additional examples follow:

- The function $f(x) = x^2 - 5x + 6$ is defined for all real numbers R; thus, $D_f = \{x \in R\}$.
- The function $g(x) = 2x + \frac{x-2}{x^2-4}$ is undefined when $x^2 - 4 = 0$; the domain should exclude all real numbers for which:

$$x^2 - 4 = 0$$
$$x^2 = 4$$
$$|x| = 2$$
$$x = \pm 2$$

  Thus, $D_g = \{x$ such that $x \in R, x \neq \pm 2\}$.
- The function $h(x) = \sqrt{3 - 2x}$ is not a real number when $3 - 2x < 0$; thus, the domain should be restricted to all real numbers for which:

$$3 - 2x \geq 0$$
$$-2x \geq -3$$
$$x \leq \frac{3}{2} \quad \text{Reverse the inequality because you divided by a negative number.}$$

  Thus, $D_h = \left\{x \text{ such that } x \in R, x \leq \frac{3}{2}\right\}$.

5.4 SKETCHING GRAPHS OF FUNCTIONS

The graph of a function f in a coordinate plane is the set of all ordered pairs (x, y) for which $x \in D_f$ and $y = f(x)$. A sketch of the graph can be made by plotting a number of points $(x, f(x))$ until a pattern becomes apparent and then drawing a smooth curve (or curves, if necessary) through the points. The sketch of the graph of the function gives some indication of the behavior of the function.

Suppose we wish to sketch the function defined by $f(x) = 2x + 5$. We can begin by finding points that satisfy the equation. For convenience, we can list the points in tabular form:

x	$f(x) = 2x + 5$
-3	$f(-3) = 2(-3) + 5 = -6 + 5 = -1$
3	$f(3) = 2(3) + 5 = 6 + 5 = 11$
.	.
.	.
.	.

Zeros of a Function

When sketching a function, it is usually helpful to find the point (or points) at which the graph of the function intersects the x-axis. These points are called the real **zeros** of the function and are determined by finding all values x for which $f(x) = 0$. For instance, $f(x) = 2x + 5$ has x-intercept $\frac{-5}{2}$. This value is obtained by solving the linear equation $2x + 5 = 0$. The value $\frac{-5}{2}$ is a zero of $f(x) = 2x + 5$.

We can say three things about a real zero of a function:

A real zero of a function is

1. an x-value for which $f(x) = 0$.
2. a real root of the equation $f(x) = 0$.
3. an x-intercept for the graph of $y = f(x)$.

Y-intercept

If zero is in the domain of f, then $f(0)$ is the y-intercept of the graph. The y-intercept for $f(x) = 2x + 5$ is $f(0) = 2(0) + 5 = 5$.

Since each x in the domain of a function corresponds to one and only one $f(x)$ in the range, a function cannot have more than one y-intercept.

We now have four points that satisfy the equation:

x	$f(x) = 2x + 5$
-3	$f(-3) = 2(-3) + 5 = -6 + 5 = -1$
3	$f(3) = \quad 2(3) + 5 = \quad 6 + 5 = 11$
$-\frac{5}{2}$	$f(-\frac{5}{2}) = 2(-\frac{5}{2}) + 5 = -5 + 5 = 0$
0	$f(0) = \quad 2(0) + 5 = \quad 0 + 5 = 5$

As Figure 5.14 illustrates, when we plot the points and connect them, the graph is a straight line.

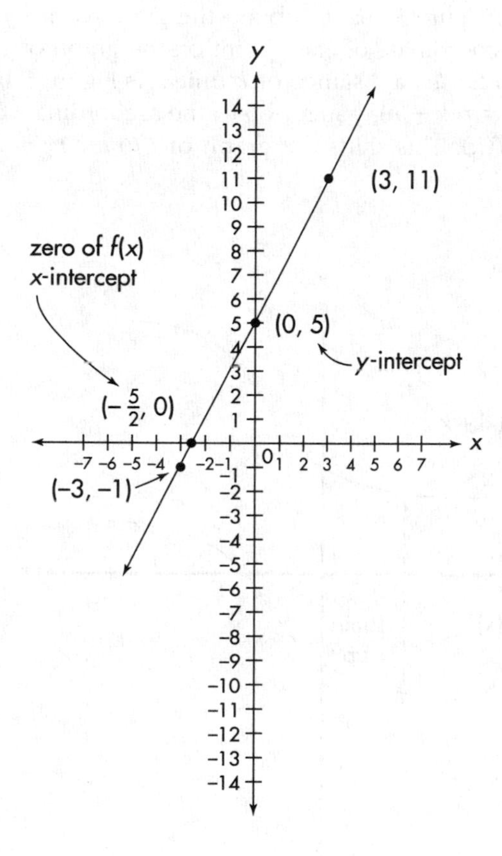

FIGURE 5.14

By looking at the graph you can see that the range of the function defined by $f(x) = 2x + 5$ is the set of all real numbers. This is often the most convenient way to determine the range of a function. You will find a graphing calculator very useful for this purpose.

Vertical and Horizontal Shifts

If f is a function and you know the graph of $y = f(x)$, the effect of adding or subtracting a positive constant k to or from $f(x)$ is called a **vertical shift.** As illustrated in Figure 5.15a, to obtain the graph of $y = f(x) + k$, you add k to the y-coordinate of each point on the graph of $y = f(x)$. This shifts the graph of f *up* a distance of k units. As Figure 5.15b shows, to obtain the graph of $y = f(x) - k$, you subtract k from the y-coordinate of each point on the graph of $y = f(x)$. This shifts the graph of f *down* a distance of k units.

Similarly, if f is a function and you know the graph of $y = f(x)$, the effect of adding or subtracting a positive constant h to or from x is called a **horizontal shift.** As illustrated in Figure 5.16a, to obtain the graph of $y = f(x + h)$, you subtract h from the x-coordinate of each point on the graph of $y = f(x)$. This shifts the graph of f *to the left* a distance of h units. As Figure 5.16b shows, to obtain the graph of $y = f(x - h)$, you add h to the x-coordinate of each point on the graph of $y = f(x)$. This shifts the graph of f *to the right* a distance of h units.

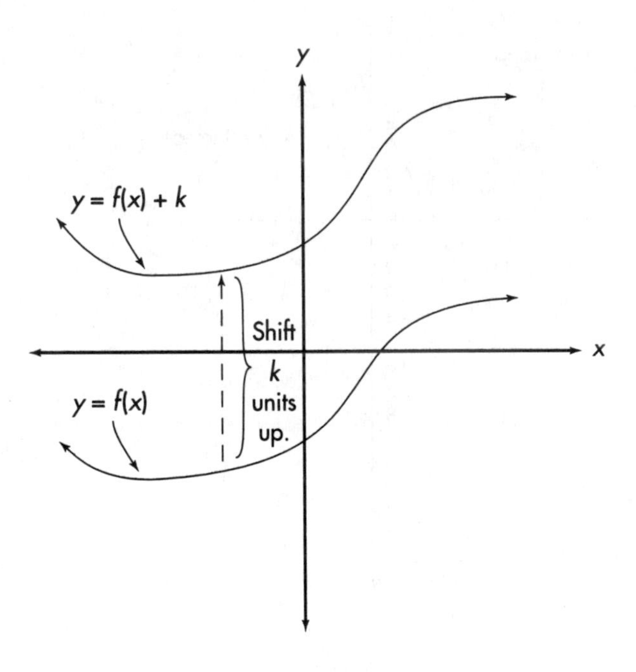

FIGURE 5.15a

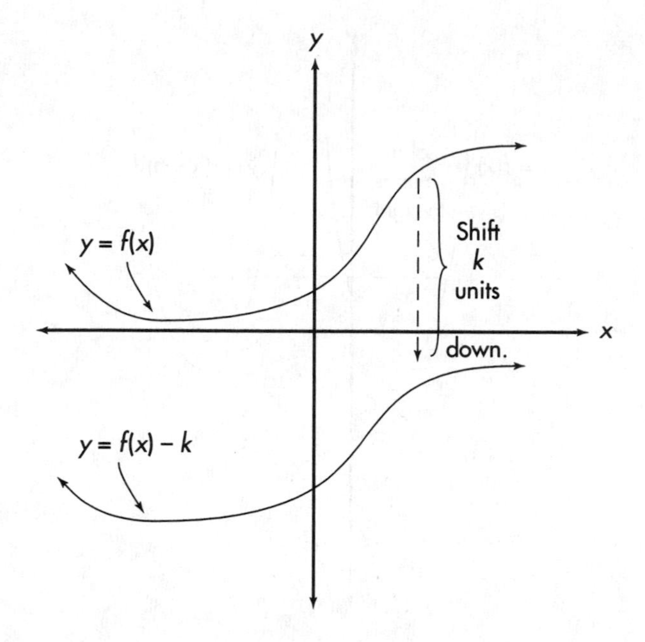

FIGURE 5.15b

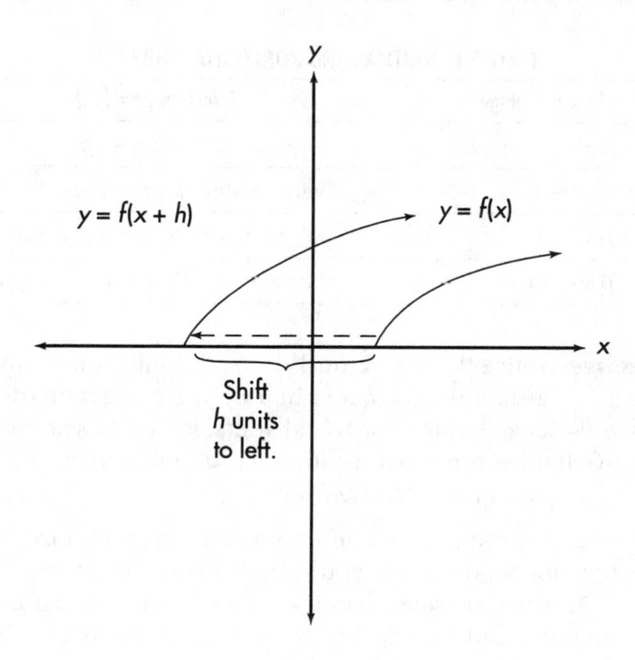

FIGURE 5.16a

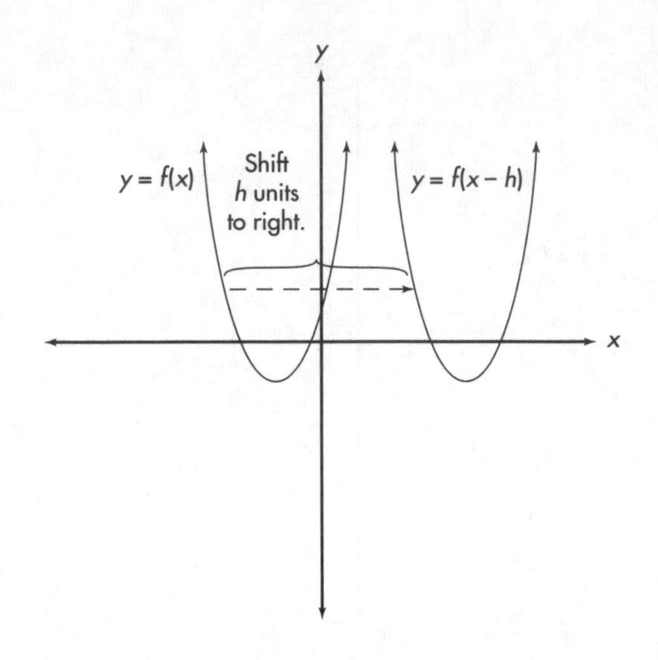

FIGURE 5.16b

Vertical and horizontal shifts are summarized in Table 5.1.

TABLE 5.1. VERTICAL AND HORIZONTAL SHIFTS

Type of change		Effect on $y = f(x)$
$y = f(x) + k$	$k > 0$	vertical shift: k units *up*
$y = f(x) - k$	$k > 0$	vertical shift: k units *down*
$y = f(x + h)$	$h > 0$	horizontal shift: h units to *left*
$y = f(x - h)$	$h > 0$	horizontal shift: h units to *right*

A helpful message: Notice that for vertical shifts, the shift is in a *positive* direction (up) when a constant is *added,* and in a *negative* direction (down) when a constant is *subtracted;* but for horizontal shifts, it switches around: the shift is in a *negative* (left) direction when a constant is *added,* and in a *positive* (right) direction when a constant is *subtracted.*

The graph of a function can contain a combination of a horizontal and vertical shift of a function whose graph you already know. For example, the graph of $f(x) = 2(x - 3) + 9$, represents a vertical shift of 9 units up, a horizontal shift of 3 units to the right, and a vertical increase of magnitude 2 of the graph of the function $f(x) = x$ (Figure 5.17). It is customary to call the function $f(x) = x$ the **parent** function for the function $f(x) = 2(x - 3) + 9$.

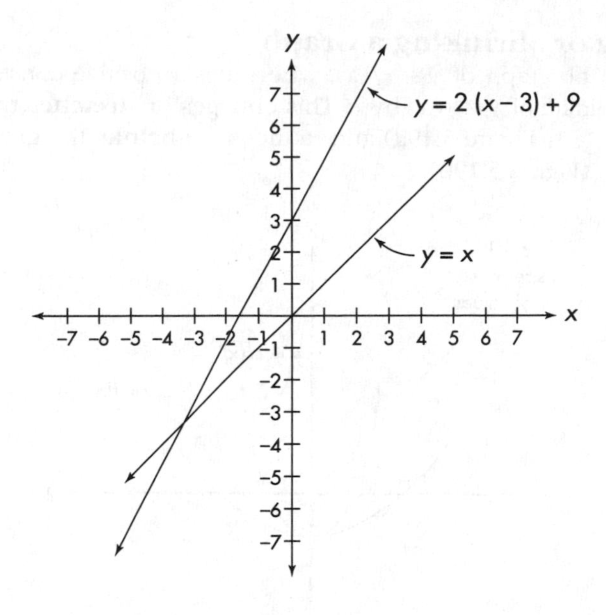

FIGURE 5.17

Reflection

The graph of $y = -f(x)$ is a **reflection** of $y = f(x)$ in the x-axis (Figure 5.18). It can be obtained by multiplying each y-coordinate by -1. Thus, if (a, b) is on the graph of $y = f(x)$ then $(a, -b)$ will be on the graph of $-f(x)$.

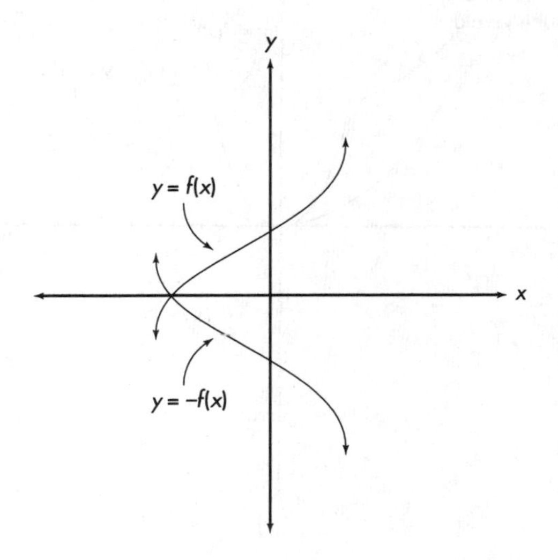

FIGURE 5.18

Stretching or Shrinking a Graph

To obtain the graph of $y = cf(x)$, where c is a positive constant, multiply each y-coordinate of $y = f(x)$ by c. This enlarges or **stretches** the graph of f *vertically* if $c > 1$ (Figure 5.19a) and reduces or **shrinks** the graph of f *vertically* if $c < 1$. (Figure 5.19b).

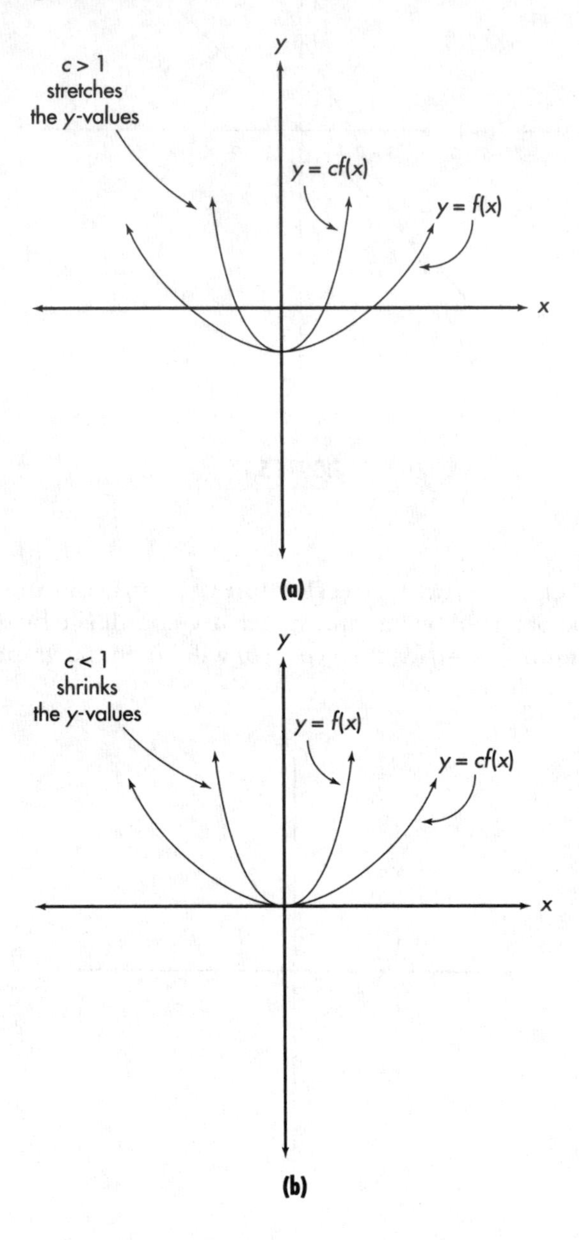

FIGURE 5.19

Asymptotes

The graph of a function sometimes approaches lines containing values for which the function is undefined. These lines are called **asymptotes.** There are three types: *vertical, horizontal,* or *oblique.*

A line $x = k$ is a **vertical asymptote** if the function approaches $-\infty$ or ∞ as x approaches k from the left or right. A vertical asymptote for $y = f(x)$ tells you about the behavior of the function as x approaches a point k not in the domain of f. For example, the graph of $f(x) = \frac{1}{x+2}$ (Figure 5.20) has a vertical asymptote at $x = -2$ because as x gets close to -2 from the left, $f(x)$ approaches $-\infty$, and as x gets close to -2 from the right, $f(x)$ approaches ∞. Observe that the point -2 is *not* in the domain of f.

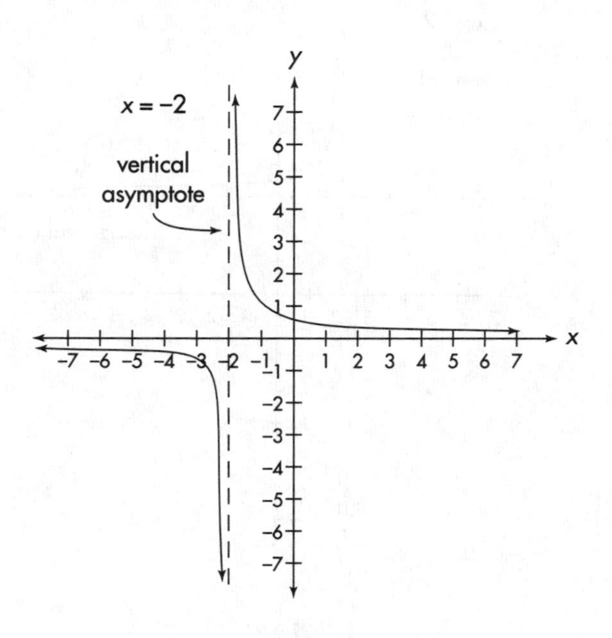

FIGURE 5.20

A line $y = b$ is a **horizontal asymptote** if the function approaches b as x increases or decreases without bound. A horizontal asymptote for $y = f(x)$ tells you about the behavior of the graph of f as x gets extremely large or extremely small. For example, $f(x) = 3 + \frac{1}{x+2}$ (Figure 5.21) has a horizontal asymptote at $y = 3$ because as x gets extremely large or extremely small, $\frac{1}{x+2}$ approaches zero, and therefore the graph of f approaches the line $y = 3$. Of course, as in the previous example, the graph of f also has a vertical asymptote at $x = -2$.

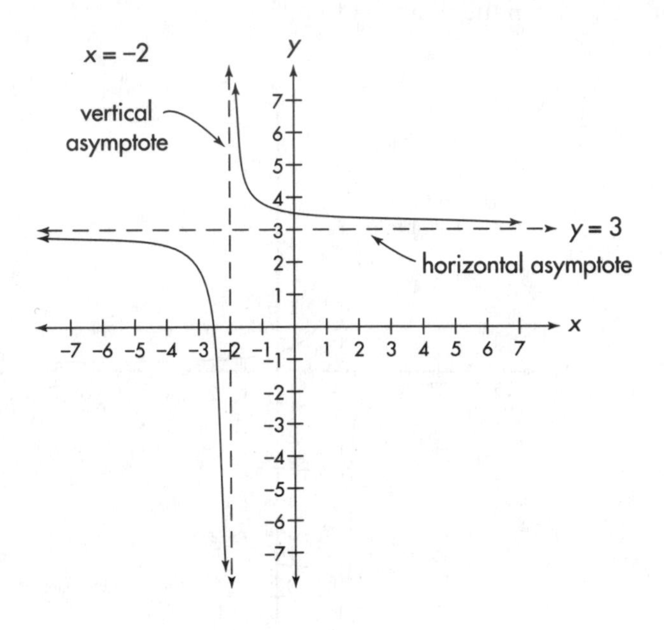

FIGURE 5.21

A function $y = g(x)$ is an **oblique asymptote** or **slant asymptote** if the function approaches $g(x)$ as x increases or decreases without bound. An oblique asymptote, like horizontal asymptotes, for $y = f(x)$ tells you about the behavior of the graph of f as x gets extremely large or extremely small. For example, the graph of $f(x) = x + 3 + \frac{1}{x+2}$ (Figure 5.22) has an oblique asymptote $y = x + 3$ because as x gets extremely large or extremely small, $\frac{1}{x+2}$ approaches zero, and therefore the graph of f approaches the line $y = x + 3$. As in the previous examples, the graph of f also has a vertical asymptote at $x = -2$.

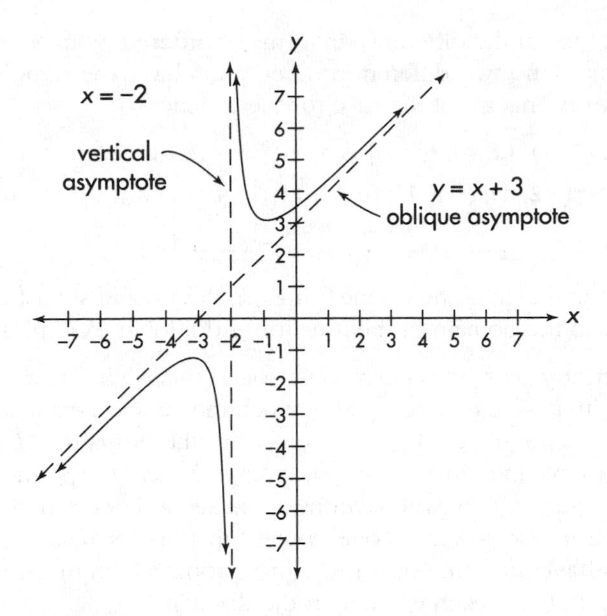

FIGURE 5.22

5.5 SPECIAL FUNCTION CHARACTERISTICS

One-to-One

A function is **one-to-one** if no two different ordered pairs have the same second component. Thus, we have the following:

ONE-TO-ONE FUNCTION

A function f is a one-to-one function

if and only if $f(a) = f(b)$ implies that $a = b$; that is,
if $(a, c) \in f$ and $(b, c) \in f$, then $a = b$.

For example,

- $S = \{(-3, 9), (-2, 4), (-1, 1), (0, 0), (1, 1), (2, 4), (3, 9)\}$ is a function, but it is *not* a one-to-one function.
- $Q = \{(0, 0), (1, 1), (2, 4), (3, 9)\}$ is a one-to-one function.

When you look at the difference in terms of ordered pairs, you can see that in the function Q, no two different ordered pairs have the same second component; however, this is not the case for the function S:

$Q = \{(0, 0), (1, 1), (2, 4), (3, 9)\}$ is a one-to-one function.

$S = \{(-3, 9), (-2, 4), (-1, 1), (0, 0), (1, 1), (2, 4), (3, 9)\}$ is *not* one-to-one.

same second components

Keep in mind that in a one-to-one function, the y-values must be unique for *each* element in the domain of the function. Additional examples follow:

- The function $f(x) = x^2$ is not a one-to-one function because a number and its opposite have the same square, which means the y-values for the function will *not* be unique for each element in the domain of f. For instance, -3 and 3 have the same square, so their y-values, $f(-3)$ and $f(3)$, will be the same; that is, they will have the same second component, namely 9.
- The function $f(x) = \sqrt{x}$ is a one-to-one function because no two different numbers have the same principal square root, which means the y-values will be unique for each element in the domain.

The definition for one-to-one functions means that in a one-to-one function each first component is paired with exactly one second component *and* each second component is paired with exactly one first component. Therefore, any horizontal line in the plane will intersect the function no more than once. This is known as the **horizontal line test.** For example, in Figure 5.23, both graphs (a) and (b) are functions, but graph (a) is a one-to-one function, and graph (b) is not.

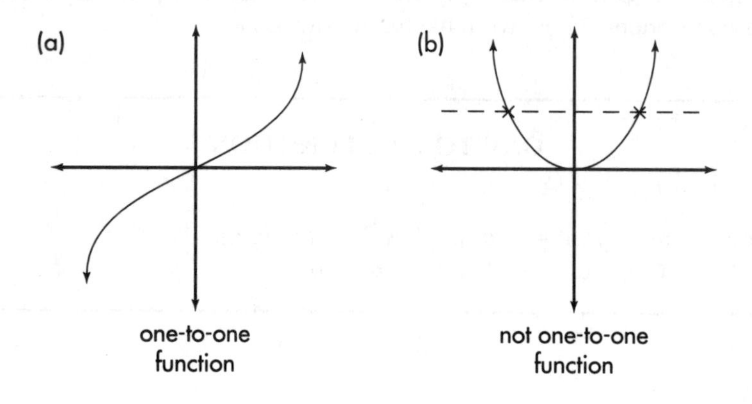

FIGURE 5.23

Notice that a horizontal line would cut the curve in more than one place in (b). Thus, we can state the horizontal line test as follows:

> **HORIZONTAL LINE TEST FOR ONE-TO-ONE FUNCTIONS**
> If any horizontal line can be drawn so that it cuts the graph of a function in more than one point, the function is *not* a one-to-one function.

A helpful message: You should note that one-to-one functions must pass both the vertical line test and the horizontal line test.

Increasing, Decreasing, Constant, and Monotone

> Suppose x_1 and x_2, with $x_1 < x_2$, are any two points in an interval. A function f is
>
> *increasing* in the interval if $f(x_1) < f(x_2)$;
> *decreasing* in the interval if $f(x_1) > f(x_2)$;
> *constant* in the interval if $f(x_1) = f(x_2)$.

See Figure 5.24 for an illustration.

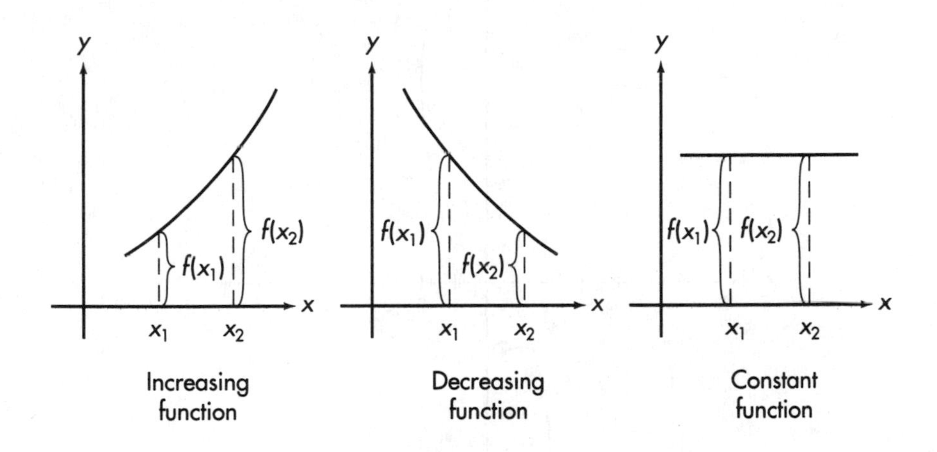

FIGURE 5.24

A function is **monotone** if it is increasing or decreasing on its entire domain. For example,

- The function $f(x) = -3x$ (Figure 5.25) is monotone decreasing on its domain.

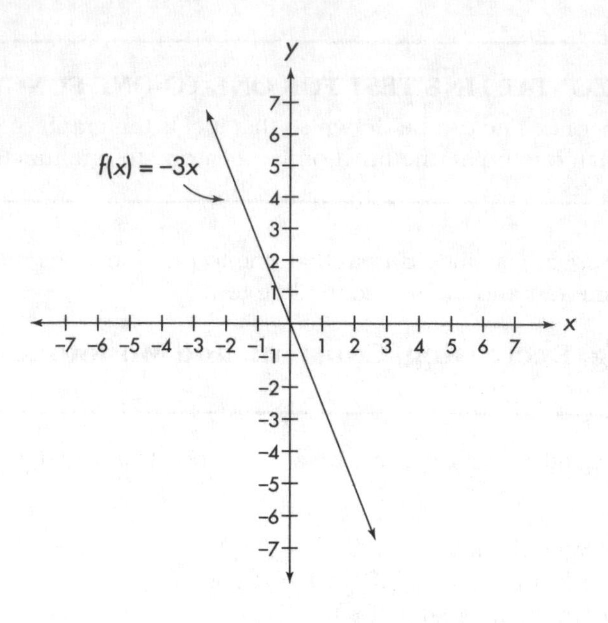

FIGURE 5.25

- The function $f(x) = x^3$ (Figure 5.26) is monotone increasing on its domain.

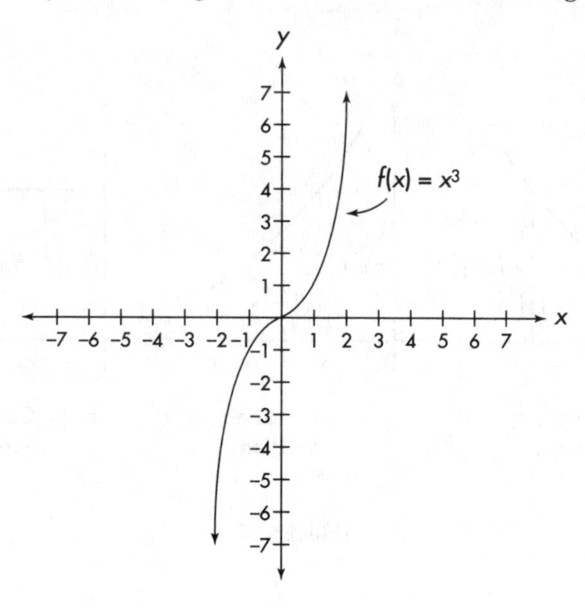

FIGURE 5.26

As x goes from left to right, the graph of a monotone increasing function rises and that of a monotone decreasing function falls.

There is a connection between the characteristics of monotone and one-to-one, namely,

> A monotone increasing or decreasing function is one-to-one.

Even and Odd

> A function is **even** if for every x in the domain of f, $-x$ is in the domain of f and $f(-x) = f(x)$.

For example, $f(x) = x^2$ is even because $f(-x) = (-x)^2 = x^2 = f(x)$ for every real number x.

> A function is **odd** if for every x in the domain of f, $-x$ is in the domain of f and $f(-x) = -f(x)$.

For example, $f(x) = x^3$ is odd because $f(-x) = (-x)^3 = -x^3 = -f(x)$ for every real number x.

The graph of an even function is symmetric with respect to the y-axis (Figure 5.27a). Pictorially, this means both sides would match if you folded the

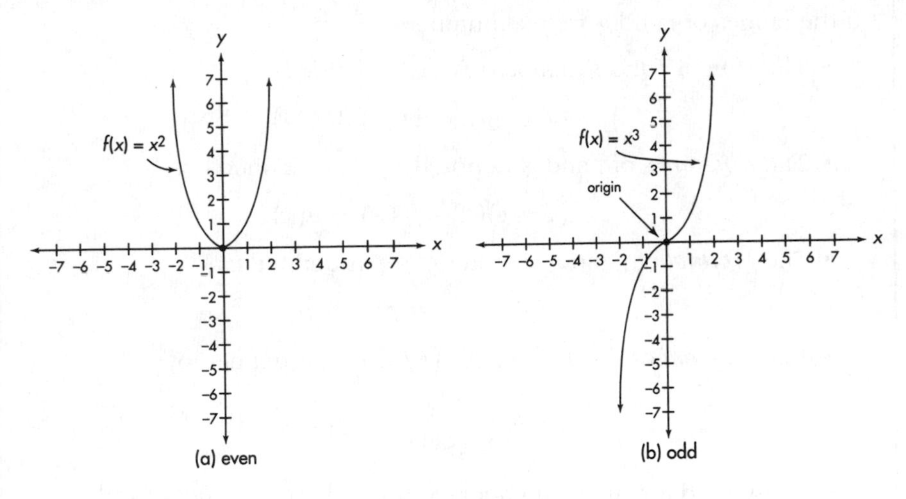

FIGURE 5.27

graph along the y-axis. The graph of an odd function is symmetric with respect to the origin (Figure 5.27b). Pictorially, this means the portion of the graph in one quandrant is an upside down copy of the portion of the graph in the diagonally opposite quadrant.

In terms of coordinates, symmetry with respect to the y-axis means that when x is on the graph, $-x$ is also. Symmetric with respect to the origin means that when (x, y) is on the graph, so is $(-x, -y)$.

5.6 THE ARITHMETIC OF FUNCTIONS

Equality of Functions

EQUALITY OF FUNCTIONS

Two functions f and g are *equal,* written $f = g$, if and only if their domains are equal and $f(x) = g(x)$ for all x in their common domain.

Arithmetic of Functions

We can add, subtract, multiply, and divide functions by performing the algebra on the equations that define the functions. The following definitions for this "arithmetic" of functions hold:

ARITHMETIC OF FUNCTIONS

If the ranges of f and g are real numbers,

- The *sum* of f and g, denoted $f + g$, is defined by:

$$(f + g)(x) = f(x) + g(x)$$

- The *difference* of f and g, denoted $f - g$, is defined by:

$$(f + g)(x) = f(x) - g(x)$$

- The *product* of f and g, denoted $f \cdot g$, is defined by:

$$(f \cdot g)(x) = f(x)g(x)$$

- The *quotient* of f and g, denoted $(f/g)(x)$, is defined by:

$$(f/g)(x) = \frac{f(x)}{g(x)} \qquad (g(x) \neq 0)$$

These hold for all real numbers x in the domain of *both* f and g.

- Let $f(x) = x^2 + 3$ and $g(x) = 3x$. Find the sum, difference, product, and quotient of f and g.

$$(f + g)(x) = f(x) + g(x) = (x^2 + 3) + 3x = x^2 + 3x + 3$$
$$(f - g)(x) = f(x) - g(x) = (x^2 + 3) - 3x = x^2 - 3x + 3$$
$$(f \cdot g)(x) = f(x)g(x) = (x^2 + 3)(3x) = 3x^3 + 9x$$
$$(f/g)(x) = \frac{f(x)}{g(x)} = \frac{x^2 + 3}{3x} \qquad (x \neq 0)$$

- Let $f(x) = x^2 - 4$ and $g(x) = x - 2$. Find the quotient of f and g.

$$(f/g)(x) = \frac{f(x)}{g(x)} = \frac{x^2 - 4}{x - 2} = \frac{(x + 2)\cancel{(x - 2)}}{\cancel{(x - 2)}} = x + 2 \qquad (x \neq 2)$$

A helpful message: You *must* keep the restriction $x \neq 2$ for the domain in the above example. The function defined by $y = x + 2$ is a polynomial whose domain is the set of all real numbers. It has no excluded values, so it is not the same as the function defined by $(f/g)(x) = x + 2$, $x \neq 2$, whose domain is the set of all real numbers except 2. Remember that in order for two functions to be equal, they must have *exactly* the same domains.

- Let $f(x) = x^2 - 9$ and $g(x) = x + 3$. Find $(f + g)(x)$, $(f - g)(x)$, $(f \cdot g)(x)$, and $(f/g)(x)$. Then use the results to find $(f + g)(2)$, $(f - g)(-1)$, $(f \cdot g)(0)$, and $(f/g)(-3)$.

$$(f + g)(x) = f(x) + g(x) = (x^2 - 9) + (x + 3) = x^2 + x - 6$$
$$(f - g)(x) = f(x) - g(x) = (x^2 - 9) - (x + 3) = x^2 - x - 12$$
$$(f \cdot g)(x) = f(x)g(x) = (x^2 - 9)(x + 3) = x^3 + 3x^2 - 9x - 27$$
$$(f/g)(x) = \frac{f(x)}{g(x)} = \frac{x^2 - 9}{x + 3} = \frac{\cancel{(x + 3)}(x - 3)}{\cancel{x + 3}} = x - 3 \quad (x \neq -3)$$
$$(f + g)(2) = (2)^2 + 2 - 6 = 0$$
$$(f - g)(-1) = (-1)^2 - (-1) - 12 = -10$$
$$(f \cdot g)(0) = (0)^3 + 3(0)^2 - 9(0) - 27 = -27$$

$(f/g)(-3)$ is undefined because -3 is not in the domain of $(f/g)(x)$.

5.7 COMPOSITION OF FUNCTIONS

COMPOSITION OF FUNCTIONS

The *composite* of f and g, denoted $f \circ g$, is the function defined by

$$(f \circ g)(x) = f(g(x)),$$

where $R_g \subseteq D_f$; that is, when x is in the domain of g, $g(x)$ is in the domain of f. Forming the composite of functions is called *composition of functions*.

The expression $(f \circ g)(x)$ is read "f of g of x" or as "f composition g." You can think of $f \circ g$ as a first process, called g, that takes a number x and produces from it the number $g(x)$, which then goes into a second process, called f, that uses $g(x)$ to produce the number $f(g(x))$.

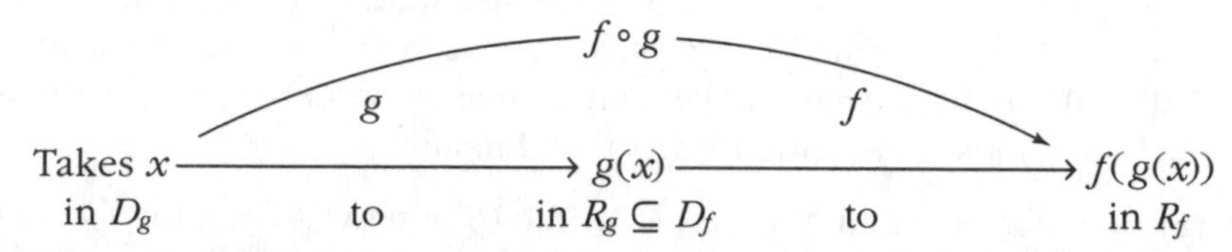

Notice that it is important that the number $g(x)$ be in the domain of f because f uses it to produce $f(g(x))$.

Figure 5.28 shows a helpful visualization of the composition of f and g as two machines lined up so that the output from the first machine (function g) feeds in as input to the second machine (function f).

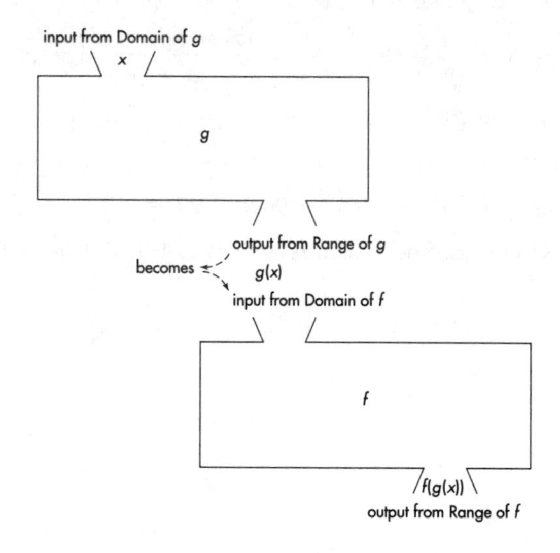

FIGURE 5.28

A helpful message: Notice that for $f \circ g$ the function g is performed *before* the function f.

Let $f(x) = x + 5$ and $g(x) = 3x$. Find $(f \circ g)(2)$ and $(g \circ f)(2)$.

$$(f \circ g)(2) = f(g(2)) = f(3(2)) = f(6) = 6 + 5 = 11$$

Evaluate g at 2. Evaluate f at 6.

$$(g \circ f)(2) = g(f(2)) = g(2 + 5) = g(7) = 3(7) = 21$$

Evaluate f at 2. Evaluate g at 7.

From this example, you can conclude that, as a general rule, the following is true:

> Composition of functions is *not* a commutative process; that is, in general,
>
> $$(f \circ g)(x) \neq (g \circ f)(x)$$

- Let $f(x) = x^2 + 3$ and $g(x) = 3x$. Find $(f \circ g)(x)$ and $(g \circ f)(x)$.

$$(f \circ g)(x) = f(g(x)) = f(3x)) = (3x)^2 + 3 = 9x^2 + 3$$

Evaluate g at x. Evaluate f at $g(x) = 3x$.

$$(g \circ f)(x) = g(f(x)) = g(x^2 + 3)) = 3(x^2 + 3) = 3x^2 + 9$$

Evaluate f at x. Evaluate g at $f(x) = x^2 + 3$.

When simplifying expressions containing functional notation, you should evaluate the functions *before* performing other operations. For example:

- If $g(x) = x + 5$, evaluate $g(5) - g(-3)$.

Perform g first, then subtract.

$$g(5) - g(-3) = (5 + 5) - (-3 + 5) = 10 - 2 = 8$$

- If $f(x) = x + 1$, evaluate $\dfrac{f(x + h) - f(x)}{h}$.

Perform f first, then simplify.

$$\frac{f(x+h) - f(x)}{h} = \frac{((x+h)+1) - (x+1)}{h} = \frac{x+h+1-x-1}{h} = \frac{h}{h} = 1.$$

A helpful message: Do not confuse the composition $(f \circ g)(x)$ of two functions with the product $(f \cdot g)(x)$ of two functions. They are *not* the same.

- Let $f(x) = x^2$ and $g(x) = 2x$. Find $(f \cdot g)(x)$ and $(f \circ g)(x)$. Then use the results to find $(f \cdot g)(5)$ and $(f \circ g)(5)$.

$$(f \cdot g)(x) = f(x)g(x) = (x^2)(2x) = 2x^3$$
$$(f \circ g)(x) = f(g(x)) = f(2x) = (2x)^2 = 4x^2$$
$$(f \cdot g)(5) = 2(5)^3 = 250$$
$$(f \circ g)(x) = 4(5)^2 = 100$$

5.8 INVERSE OF FUNCTIONS

Let's play a game:

- I have a number. When I multiply it by 2 and add 5, I get 11. Can you guess my number? Of course you can; the number is 3.
- More generally, I have a number, call it x. When I multiply x by 2 and add 5, I get $f(x) = 2x + 5$. Can you tell me a function that will start with $f(x)$ and take me back to x? You likely obtained a function that looks like $g(x) = \frac{x-5}{2}$. We can check that g "undoes" f by evaluating $(g \circ f)(x)$:

$$(g \circ f)(x) = g(f(x)) = g(2x + 5) = \frac{(2x + 5) - 5}{2} = \frac{2x}{2} = x$$

Start with x. Evaluate f at x. Evaluate g at $f(x) = 2x + 5$. Simplify. You get x back.

As a matter of fact, it works both ways; that is, f "undoes" g:

$$f \circ g(x) = f(g(x)) = f\left(\frac{x-5}{2}\right) = 2\left(\frac{x-5}{2}\right) + 5 = x - 5 + 5 = x$$

Start with x. Evaluate g at x. Evaluate f at $g(x) = \left(\frac{x-5}{2}\right)$. Simplify. You get x back.

Pairs of functions, such as f and g, that undo what the other does are called **inverses** of each other. We have the following definition.

> Two functions f and g are called inverses of each other if and only if
>
> $$(f \circ g)(x) = (g \circ f)(x) = x,$$
>
> where $R_g \subseteq D_f$ and $R_f \subseteq D_g$.

So can we always do this? Can we always find a function that takes us back from a given function—that "undoes" the given function? Let's play our game again:

- I have a number. When I square it, I get 9. Can you guess my number? This time you would not be sure of your answer, since there are two numbers, 3 and -3, that give 9 when squared.

- More generally, I have a number, call it x. When I square x, I get $f(x) = x^2$. Can you tell me a function that will start with $f(x)$ and take me back to x? Your best guess would be a function like $g(x) = \sqrt{x}$. Check it by evaluating $(g \circ f)(x)$:

$$(g \circ f)(x) = g(f(x)) = g(x^2) = \sqrt{x^2} = |x| \neq x,$$

Start with x. Evaluate f at x. Evaluate g at $f(x) = x^2$. Simplify. You do *not* get x back.

So $g(x) = \sqrt{x}$ didn't work. What we need is $g(x) = \pm\sqrt{x}$, but this is not a function.

The problem is that the ordered pairs for $f(x) = x^2$ have second elements that are paired with *more than one* different first element; for instance, $(3, 9) \in f$ and $(-3, 9) \in f$ have the same second element. For these ordered pairs, when you try to find a function that will take 9 back, there is not a single number to go back to. Which should it be—3 or -3? You can't use both. The ordered pairs (9, 3) and (9, -3) cannot be in the same function, since they have the same first element.

What is the difference between $f(x) = 2x + 5$ and $f(x) = x^2$? The function $f(x) = 2x + 5$ is a one-to-one function; the function $f(x) = x^2$ is not. We have the following:

The inverse of a function f is usually denoted f^{-1} (read "f inverse.") If a function is one-to-one, then f^{-1} may be found by interchanging the first and second components in the ordered pairs of f. This leads to an alternate definition for the inverse of a function.

INVERSE OF A FUNCTION

If f is a one-to-one function, f^{-1} is the set of ordered pairs obtained from f by interchanging the first and second components of each of its ordered pairs.

A helpful message: The notation f^{-1} does not mean "reciprocal." The $^{-1}$ is *not* an exponent. Do not write $f^{-1} = \frac{1}{f}$!

For example,

$$Q = \{(0, 0), (1, 1), (2, 4), (3, 9)\} \text{ is a one-to-one function and}$$
$$Q^{-1} = \{(0, 0), (1, 1), (4, 2), (9, 3)\} \text{ is its inverse.}$$

A helpful message: Some authors define f^{-1} to be the *relation* obtained from f by interchanging the first and second components of each ordered pair *without restricting f to be a one-to-one function*. In that case, f^{-1} may or may not be a function.

There is an interesting relationship between the graph of a one-to-one function and the graph of its inverse. Each is the mirror image reflection of the other in the line $y = x$. This occurs because when (x, y) is in the graph of f, (y, x) is in the graph of f^{-1}. Thus, we have the following:

REFLECTION PROPERTY FOR INVERSE FUNCTIONS

The graphs of f and f^{-1} are reflections of one another about the line $y = x$.

This property is illustrated in Figure 5.29.

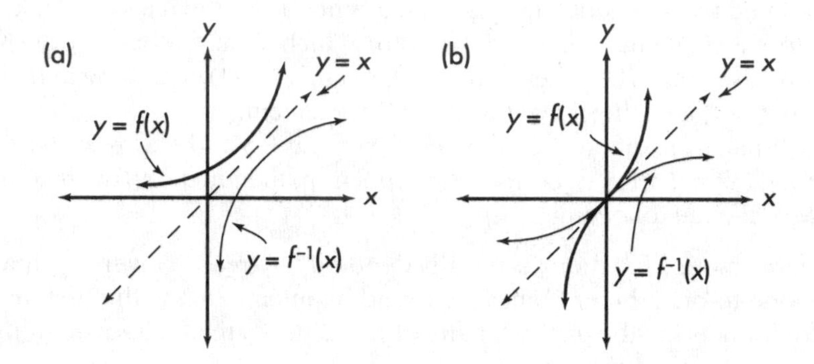

FIGURE 5.29

Two common methods for determining the equation that defines the inverse of a function are shown below.

Method 1. Set $(f \circ f^{-1})(x) = x$ and solve for $f^{-1}(x)$.

- Find $f^{-1}(x)$ for $y = f(x) = 2x + 5$.
  Set $(f \circ f^{-1})(x) = x$ and solve for $f^{-1}(x)$:

$$\begin{aligned} (f \circ f^{-1})(x) &= x \\ f(f^{-1}(x)) &= x \\ 2f^{-1}(x) + 5 &= x \\ 2f^{-1}(x) &= x - 5 \\ f^{-1}(x) &= \frac{x - 5}{2} \end{aligned}$$

Method 2. First, interchange x and y in $y = f(x)$, then solve $x = f(y)$ for y.

- Find $f^{-1}(x)$ for $y = f(x) = 2x + 5$.

Interchange x and y:

$$y = 2x + 5$$

$$x = 2y + 5$$

Solve $x = 2y + 5$ for y (you may find it helpful to exchange the sides first):

$$2y + 5 = x$$

$$2y = x - 5$$

$$y = f^{-1}(x) = \frac{x - 5}{2}$$

A helpful message: When it is clear no confusion will occur, it is customary to designate y as the value of f^{-1} at x. Thus, when $y = f(x)$ defines f, we say $y = f^{-1}(x)$ defines the inverse of f, provided the inverse exists. As in the previous examples, we started with $y = f(x) = 2x + 5$ and obtained $y = f^{-1}(x) = \frac{x-5}{2}$.

Now let's reconsider $f(x) = x^2$. We know that f is not a one-to-one function. However, if the domain of f is restricted to the set of *nonnegative* real numbers, then f will be a one-to-one function. We will not have the problem we encountered earlier, since (3, 9) will be in f, but (−3, 9) will not (because −3 is not in the restricted domain of f). Thus, the function $g(x) = \sqrt{x}$ will take 9 back to 3; and, in general, g will "undo" f by returning all the squares back to their *nonnegative* roots in the restricted domain of f. That is, let $x \geq 0$; then:

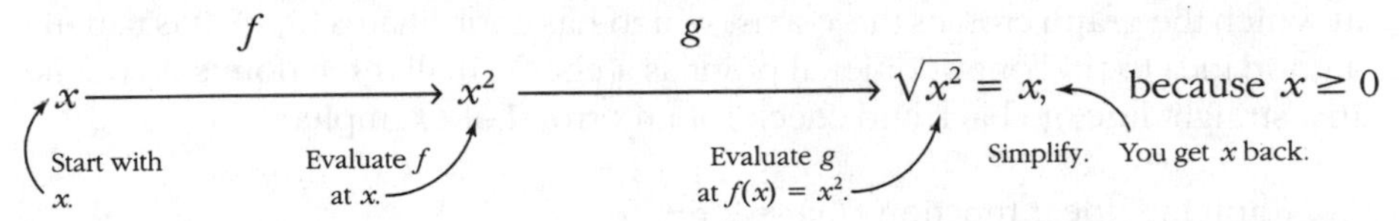

From the above discussion, we conclude that if a function is not a one-to-one function, we can restrict the domain so that the function is one-to-one in the restricted domain. The inverse of the function will be defined in the restricted domain of the function.

5.9 FUNCTIONS AND THEIR GRAPHS

The concept of function is fundamental to all of mathematics. In this section, we discuss functions frequently encountered in mathematics and their special names.

Linear Functions

> A linear function is defined by:
>
> $$y = f(x) = mx + b \quad (m \neq 0)$$

The domain and range are both equal to the set of real numbers.

The zeros are found by solving the equation $mx + b = 0$. Thus, $x = -\frac{b}{m}$ is the only zero of the function. The graph crosses the x-axis at the point $\left(-\frac{b}{m}, 0\right)$.

It can be shown that the graph of an equation of the form $y = mx + b$ is a straight line with slope m and y-intercept b. This is called the **slope-intercept** form of a line.

Linear equations of the form $Ax + By = C$, where $B \neq 0$, can be transformed into an equivalent equation having the form $y = mx + b$ as follows:

$$Ax + By = C$$

$$By = -Ax + C$$

$$y = -\frac{A}{B}x + \frac{C}{B}$$

Thus, every linear equation $Ax + By = C$ with $B \neq 0$ determines a linear function. The line determined by the equation has slope $-\frac{A}{B}$ and y-intercept $\frac{C}{B}$.

Since any two distinct points determine a line, we can graph a linear function by determining two points that satisfy its equation. In practice, the two points that usually work best are the point at which the graph crosses the x-axis, which for equations in the form $y = mx + b$ is given by $\left(-\frac{b}{m}, 0\right)$, and the point at which the graph crosses the y-axis, which has coordinates $(0, b)$. It is usually a good idea to find one additional point as a check. If all three points do not lie in a straight line, go back and check for an error. For example:

Graph the linear function $f(x) = 2x + 5$.

1. Solve the equation $2x + 5 = 0$ to determine the point at which the graph crosses the x-axis:

$$2x + 5 = 0$$

$$x = -\frac{5}{2}$$

   The graph crosses the x-axis at $\left(-\frac{5}{2}, 0\right)$.
2. Since the equation is in slope-intercept form, the graph will cross the y-axis at $(0, 5)$.
3. Determine a third point:

$$f(2) = 2(2) + 5 = 9$$

   The point $(2, 9)$ satisfies the equation so it lies on the graph of f.
4. Plot the points and sketch the graph (Figure 5.30):

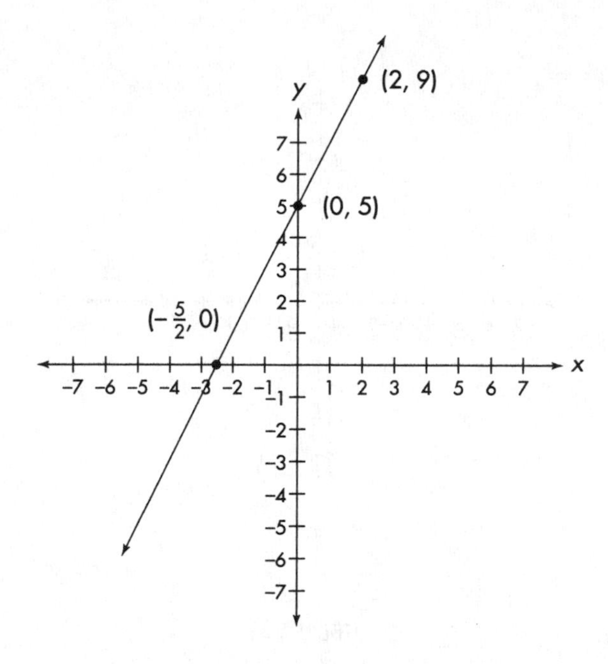

FIGURE 5.30

Since linear functions are an important type of function and their graphs are lines, we will discuss some basic concepts about the equations of lines before going on.

The equation of a line can be determined using one of the following:

1. The **slope-intercept form:** $y = mx + b$, where m is the slope of the line and b is its y-intercept
2. The **standard form:** $Ax + By = C$, where A and B are not both zero
3. The **point-slope form:** $y - y_1 = m(x - x_1)$, where m is the slope of the line and (x_1, y_1) is a point on the line

Which form is best to use in a particular problem can be decided based on the information about the line given.

Use the slope-intercept form, $y = mx + b$, when the slope m and y-intercept b are given. For example, to determine the equation of the line that has slope -4 and y-intercept 3 (Figure 5.31):

Substitute -4 for m and 3 for b:

$$y = -4x + 3$$

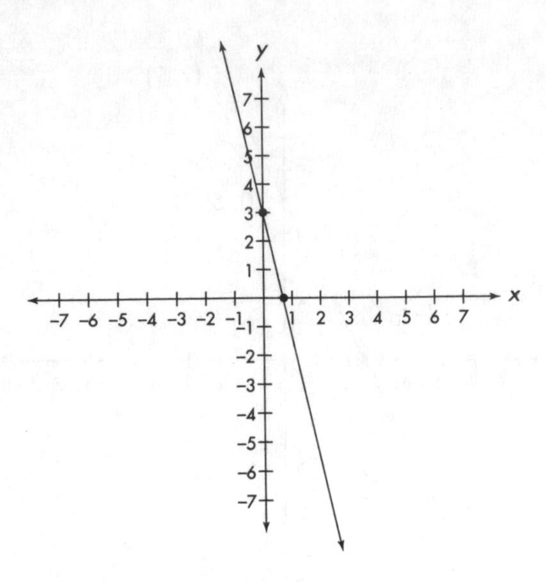

FIGURE 5.31

Use the standard form $Ax + By = C$ for horizontal lines located at the point $(0, k)$. Substitute 0 for A, 1 for B, and k for C. For example, the equation of the horizontal line located at the point $(0, -5)$ is $y = -5$ (Figure 5.32). The

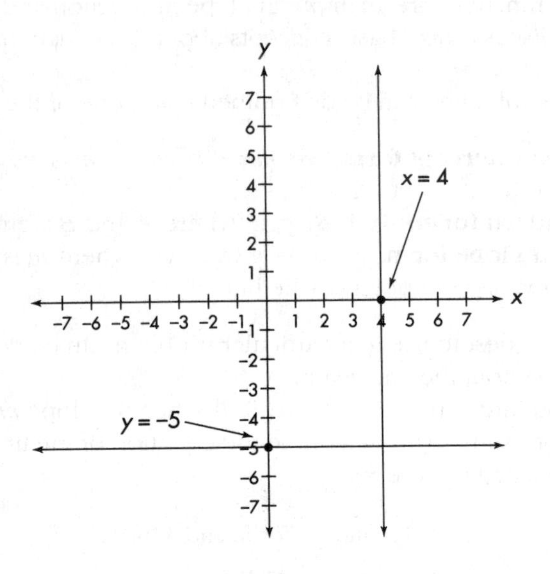

FIGURE 5.32

equation of a vertical line located at $(h, 0)$, *which is not a linear function,* can be obtained by substituting 0 for B, 1 for A, and h for C. For example, the equation of the vertical line located at the point (4, 0), is $x = 4$ (Figure 5.32).

Use the point-slope form, $y - y_1 = m(x - x_1)$, when the slope and the coordinates of a point on the line are given. Substitute the slope for m, the x-coordinate of the point for x_1, and the y-coordinate of the point for y_1. For example, to determine the equation of the line with slope $\frac{2}{3}$ that contains the point $(-1, -4)$ (Figure 5.33):

Substitute $\frac{2}{3}$ for m, -1 for x_1, and -4 for y_1:

$$y - (-4) = \frac{2}{3}(x - (-1))$$

$$y + 4 = \frac{2}{3}(x + 1)$$

$$y + 4 = \frac{2}{3}x + \frac{2}{3}$$

$$y = \frac{2}{3}x - \frac{10}{3}$$

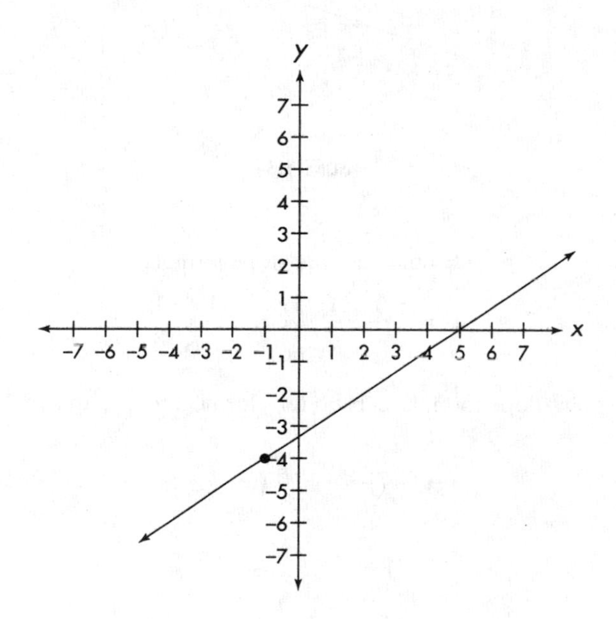

FIGURE 5.33

Also use the point-slope form, $y - y_1 = m(x - x_1)$, when the coordinates of two points on the line are given. Substitute the coordinates of the two points into the slope formula to determine m. Then substitute the computed slope m, the x-coordinate of either one of the points for x_1, and the y-coordinate of that same point for y_1. For example, to determine the equation of the line through $(2, -5)$ and $(-1, -6)$ (Figure 5.34):

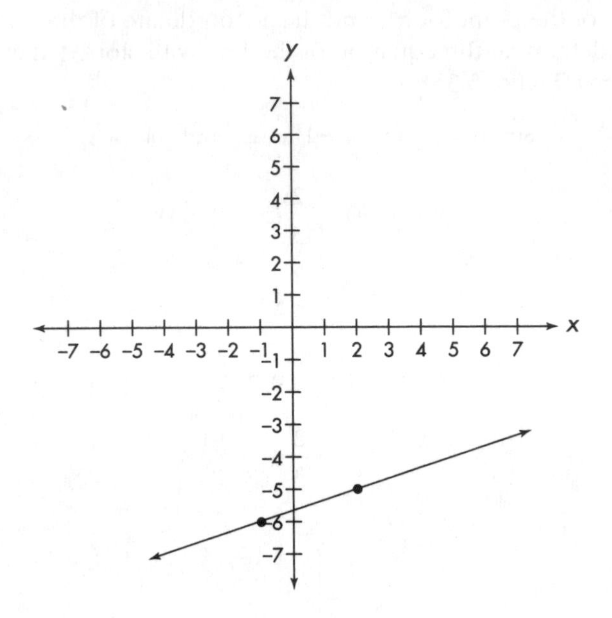

FIGURE 5.34

Substitute into the slope formula:

$$m = \frac{-6 - (-5)}{-1 - 2} = \frac{-1}{-3} = \frac{1}{3}$$

In the point-slope formula, substitute $\frac{1}{3}$ for m, 2 for x_1, and -5 for y_1:

$$y - (-5) = \frac{1}{3}(x - 2)$$

$$y + 5 = \frac{1}{3}x - \frac{2}{3}$$

$$y = \frac{1}{3}x - \frac{17}{3}$$

From the earlier discussion on slope of a line, we know that if two lines are parallel their slopes are equal; and if two lines are perpendicular, their slopes are negative reciprocals of each other.

Thus, as shown in Figure 5.35, the line with y-intercept -5 that is parallel to the line $y = -4x + 3$ has equation $y = -4x - 5$; and similarly, the equation of the line with y-intercept -5 and perpendicular to $y = -4x + 3$ has equation $y = \frac{1}{4}x - 5$.

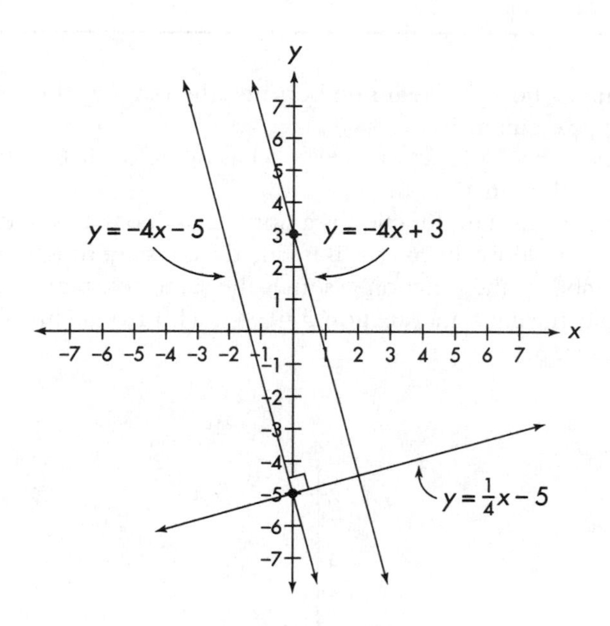

FIGURE 5.35

The following summary of linear equations should be helpful when you are working with linear functions and their graphs.

TABLE 5.2. SUMMARY OF LINEAR EQUATIONS

Slope-intercept form (functional form)	$y = mx + b$
Point-slope form	$y - y_1 = m(x - x_1)$
Standard form	$Ax + By = C$
Horizontal line	$y = k$ for any constant k
Vertical line (not a function)	$x = h$ for any constant h

Constant Functions

A constant function is defined by:

$$y = f(x) = b$$

The domain is the set of real numbers and the range is the set $\{b\}$ consisting of the single element b.

If $b \neq 0$, the constant function $f(x) = b$ has no zeros; if $b = 0$, every value of x is a zero of the function.

The graph of a constant function is a horizontal line (has slope zero) that is $|b|$ units above or below the x-axis. It is called a constant function because for every real number x, the function assumes the same constant value b.

For example, the constant function $y = -4$ is a horizontal line 4 units below the x-axis (Figure 5.36).

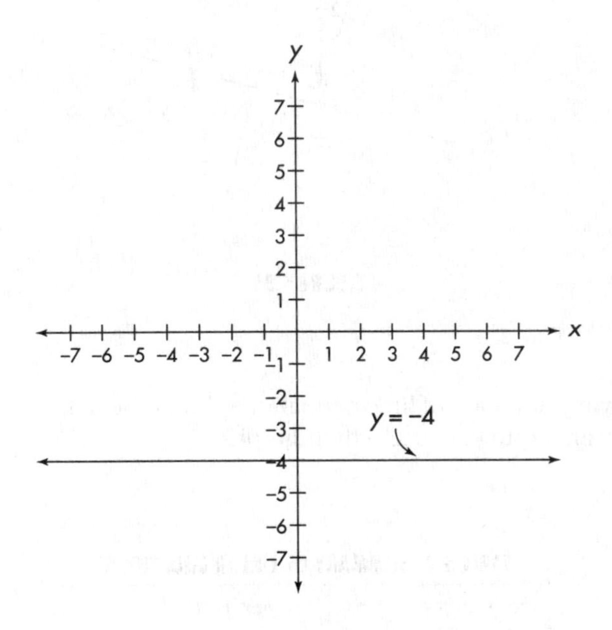

FIGURE 5.36

A helpful message: The equation $x = b$ is *not* a constant function. In fact, it is not a function at all because the ordered pairs look like this: $(b, 0)$, $(b, 1)$, $(b, 2)$, and so forth, for *every* y-value. Since a function cannot have different ordered pairs that have the same first element, this is not a function.

Quadratic Functions

A quadratic function is defined by:

$$y = f(x) = ax^2 + bx + c \quad (a \neq 0)$$

The domain is the set of all real numbers and the range is a subset of the real numbers.

The zeros are determined by finding the roots of the quadratic equation $ax^2 + bx + c = 0$. By the quadratic formula, if $b^2 - 4ac > 0$, there are two real unequal zeros. If $b^2 - 4ac = 0$, there is one real zero (double root). If $b^2 - 4ac < 0$, there are no real zeros for the function.

The graph of a quadratic function defined by $f(x) = ax^2 + bx + c$ is always a **parabola.** If $a > 0$, the parabola opens upward and has a low point (Figure 5.37a). If $a < 0$, the parabola opens downward and has a high point (Figure 5.37b). The high or low point is called the **vertex** of the parabola. When the vertex is a low point of the parabola, the y-coordinate is a **minimum** value for the function. When the vertex is a high point of the parabola, the y-coordinate is a **maximum** value for the function. The parabola is symmetric about a vertical line through its vertex. This line is called its **axis of symmetry.**

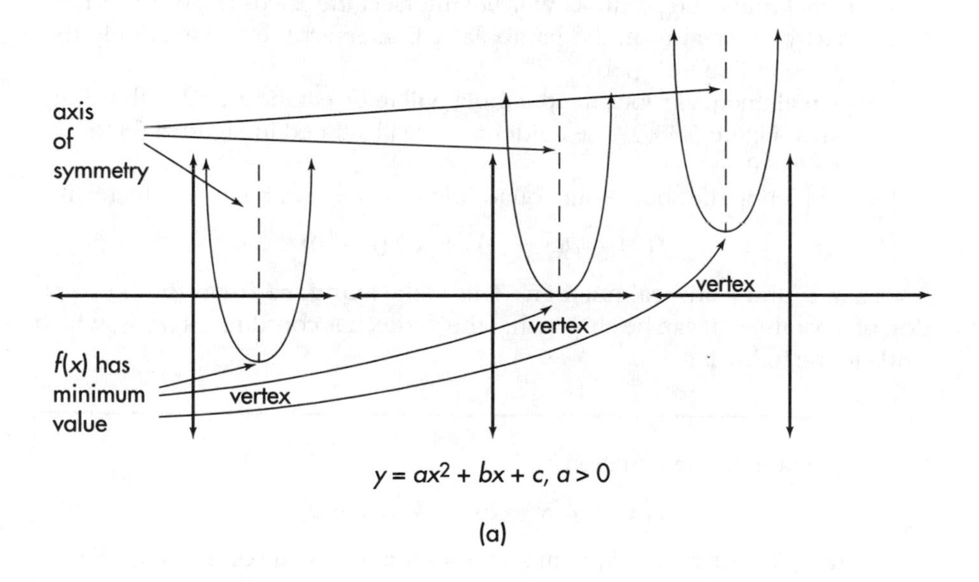

FIGURE 5.37a

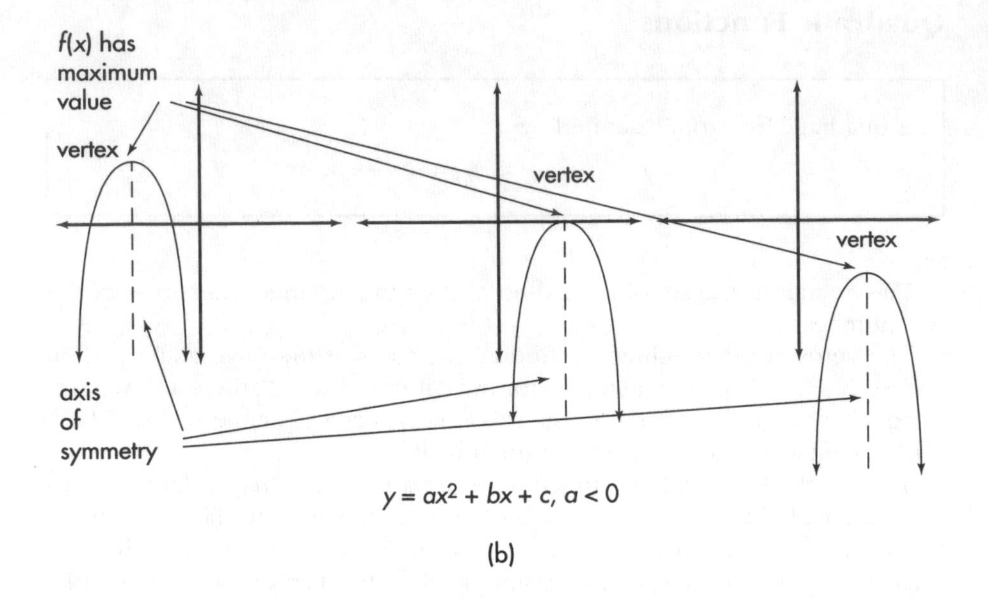

FIGURE 5.37b

The graph of the function may or may not intersect the x-axis, depending on the solution set of $ax^2 + bx + c = 0$. The following guidelines hold:

If $ax^2 + bx + c = 0$ has

a. *no* real roots, the parabola will *not* intersect the x-axis (Figure 5.38a).

b. exactly *one* real root, the parabola will intersect the x-axis at only that *one* point (Figure 5.38b).

c. *two* real unequal roots, the parabola will intersect the x-axis at those *two* points (Figure 5.38c). The guidelines are illustrated in Figure 5.38 for the case $a > 0$.

By completing the square, any quadratic function can be put in the form

$$f(x) = a(x - h)^2 + k \quad (a \neq 0)$$

where a, h, and k are real numbers. This is the **standard form** for the equation of a parabola. It can be shown that the vertex has coordinates (h, k), which leads to the following:

> The graph of the equation
>
> $$f(x) = a(x - h)^2 + k \quad (a \neq 0)$$
>
> where a, h, and k are real numbers is a parabola with vertex (h, k). The parabola opens upward if $a > 0$ and downward if $a < 0$.

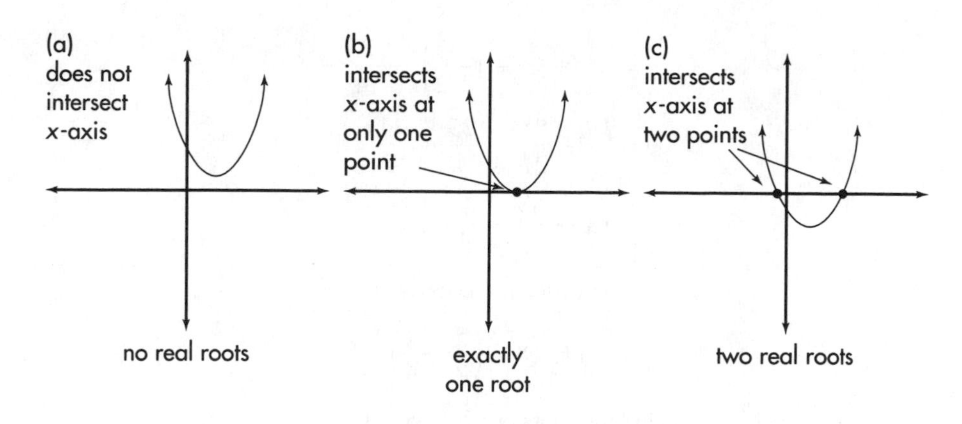

FIGURE 5.38

A helpful message: In some texts, the standard form of a parabola with vertex at (h, k) that opens upward is given by

$$(x - h)^2 = 4p(y - k) \quad p > 0$$

and one that opens downward by

$$(x - h)^2 = -4p(y - k) \quad p > 0.$$

These forms are not expressed explicitly in functional form. For the purposes of investigating functional characteristics (especially through the use of a graphics utility), we prefer the form

$$y = f(x) = a(x - h)^2 + k \quad (a \neq 0).$$

To graph $y = ax^2 + bx + c$:

1. Put the equation in standard form by completing the square.
2. Find and plot the vertex.
3. Determine the real roots, if any, of $ax^2 + bx + c = 0$ and/or determine two or three other points on the graph (being sure to select points on either side of the vertex).
4. Sketch the graph opening upward ($a > 0$) or opening downward ($a < 0$), keeping it symmetric about a vertical line through the vertex.

For example, graph $y = x^2 - 5x + 6$.

Complete the square to write the equation in standard form:

$$y = x^2 - 5x \qquad + 6$$

Add and subtract half the square of −5 on the right side:

$$y = x^2 - 5x + \left(-\frac{5}{2}\right)^2 - \left(-\frac{5}{2}\right)^2 + 6$$

This is zero, so the right side is unchanged.

$$y = x^2 - 5x + \left(-\frac{5}{2}\right)^2 - \frac{25}{4} + 6$$

$$y = \left(x - \frac{5}{2}\right)^2 - \frac{25}{4} + \frac{24}{4}$$

$$y = \left(x - \frac{5}{2}\right)^2 - \frac{1}{4}$$

The vertex has coordinates:

$$(h, k) = \left(\frac{5}{2}, -\frac{1}{4}\right)$$

Find the roots of $x^2 - 5x + 6 = 0$:

$$x^2 - 5x + 6 = 0$$

$$(x - 3)(x - 2) = 0$$

$$x = 3 \text{ or } 2$$

Find other points on the graph:

$$f(0) = (0)^2 - 5(0) + 6 = 6 \rightarrow (0, 6) \in f$$

$$f(1) = (1)^2 - 5(1) + 6 = 2 \rightarrow (1, 2) \in f$$

$$f(4) = (4)^2 - 5(4) + 6 = 2 \rightarrow (4, 2) \in f$$

$$f(5) = (5)^2 - 5(5) + 6 = 6 \rightarrow (5, 6) \in f$$

Since $a = 1 > 0$, the parabola turns upward and is symmetric about a line through $\left(\frac{5}{2}, -\frac{1}{4}\right)$; the vertex is a low point where f has minimum value $-\frac{1}{4}$ (Figure 5.39). Therefore, the range of the function is all real numbers $\geq -\frac{1}{4}$.

In Section 5.4 you learned about horizontal and vertical shifts, reflections, and stretching or shrinking. These tools used in sketching graphs of functions are called **transformations.** Observe that if we consider

$$y = f(x) = x^2$$

as the parent function, then

$$y = f(x) = \left(x - \frac{5}{2}\right)^2 - \frac{1}{4} \qquad a \neq 0$$

can be obtained by shifting the graph of $y = f(x) = x^2$ horizontally to the right $\frac{5}{2}$ units and vertically down $\frac{1}{4}$ unit (Figure 5.40).

The shape of the graph of the parent function remains unchanged; only its position in the coordinate plane is different. Thus, by knowing the basic shape of $y = f(x) = x^2$, you can graph various other functions that are the results of a series of transformations applied to it.

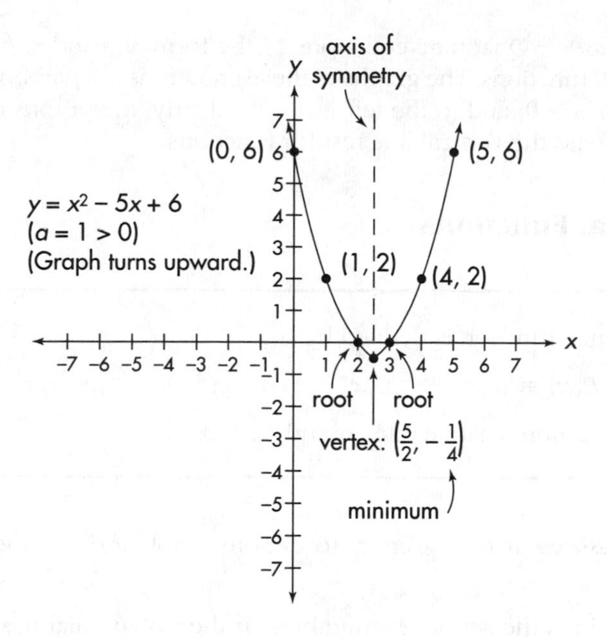

FIGURE 5.39

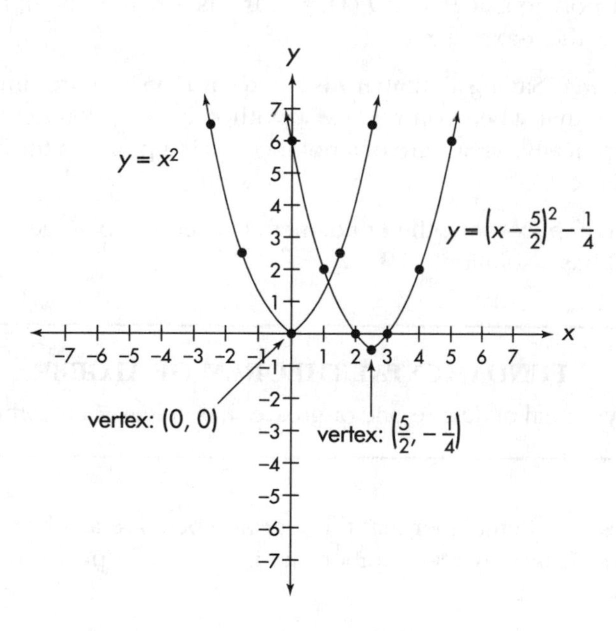

FIGURE 5.40

A helpful message: Quadratic equations of the form $x = ay^2 + by + c$, where $a \neq 0$, are *not* functions. The graphs of these equations are parabolas that open to the right if $a > 0$, and to the left if $a < 0$. Clearly, the graphs of these relations do not pass the vertical line test for functions.

Polynomial Functions

A polynomial function is defined by

$$y = P(x) = a_n x^n + a_{n-1} x^{n-1} + a_{n-2} x^{n-2} + \cdots + a_1 x^1 + a_0$$

where n is a nonnegative integer and $a_n \neq 0$.

A helpful message: It is customary to denote a polynomial function as $P(x)$ or $p(x)$.

The domain is the set of real numbers. If the polynomial has odd degree, the range is also the set of real numbers. If the polynomial has even degree, the range is a subset of the real numbers. Linear, constant, and quadratic functions are special types of polynomial functions.

The zeros are the solutions of the equation $P(x) = 0$. A number r is a **zero,** or **root,** of a polynomial $P(x)$ if $P(r) = 0$. If r is a real number, the graph of $P(x)$ intersects the x-axis at r.

A helpful message: Saying a number r is a zero of $P(x)$ is not saying that r must be 0. It means that when you replace x with r in $P(x)$, you get 0 when you simplify. Graphically, a real zero of a polynomial is a point on the x-axis where the y-value is zero.

Allowing complex roots, the Fundamental Theorem of Algebra assures us that $P(x) = 0$ has a solution.

FUNDAMENTAL THEOREM OF ALGEBRA

Every polynomial of degree one or greater has at least one complex root.

A helpful message: Remember that the real numbers are a subset of the complex numbers. Thus, any real number that is a root of a polynomial is a complex root also.

The remainder theorem provides a method for evaluating a polynomial.

THE REMAINDER THEOREM

If a polynomial $P(x)$ is divided by $x - a$, the remainder is $P(a)$.

This means you can evaluate $P(x)$ at the number a either by direct substitution, that is, replacing x with a in $P(x)$, or by finding the remainder when $P(x)$ is divided by $x - a$. The latter approach can be accomplished efficiently by using synthetic division (see Appendix D).

A root r of $P(x)$ has **multiplicity** k if $(x - r)^k$ is a factor of $P(x)$. Allowing complex roots and counting multiplicities as individual roots, we have the following:

Every polynomial $P(x)$ of positive degree n has exactly n roots (or zeros).

This means that if you allow complex roots and count a multiple root again each time it occurs more than once, every linear function has exactly one root; every quadratic function has exactly two roots; and so on. The roots (or zeros) are guaranteed to be there, although they are sometimes difficult to determine.

For example, -3 and 4 are zeros of $P(x) = x^2 - x - 12$ because the following are true:

$$P(-3) = (-3)^2 - (-3) - 12 = 9 + 3 - 12 = 0$$

$$P(4) = (4)^2 - (4) - 12 = 16 - 4 - 12 = 0$$

The graph (Figure 5.41) of $P(x) = x^2 - x - 12$ intersects the x-axis at -3 and 4.

The remainder theorem can be used to obtain the following:

THE FACTOR THEOREM

$P(r) = 0$ if and only if $x - r$ is a factor of $P(x)$.

For instance, since for $P(x) = x^2 - x - 12$, $P(-3) = 0$ and $P(4) = 0$, $(x + 3)$ and $(x - 4)$ are factors of $P(x) = x^2 - x - 12$, and conversely.

Thus, you can factor $P(x)$ by finding its zeros, and you can determine if r is a zero by dividing $P(x)$ by $x - r$. If the remainder is zero, r is a zero of $P(x)$.

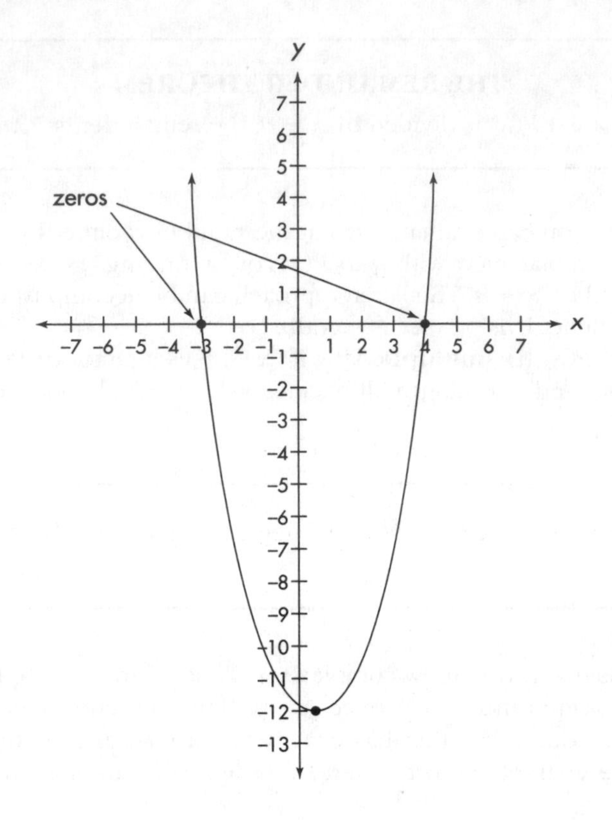

FIGURE 5.41

The remainder and factor theorems provide ways to test whether a number is a root of $P(x)$. They do not provide clues as to how to pick the numbers you should test. The following theorem may prove helpful.

> **RATIONAL ROOTS OF A POLYNOMIAL**
>
> If $P(x) = a_n x^n + a_{n-1}x^{n-1} + a_{n-2}x^{n-2} + \ldots + a_1 x^1 + a_0$ is a polynomial function with integral coefficients and $\frac{p}{q}$ is a rational root of $P(x)$ in lowest terms, then p is a factor of a_0 and q is a factor of a_n.

This theorem can be used to obtain some possible roots of $P(x)$ that you can test by using the remainder or factor theorems. For example:

Factor and find the zeros for $P(x) = 4x^3 - 3x + 1$.

Possible numerators for a rational zero are factors of 1: ± 1. Possible denominators are factors of 4: ± 1, ± 2, and ± 4.

Possible rational zeros of $4x^3 - 3x + 1$ are ± 1, $\pm\frac{1}{2}$, $\pm\frac{1}{4}$. Using synthetic division to test the possible zeros:

Test $x = 1$:

$$\begin{array}{r|rrrr} 1 & 4 + & 0 - & 3 + & 1 \\ & + & 4 + & 4 + & 1 \\ \hline & 4 + & 4 + & 1 & 2 \end{array} \leftarrow \text{remainder; 1 is not a zero.}$$

Test $x = -1$:

$$\begin{array}{r|rrrr} -1 & 4 + & 0 - & 3 + & 1 \\ & & -4 + & 4 & -1 \\ \hline & 4 & -4 + & 1 & 0 \end{array} \leftarrow \text{remainder; } -1 \text{ is a zero.}$$

Use the factor theorem to express $P(x) = 4x^3 - 3x + 1$ as

$$P(x) = 4x^3 - 3x + 1 = (x + 1)(4x^2 - 4x + 1)$$

You can continue testing the other possible rational roots or you can factor the quadratic trinomial in the above expression:

$$4x^2 - 4x + 1 = (2x - 1)^2$$

Solving $2x - 1 = 0$ gives $x = \frac{1}{2}$ (root of multiplicity 2).

Thus, $P(x) = (x + 1)(2x - 1)^2$, and has roots -1 and $\frac{1}{2}$ (multiplicity 2).

If $P(x)$ has a complex root, you will find the following theorem useful:

> Let $P(x)$ be a polynomial function with real coefficients. If $a + bi$ is a complex root of $P(x)$, then its complex conjugate $a - bi$ is also a root of $P(x)$.

In simple terms, this means complex roots of polynomials that have real coefficients occur as *conjugate pairs*. Before showing an example, we must go over some basic concepts for complex numbers. Performing operations with complex numbers is essentially like working with binomials. The rules follow:

COMPLEX NUMBERS

$i^2 = -1$
Ex: $(2i)(2i) = 4i^2 = 4(-1) = -4$
$\sqrt{-b^2} = bi \ (b \geq 0)$
Ex: $\sqrt{-36} = 6i$
$\sqrt{-a} = i\sqrt{a}$
Ex: $\sqrt{-3} = i\sqrt{3}$ or $\sqrt{3}i$
The conjugate of $a + bi$ is $a - bi$.
Ex: $2 - 3i$ is the conjugate of $2 + 3i$.

$(a + bi)(a - bi) = a^2 + b^2$
Ex: $(2 + 3i)(2 - 3i) = 4 - 9i^2 = 4 + 9 = 13$
$(a + bi) + (c + di) = (a + c) + (b + d)i$
Ex: $(2 + 3i) + (1 + 5i) = (2 + 1) + (3 + 5)i = 3 + 8i$
$(a + bi)(c + di) = (ac - bd) + (ad + bc)i$
Ex: $(2 + 3i) + (1 + 5i) = (2 - 15) + (10 + 3)i = -13 + 13i$

A helpful message: Be careful when writing complex numbers containing radicals. For instance, $\sqrt{-3} = \sqrt{3}i$, not $\sqrt{3i}$.

Now for an example, find the zeros for $P(x) = x^3 - 1$.

This means to solve $x^3 - 1 = 0$, which you know has exactly three roots.

You should recognize that you can factor the left side as the difference of two cubes:

$$x^3 - 1 = 0$$
$$(x - 1)(x^2 + x + 1) = 0$$

From the first factor, we have that $x = 1$ is a zero of $P(x)$.

Using the quadratic formula, we can determine the other two zeros:

$$x = \frac{-1 \pm \sqrt{1 - 4}}{2} = \frac{-1 \pm \sqrt{-3}}{2} = \frac{-1 \pm i\sqrt{3}}{2} = -\frac{1}{2} \pm \frac{\sqrt{3}}{2}i$$

Therefore, the zeros are 1, $-\frac{1}{2} + \frac{\sqrt{3}}{2}i$, and $-\frac{1}{2} - \frac{\sqrt{3}}{2}i$. Observe that the latter two zeros are complex conjugates of each other.

Another theorem that is useful is the following:

Let $P(x)$ be a polynomial function with real coefficients. If a and b are real numbers such that $P(a)$ and $P(b)$ have opposite signs, then $P(x)$ has at least one zero between a and b.

This means if the graph of $P(x)$ crosses the x-axis between a and b you know there is a zero between a and b (see Figure 5.42).

You can use this theorem to help you locate irrational roots.

A helpful message: In most college algebra courses today, students are encouraged to use computers or graphing calculators to investigate the behavior of polynomial functions, including determining the zeros.

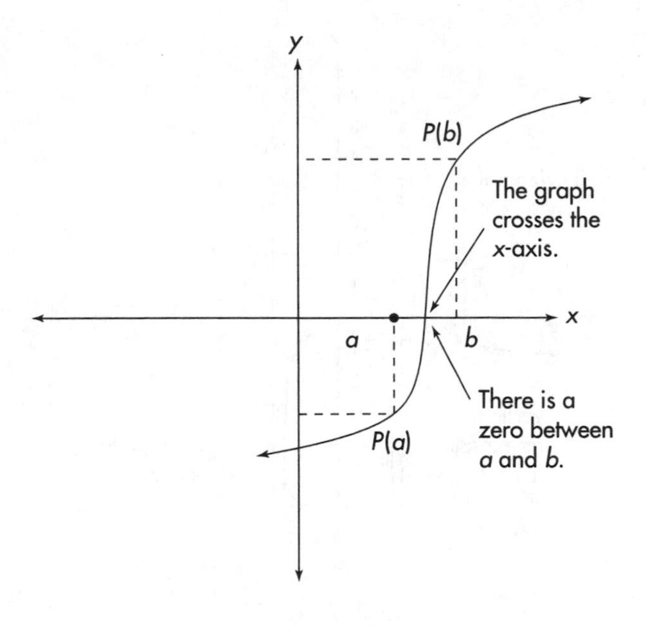

FIGURE 5.42

Rational Functions

A rational function is defined by

$$y = f(x) = \frac{p(x)}{q(x)},$$

where $p(x)$ and $q(x)$ are polynomials ($q(x) \neq 0$).

The domain excludes any real number for which $q(x) = 0$. The range will be a subset of the real numbers.

The zeros, if any, are the values of x for which $f(x) = 0$.

To graph rational functions, you will usually find it necessary to determine the behavior of the function at excluded values of its domain and at extremely large or extremely small values for x. In other words, you must look for **asymptotes** associated with the function.

For example, the graph of $f(x) = \frac{x+3}{x^2-4}$ (Figure 5.43) has a zero at $x = -3$, two vertical asymptotes, $x = 2$ and $x = -2$, and a horizontal asymptote at $y = 0$ (since $f(x)$ approaches 0 as x gets extremely large or extremely small).

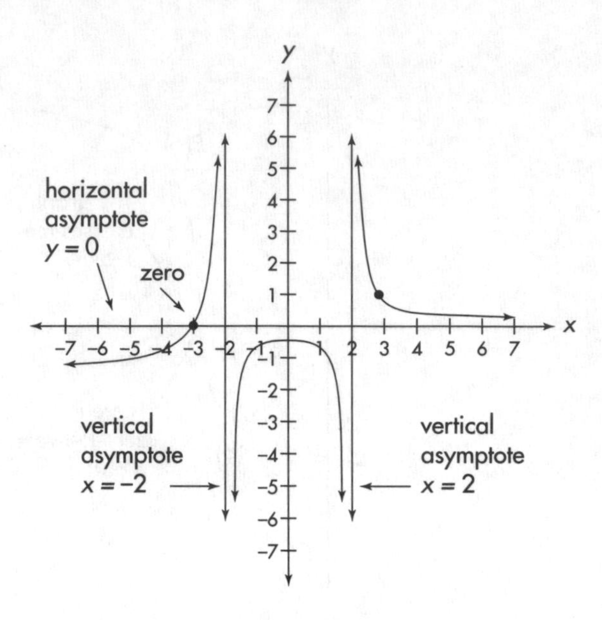

FIGURE 5.43

Absolute Value Function

> An absolute value function is defined by
>
> $$y = f(x) = |ax + b|$$

The domain is the set of all real numbers and the range is restricted to nonnegative values.

The function has a zero at $x = -\frac{b}{a}$.

The graph of $y = |2x - 8|$ is shown in Figure 5.44.

Square Root Functions

> A square root function is defined by:
>
> $$y = f(x) = \sqrt{ax + b}$$

The domain is the set of all real numbers such that $ax + b \geq 0$. The range is the set of nonnegative real numbers $[0, \infty)$.

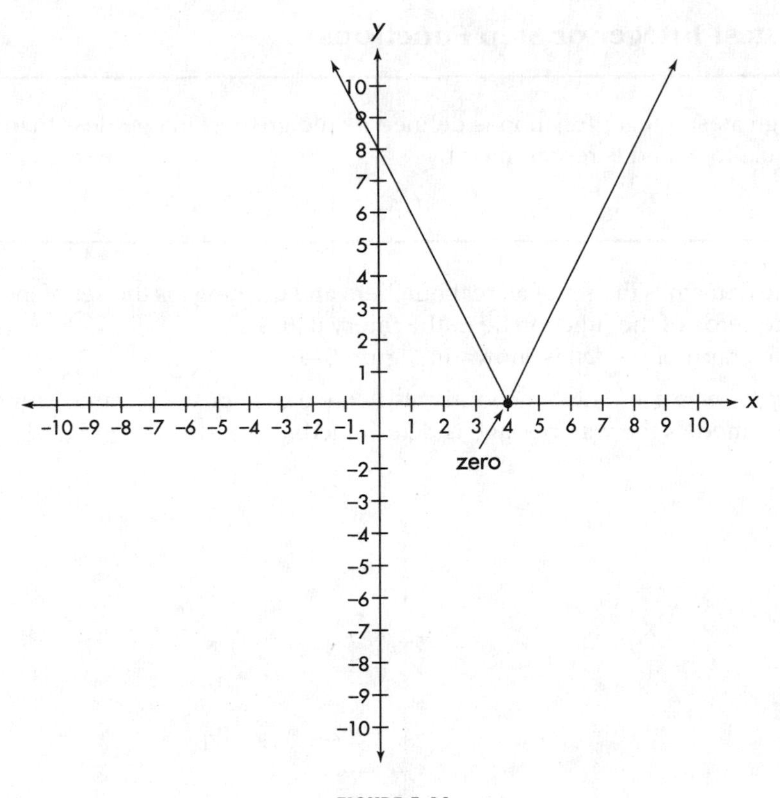

FIGURE 5.44

The function has a zero at $x = -\frac{b}{a}$.
The graph of the function is to the right of $x = -\frac{b}{a}$ and above the x-axis.
The graph of $y = \sqrt{x + 3}$ is shown in Figure 5.45.

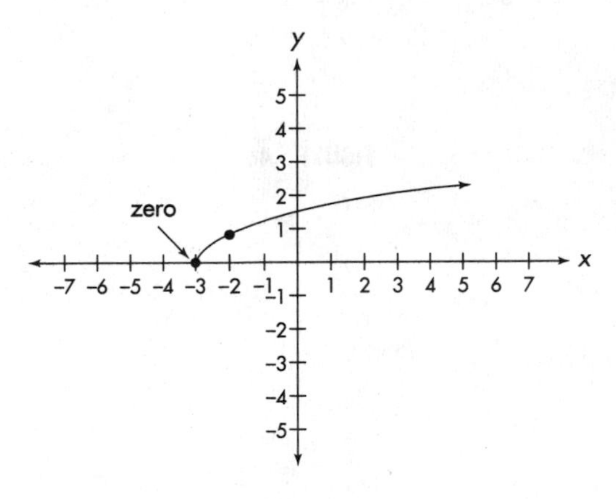

FIGURE 5.45

Greatest Integer or Step Functions

A greatest integer function is defined as the greatest integer less than or equal to x, and is represented by:

$$y = f(x) = [x]$$

The domain is the set of all real numbers and the range is the set of integers. The zeros of the function lie in the interval [0, 1).

The graph of $y = [x]$ is shown in Figure 5.46.

A helpful message: This function is also called the "postage stamp" function, since it models the way postage is determined.

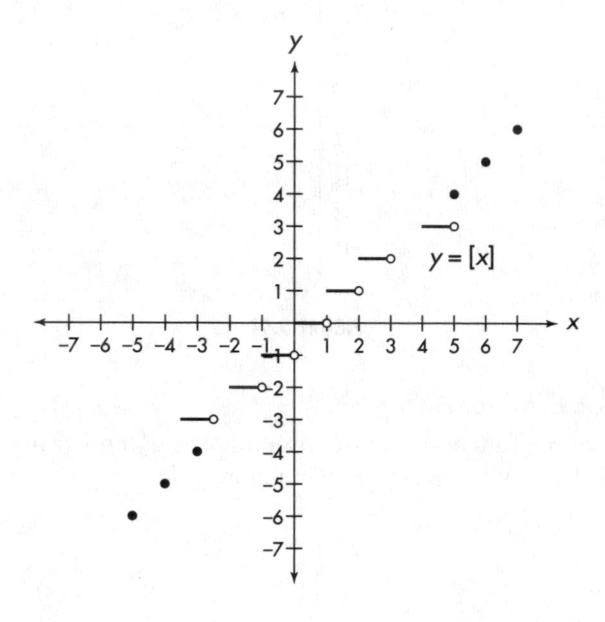

FIGURE 5.46

Exponential Functions

An exponential function is defined by

$$y = f(x) = b^x \qquad (b \neq 1,\ b > 0)$$

where b is called the *base* of the exponential function.

The domain is the set of real numbers and the range is the set of positive real numbers.

The function has no zeros.

Exponential functions have the following properties:

PROPERTIES FOR EXPONENTIAL FUNCTIONS

For real numbers $b (b \neq 1,\ b > 0)$ and x, y:

$$b^x > 0 \text{ for all real numbers } x$$

$$b^0 = 1$$

$$b^1 = b$$

$$\text{If } x = y, \text{ then } b^x = b^y$$

$$b^{-x} = \frac{1}{b^x}$$

$$b^x b^y = b^{x+y}$$

$$\frac{b^x}{b^y} = b^{x-y}$$

$$(b^x)^y = b^{xy}$$

The graph of $y = f(x) = b^x$

- passes through the points (0, 1) and (1, b) and is asymptotic to the x-axis;
- is an increasing function if $b > 1$;
- is a decreasing function if $0 < b < 1$.

The graph is located in the first and second quadrants only.

Two exponential functions are of particular importance:

- $y = f(x) = 10^x$ (Figure 5.47a, page 230)
- $y = f(x) = e^x$ (Figure 5.47b, page 231)

The latter is the **natural exponential function,** where e is the irrational number approximately equal to 2.71828.

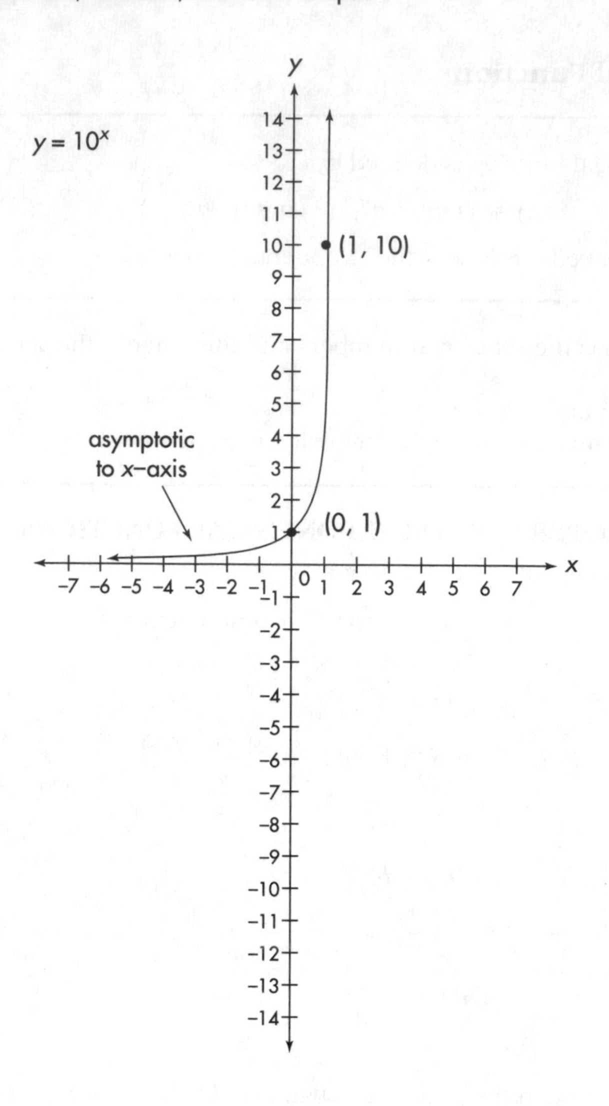

FIGURE 5.47a

Logarithmic Functions

> A logarithmic function is defined by
>
> $$y = f(x) = \log_b x \Leftrightarrow b^y = x \qquad (x > 0)$$
>
> where b is called the *base* of the logarithmic function, with $b \neq 1$, $b > 0$.

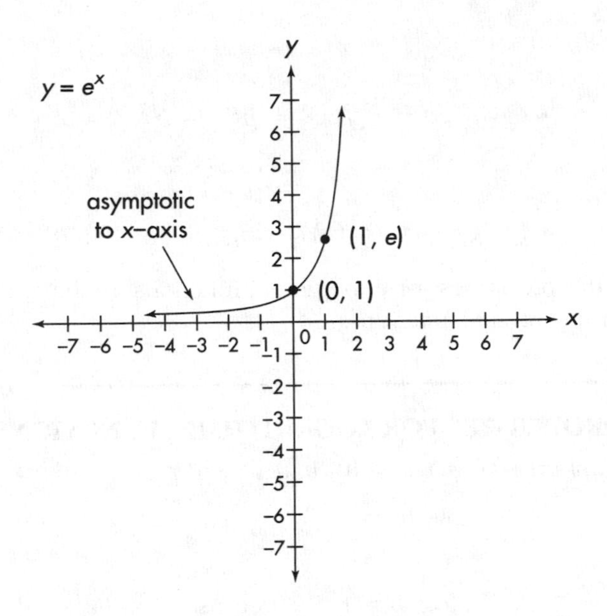

FIGURE 5.47b

The domain is the set of positive real numbers and the range is the set of real numbers.

The function has a zero at $x = 1$.

The graph of the function is located in the first and fourth quadrants only.

Logarithms are *exponents*. If $y = \log_b x$, then y is the *exponent* that is used on b to get x; that is, $b^y = x$. For example:

- $\log_2 8 = 3$
  This is true because 3 is the exponent you can use on 2 to get 8; that is, $2^3 = 8$.
- $\log_2 \left(\frac{1}{8}\right) = -3$
  This is true because -3 is the exponent you use on 2 to get $\frac{1}{8}$; that is, $2^{-3} = \frac{1}{8}$.
- $\log_{10} 100 = 2$
  This is true because 2 is the exponent you use on 10 to get 100; that is, $10^2 = 100$.
- $\log_{10} (0.001) = -3$
  This is true because -3 is the exponent you use on 10 to get (0.001); that is, $10^{-3} = 0.001$.

The logarithm function is the *inverse* of the exponential function, and vice versa. To show this, we find $(g \circ f)(x)$ and $(f \circ g)(x)$ for $f(x) = \log_b x$ and $g(x) = b^x$ $(b \neq 1,\ b > 0)$.

$\log_b x$ is the exponent you use on b to get x

- $(g \circ f)(x) = g(f(x)) = b^{f(x)} = b^{\log_b x} = x$

x is the exponent you use on b to get b^x

- $(f \circ g)(x) = f(g(x)) = \log_b g(x) = \log_b b^x = x$

Based on the properties for exponential functions we have the following properties for logarithmic functions:

PROPERTIES FOR LOGARITHMIC FUNCTIONS

For real numbers b ($b \neq 1$, $b > 0$), u, v, w, and p:

$$\log_b b^x = x$$

$$b^{\log_b x} = x$$

$$\log_b b = 1 \qquad \text{(because } b^1 = b\text{)}$$

$$\log_b 1 = 0 \qquad \text{(because } b^0 = 1\text{)}$$

If $\log_b u = \log_b v$, then $u = v$.

$$\log_b \frac{1}{u} = -\log_b u$$

$$\log_b uv = \log_b u + \log_b v$$

$$\log_b \frac{u}{v} = \log_b u - \log_b v$$

$$\log_b u^p = p \log_b u$$

For real numbers b ($b \neq 1$, $b > 0$), x, and y:
The graph of $y = f(x) = \log_b x$

- passes through (1, 0) and (b, 1) and is asymptotic to the y-axis;
- is an increasing function if $b > 1$;
- is a decreasing function if $0 < b < 1$.

Two logarithmic functions are of special importance:

- The logarithmic function $y = f(x) = \log_{10} x$ (the **common logarithmic function**) is the inverse of the exponential function $y = f(x) = 10^x$ (Figure 5.48a).
- The logarithm function $y = f(x) = \log_e x$ (the **natural logarithmic function**) is the inverse of the exponential function $y = f(x) = e^x$ (Figure 5.48b). This function is usually denoted $y = f(x) = 1\text{n } x$.

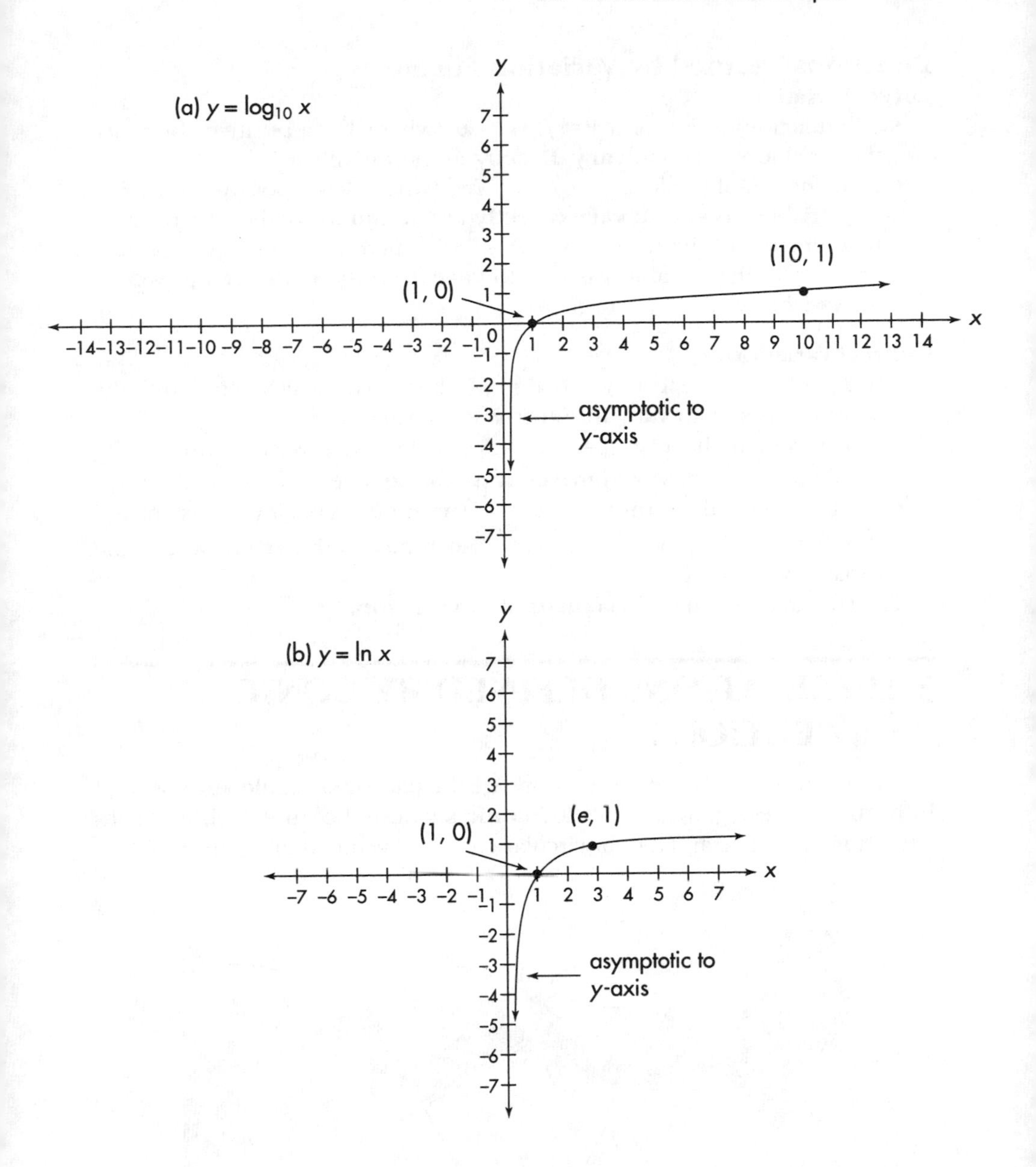

FIGURE 5.48

A helpful message: Common (base 10) logarithms were once used extensively in computations, particularly for problems involving exponents. Today, it is more efficient to use a scientific calculator to perform these calculations.

Functions Defined by Variation Formulas

Direct Variation

- In a function of the form $y = f(x) = kx$ (where k is a positive constant), the variable y is said to **vary directly** as the variable x.
- In a function of the form $y = f(x) = kx^2$ (where k is a positive constant), the variable y is said to **vary directly as the square** of the variable x.
- In a function of the form $y = f(x) = kx^n$ (where k is a positive constant and $n > 0$), the variable y is said to **vary directly as the *n*th power** of the variable x.

Indirect Variation

- In a function of the form $y = f(x) = \frac{k}{x}$ (where k is a positive constant), the variable y is said to **vary indirectly** as the variable x.
- In a function of the form $y = f(x) = \frac{k}{x^2}$ (where k is a positive constant), the variable y is said to **vary indirectly as the square** of the variable x.
- In a function of the form $y = f(x) = \frac{k}{x^n}$ (where k is a positive constant and $n > 0$), the variable y is said to **vary indirectly as the *n*th power** of the variable x.

The constant k is called the **constant of variation.**

5.10 RELATIONS DEFINED BY CONIC SECTIONS

The four basic kinds of conic sections are the **parabola, circle, ellipse,** and **hyperbola.** These graphs are called **conic sections** because each is the intersection of a plane and a right-circular cone as illustrated in Figure 5.49.

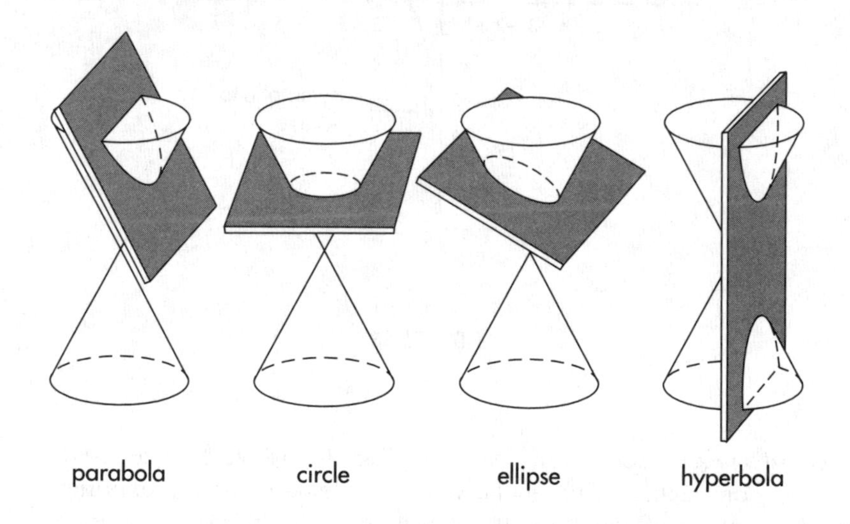

FIGURE 5.49

In the section on quadratic functions, we discussed the parabola. This section will focus on the circle, ellipse, and hyperbola.

The equation of a conic can be written as follows:

$$Ax^2 + By^2 + Cx + Dy + F = 0 \qquad (A \neq 0 \text{ and } B \neq 0)$$

This equation defines a relation that is *not* a function and has different graphs depending on the values of the coefficients A, B, and C. If the graph is not *degenerate* (a point, two intersecting lines, and so on), we have the following:

The graph of the equation $Ax^2 + By^2 + Cx + Dy + F = 0$ ($A \neq 0$ and $B \neq 0$) takes the following form:

- a circle if $A = B$, with standard form:

$$(x - h)^2 + (y - k)^2 = r^2$$

- an ellipse if $AB > 0$, with standard form:

$$\frac{(x-h)^2}{a^2} + \frac{(y-k)^2}{b^2} = 1$$

- a hyperbola if $AB < 0$, with standard forms:

Form (1) $$\frac{(x-h)^2}{a^2} - \frac{(y-k)^2}{b^2} = 1$$

Form (2) $$\frac{(y-k)^2}{b^2} - \frac{(x-h)^2}{a^2} = 1$$

Circle

The standard form of the equation of a circle is given by

$$(x - h)^2 + (y - k)^2 = r^2$$

Here (h, k) is the center and the radius is $|r|$ units. For example, consider the equation:

$$(x - 1)^2 + (y + 2)^2 = 9$$

The graph is a circle centered at $(1, -2)$, with radius 3 units (Figure 5.50, page 236).

Ellipse

The standard form of the equation of an ellipse is given by

$$\frac{(x-h)^2}{a^2} + \frac{(y-k)^2}{b^2} = 1$$

The center is at (h, k). The ellipse has **vertices** $(h - a, k)$, $(h + a, k)$, $(h, k - b)$, and $(h, k + b)$. The line segment joining the vertices $(h - a, k)$ and $(h + a, k)$ is a **horizontal axis** of symmetry, and the line segment joining the vertices $(h, k - b)$ and $(h, k + b)$ is a **vertical axis** of symmetry. The longer axis is

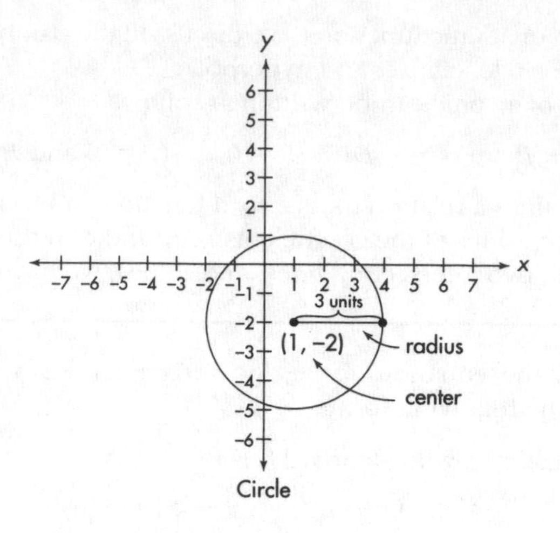

FIGURE 5.50

called the **major axis** and the shorter axis is called the **minor axis.** The lengths of the two axes are $2|a|$ and $2|b|$. (When the two axes are equal, the ellipse is a circle.) For example, consider the equation:

$$\frac{(x-1)^2}{16} + \frac{(y+2)^2}{9} = 1$$

The graph is an ellipse with center at $(1, -2)$ and vertices $(-3, -2)$, $(5, -2)$, $(1, -5)$, and $(1, 1)$ (Figure 5.51). The length of the major axis is 8 units and the length of the minor axis is 6 units.

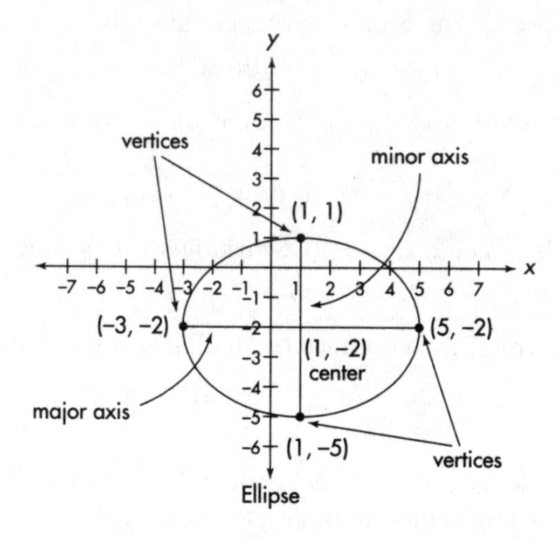

FIGURE 5.51

Hyperbola

The first standard form for the equation of a hyperbola is given by

$$\frac{(x-h)^2}{a^2} - \frac{(y-k)^2}{b^2} = 1$$

The center is at (h, k). The hyperbola opens left and right along the line $y = k$, and it passes through the **vertices** $(h - a, k)$ and $(h + a, k)$. It has the intersecting lines $y = k + \frac{b}{a}(x - h)$ and $y = k - \frac{b}{a}(x - h)$ as (slanting) **asymptotes.** The asymptotes are the diagonals of a rectangle with dimensions $2|a|$ by $2|b|$ centered at (h, k). For example, consider the equation:

$$\frac{(x-1)^2}{16} - \frac{(y+2)^2}{9} = 1$$

The graph is a hyperbola with center at $(1, -2)$ and vertices $(-3, -2)$ and $(5, -2)$ (Figure 5.52). The asymptotes are $y = -2 + \frac{3}{4}(x - 1) = \frac{3}{4}x - \frac{11}{4}$ and $y = -2 - \frac{3}{4}(x - 1) = -\frac{3}{4}x - \frac{5}{4}$. The rectangle enclosing the center has dimensions 8 by 6.

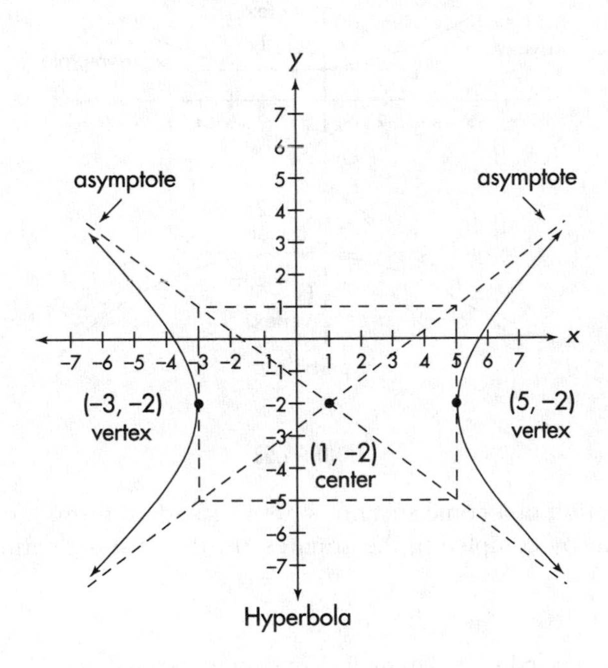

FIGURE 5.52

The second standard form for the equation of a hyperbola is given by

$$\frac{(y-k)^2}{b^2} - \frac{(x-h)^2}{a^2} = 1$$

The center is (h, k). The hyperbola opens up and down along the line $x = h$, and it passes through the **vertices** $(h, k - b)$, and $(h, k + b)$. As in the previous case, it has the intersecting lines $y = k + \frac{b}{a}(x - h)$ and $y = k - \frac{b}{a}(x - h)$ as (slanting) **asymptotes.** The asymptotes are the diagonals of a rectangle with dimensions $2|a|$ by $2|b|$ centered at (h, k). For example, consider the equation:

$$\frac{(y+2)^2}{9} - \frac{(x-1)^2}{16} = 1$$

The graph is a hyperbola with center at $(1, -2)$ and vertices $(1, -5)$, and $(1, 1)$ (Figure 5.53). The asymptotes are $y = -2 + \frac{3}{4}(x - 1) = \frac{3}{4}x - \frac{11}{4}$ and $y = -2 - \frac{3}{4}(x - 1) = -\frac{3}{4}x - \frac{5}{4}$. The rectangle enclosing the center has dimensions 8 by 6.

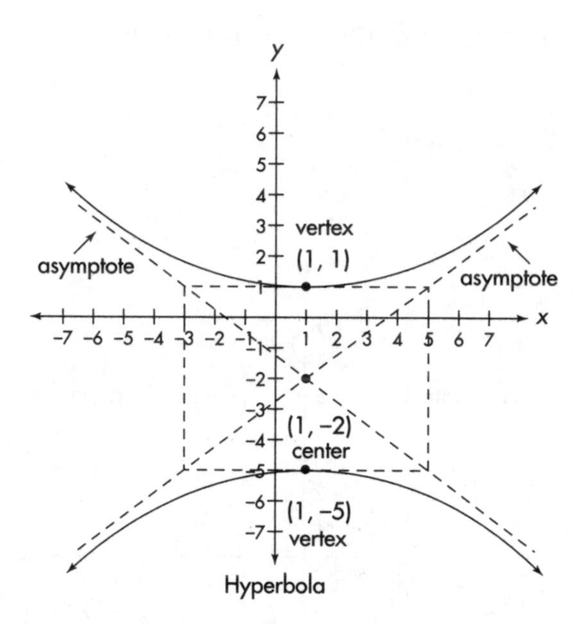

FIGURE 5.53

If the equation of a conic section is not in standard form, it can be put in standard form by completing the squares on the x and y terms. Some examples follow:

- $x^2 + y^2 + 10x - 6y + 30 = 0$

  Group the x's and y's and move the constant to the right side:

  $(x^2 + 10x) + (y^2 - 6y) = -30$

Complete the square by adding the square of half the coefficient of the first degree term to both sides:

$$(x^2 + 10x + 25) + (y^2 - 6y + 9) = -30 + 25 + 9$$

Factor the left side:

$$(x + 5)^2 + (y - 3)^2 = 4$$

This is the standard form of a circle centered at $(-5, 3)$ with radius 2.

- $25x^2 + 16y^2 - 100x - 32y - 284 = 0$

Group the x's and y's and move the constant to the right side:

$$(25x^2 - 100x) + (16y^2 - 32y) = 284$$

Factor out the coefficient of the squared term for each grouping:

$$25(x^2 - 4x) + 16(y^2 - 2y) = 284$$

Complete the square inside the parentheses on the left side. Add the same amount to the other side, being sure to multiply by the coefficient of the parentheses before you do:

$$25(x^2 - 4x + 4) + 16(y^2 - 2y + 1) = 284 + \underbrace{100}_{25(4)} + \underbrace{16}_{16(1)}$$

Factor the left side:

$$25(x - 2)^2 + 16(y - 1)^2 = 400$$

Divide both sides by 400:

$$\frac{(x-2)^2}{16} + \frac{(y-1)^2}{25} = 1$$

This is the standard form of an ellipse centered at $(2, 1)$.

5.11 RELATIONS DEFINED BY INEQUALITIES IN THE PLANE

The solution set of an inequality in two variables is the relation consisting of all ordered pairs that make the inequality a true statement. The graph of a two-variable inequality is the set of all points in the solution set of the inequality.

Two-Variable Linear Inequalities

Look at the following mathematical sentences:

- $y > x + 1$
- $2x - 3y < 10$
- $-3x + 2y \geq 4$
- $x \leq 2y + 1$

These are called **two-variable linear inequalities.** The graph of a linear inequality is a **half-plane.**

To graph a linear inequality:

1. Graph the linear equation that results when the inequality symbol is replaced with an equal sign. Use a dashed line for $<$ or $>$ inequalities and a solid line for $\leq$ or $\geq$ inequalities. This is the boundary line.
2. Select and shade the correct half-plane by testing a point that is *not* on the boundary line (the origin is usually a good choice unless the boundary passes through it). If the coordinates of the point satisfy the inequality, shade the half-plane containing the test point; if not, shade the half-plane that does *not* contain the test point.

For example, graph the inequality $y > x + 1$.

Graph $y = x + 1$ using a dashed line because the inequality contains $>$:

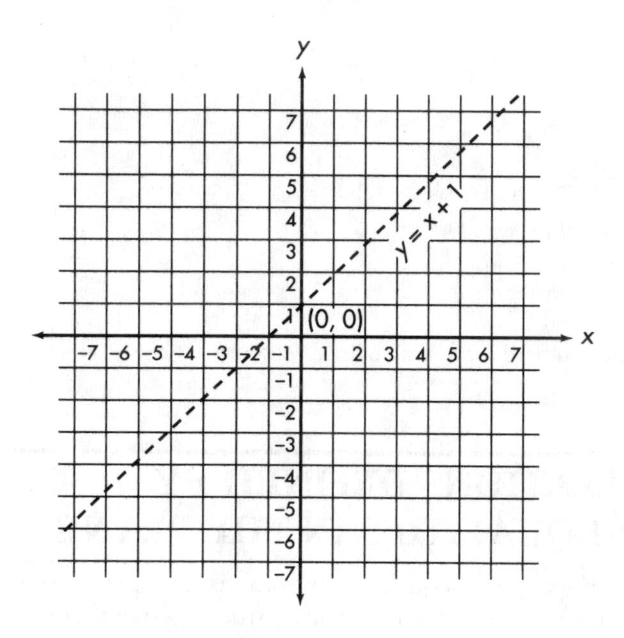

FIGURE 5.54

Test (0, 0):

$$y > x + 1$$

$$0 \overset{?}{>} 0 + 1$$

$$0 \not> 1$$

Shade the half-plane that does *not* contain (0, 0).

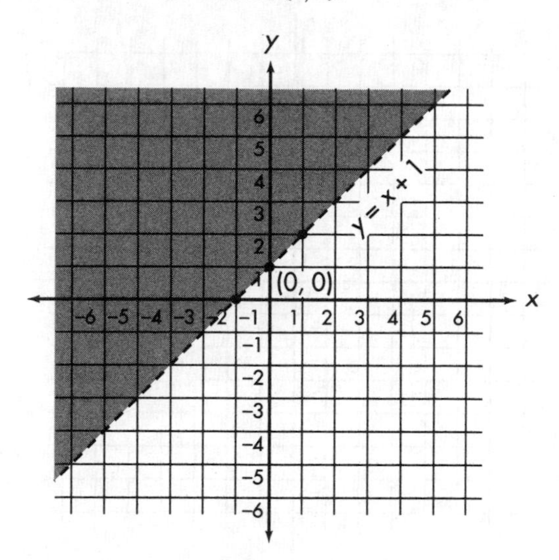

FIGURE 5.55

Two-Variable Quadratic Inequalities

The following expressions represent **two-variable quadratic inequalities:**

- $y < ax^2 + bx + c$
- $y \leq ax^2 + bx + c$
- $y > ax^2 + bx + c$
- $y \geq ax^2 + bx + c$

These can be graphed in a manner similar to that used for graphing two-variable linear inequalities:

1. Graph the equation $y = ax^2 + bx + c$. Use a dashed line for $<$ or $>$ inequalities and a solid line for $\leq$ or $\geq$ inequalities. This is the boundary curve.
2. Select and shade the correct section by testing a point that is *not* on the boundary curve (the origin is usually a good choice unless the boundary passes through it). If the coordinates of the point satisfy the inequality, shade the section containing the test point; if not, shade the section that does *not* contain the test point.

For example, graph the inequality $y \leq x^2 - 5x + 6$.

Graph $y = x^2 - 5x + 6$ using a solid line because the inequality contains $\leq$:

Test (0, 0):

$$y \leq x^2 - 5x + 6$$
$$0 \overset{?}{\leq} 0^2 - 5(0) + 6$$
$$0 \overset{\checkmark}{\leq} 6$$

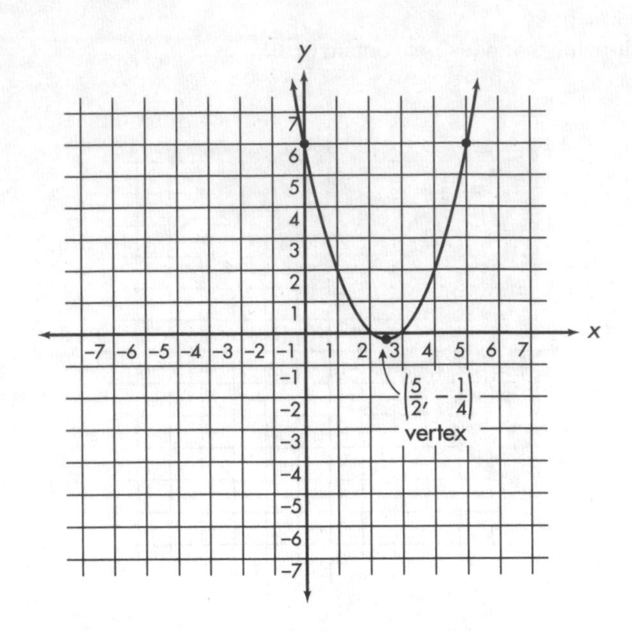

FIGURE 5.56

Shade the section that contains (0, 0).

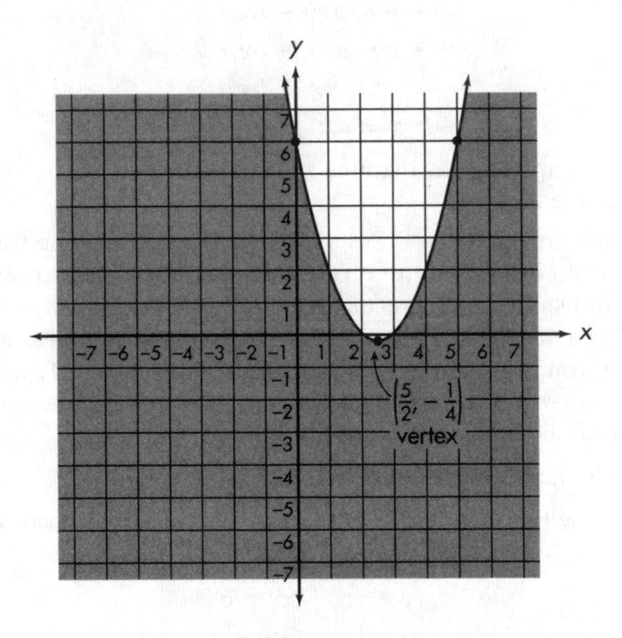

FIGURE 5.57

5.12 APPLICATIONS OF FUNCTIONS

Functions have extensive application in real-world situations. The following examples give a sample of the use of functions to model practical situations.

Solve: A fax transmission costs \$2.50 for the first page and \$2.00 for each additional page. There is no charge for the cover page. The cost C of sending a fax transmission is a function of the number n of pages transmitted. Write a function that expresses the cost C as a function of n and then determine the cost of sending an 8-page fax transmission (no cover page); that is, find $C(8)$.

Solution:

$$
\begin{aligned}
C(n) &= 2.50 + 2.00n \\
C(8) &= 2.50 + 2.00(8) \\
&= \$18.50
\end{aligned}
$$

Solution: The cost of sending a fax transmission of 8 pages is \$18.50.

Solve: A manufacturer has a total revenue function given by

$$R(q) = 0.75q$$

Here q is the number of items sold and the cost per item is 75 cents.

The total cost function is given by

$$C(q) = 2{,}500 + 0.25q$$

The profit function is given by

$$P(q) = R(q) - C(q)$$

Determine the **break-even point,** which is defined as the level of production at which profit is zero, that is, $P(q) = 0$.

Solution:

$$
\begin{aligned}
P(q) &= R(q) - C(q) \\
P(q) &= (0.75q) - (2{,}500 + 0.25q) \\
P(q) &= 0.75q - 2{,}500 - 0.25q \\
P(q) &= 0.50q - 2{,}500
\end{aligned}
$$

To find the break-even point, we solve $0.50q - 2{,}500 = 0$:

$$
\begin{aligned}
0.50q - 2{,}500 &= 0 \\
0.50q &= 2{,}500 \\
q &= 5{,}000 \text{ items}
\end{aligned}
$$

The break-even point occurs when 5,000 items are sold.

Solve: A rocket is fired into the air, following a path given by the function $f(t) = -16t^2 + 64t$, where $f(t)$ is the height in feet when time t is in seconds. Find the maximum height reached by the rocket before it returns to Earth.

Solution: The graph of the function $f(t) = -16t^2 + 64t$ is a parabola that turns down. The maximum height will be reached at the vertex of the parabola. To find the coordinates of the vertex, complete the square on the right side:

$$f(t) = -16t^2 + 64t$$
$$f(t) = -16(t^2 - 4t + 4) + 16(4)$$
$$f(t) = -16(t - 2)^2 + 64$$

The vertex is (2, 64).

Solution: The maximum height of 64 feet is reached at time $t = 2$ seconds.

Solve: If the temperature of an enclosed gas remains constant, the pressure is inversely proportional to the volume. When the pressure is 50 kg/cm^2, the volume is 84 liters. If the temperature is held constant and the pressure is increased to 60 kg/cm^2, what is the volume?

Solution: Since the pressure is inversely proportional to the volume, we can represent the volume V as a function of the pressure p as follows:

$$V(p) = \frac{k}{p} \quad \text{(where } k \text{ is the constant of variation)}$$

We want to find $V(60\text{ kg/cm}^2)$.

From the information in the problem, we know:

$$V(50\text{ kg/cm}^2) = 84\ \ell, \text{ from which we obtain } \frac{k}{50\text{ kg/cm}^2} = 84\ \ell$$

We can now solve for k:

$$k = (50\text{ kg/cm}^2)(84\ \ell) = 4{,}200\ \frac{\text{kg}\cdot\ell}{\text{cm}^2}$$

Now we can use $k = 4{,}200\ \frac{\text{kg}\cdot\ell}{\text{cm}^2}$ in the function and then determine $V(60\text{ kg/cm}^2)$:

$$V(p) = \frac{4200\ \frac{\text{kg}\cdot\ell}{\text{cm}^2}}{p} \text{ is the completely defined function.}$$

$$V(60) = \frac{4{,}200\ \frac{\cancel{\text{kg}}\cdot\ell}{\cancel{\text{cm}^2}}}{60\ \frac{\cancel{\text{kg}}}{\cancel{\text{cm}^2}}}$$

$$V(60) = 70 \text{ liters}$$

Solution: If the temperature remains constant while the pressure is increased to 60 kg/cm^2, the volume of the gas will be 70 liters.

Solve: The formula for compound interest is given by

$$S = P\left(1 + \frac{r}{m}\right)^{mt}$$

Here the interest is compounded m times per year for t years at the yearly rate of interest r. How many years will it take for \$2,500 invested at $5\frac{1}{2}\%$ compounded semiannually (twice a year) to amount to \$4,800?

Substituting into the formula and simplifying, we obtain:

$$4{,}800 = 2{,}500\left(1 + \frac{0.055}{2}\right)^{2t}$$

$$1.92 = (1.0275)^{2t}$$

Because the exponential and logarithmic functions are inverses of each other, we can undo one with the other. So, to solve for t, we will take the natural logarithm of both sides:

$$\ln (1.92) = \ln (1.0275)^{2t}$$

Using the reflexive property for equality, we obtain:

$$\ln (1.0275)^{2t} = \ln (1.92)$$

Using the properties of logarithms, we rewrite the left side and solve for t:

$2t \ln (1.0275) = \ln (1.92)$

$$t = \frac{\ln (1.92)}{2 \ln (1.0275)}$$ ← See Appendix A for guidance in computing this expression using a hand-held scientific calculator.

$$t = 12 \text{ years (approximately)}$$

Solution: It will take approximately 12 years for \$2,500 invested at $5\frac{1}{2}\%$ compounded semiannually to amount to \$4,800.

Solve: The number P of bacteria in a culture is given as a function of the time t in hours by

$$P(t) = 3{,}000e^{0.8t}$$

Approximately how many bacteria will be present after 5 hours?

Solution: We need to find $P(5)$:

$P(t) = 3{,}000e^{0.8t}$

$P(5) = 3{,}000e^{0.8(5)}$

$P(5) = 3{,}000e^{4}$ ← See Appendix A for guidance in computing this expression using a hand-held scientific calculator.

$P(5) = 3{,}000(54.59815\ldots)$
$P(5) = 163{,}794$ (approximately)

Solution: After 5 hours, the number of bacteria present will be about 163,794.

CHAPTER 5 SUMMARY

5.1 If two copies of the real number line are placed perpendicular to each other, so that they intersect at the zero point on each line, the lines form the **axes** of a rectangular coordinate system called the **Cartesian coordinate plane.** The horizontal real line with positive direction to the right is called the **horizontal axis,** or the ***x*-axis,** and the vertical real line with positive direction upward is called the **vertical axis,** or the ***y*-axis.** Their point of intersection is called the **origin.**

Each point in the plane is associated with an **ordered pair** (x, y) of real numbers x and y, called its **coordinates.** In an ordered pair the first element is called the **abscissa** and the second element, the **ordinate.** Two ordered pairs are **equal** if and only if they have *exactly* the same coordinates. The plane in which the coordinate system lies is divided into four sections called **quadrants.** They are named with the Roman numerals **I, II, III,** and **IV.** The numbering process begins in the upper right section and proceeds counterclockwise.

If (x_1, y_1) and (x_2, y_2) are two points in the plane, the **distance** d between them is given by $d = \sqrt{(x_2 - x_1)^2 + (y_2 - y_1)^2}$.

If (x_1, y_1) and (x_2, y_2) are two points in the plane, the **midpoint** of the line segment between (x_1, y_1) and (x_2, y_2) has coordinates given by $\left(\frac{x_1+x_2}{2}, \frac{y_1+y_2}{2}\right)$.

If (x_1, y_1) and (x_2, y_2) are the coordinates of two points on the line, the formula for the **slope** of the line is $m = \frac{y_2-y_1}{x_2-x_1}$ $(x_1 \neq x_2)$. When a line slopes *upward* to the right, its slope is *positive,* and when a line slopes *downward* to the right, its slope is *negative.* All **horizontal lines** have slope 0. **Vertical lines** have no slope. If two lines are **parallel** their slopes are equal; and if two lines are **perpendicular,** their slopes are negative reciprocals of each other.

5.2 The **Cartesian product** of two sets A and B, denoted $A \times B$, is the set of all ordered pairs (x, y) such that $x \in A$ and $y \in B$. The set of all possible ordered pairs of real numbers is denoted $R \times R$, or simple R^2. The set R^2 is represented by the Cartesian coordinate plane. Any subset of R^2 is a **relation** in R^2. The set of all first components in the ordered pairs in a relation is called the **domain** of the relation, and the set of all second components is called the **range** of the relation.

A **function** is a relation in which each first component is paired with *one and only one* second component. All functions are relations, but not all relations are functions. A function is a relation in which no two different ordered pairs have the same first component.

5.3 Functions are usually denoted by lowercase letters such as f, g, h, and so on. A function may be defined by listing its **ordered pairs,** a **table,** an **arrow diagram,** using two **number lines,** a **rule,** an **equation,** or a **graph.**

In the function defined by $y = f(x)$, y is called the **dependent variable** and x the **independent variable.** The notation $f(x)$ is read "f of x," and indicates you must substitute a value of x, called an **argument** of f, into the equation defining the function to find y, the **value** of f at x, or the **image** of x under f. The notation D_f is used to indicate the **domain,** and the notation R_f is used to indicate the **range** of a function f.

The ordered pairs of the function are written in the form (x, y) or $(x, f(x))$. For simplicity, we often refer to a function by the equation that describes it.

The **vertical line test** for functions states the following: If any vertical line can be drawn so that it cuts the graph of a relation in more than one point, the relation is *not* a function.

If a function f is defined by an equation and no domain is specified, the largest possible subset of the real numbers for which each first component causes the corresponding function value to be a *real* number is its **domain of definition.** When a function is discussed, it is customary to consider its domain of definition to be the domain of the function. In determining the domain of a function defined by an equation, start with the set of real numbers and eliminate any numbers that would make the equation undefined. If a rational expression is involved, exclude any numbers that would make a denominator zero. If a radical with an *even* index is involved, omit all values for which the expression under the radical is negative.

5.4 The **graph** of a function f in a coordinate plane is the set of all ordered pairs (x, y) for which $x \in D_f$ and $y = f(x)$.

Any points at which the graph of the function intersects the x-axis are called the real **zeros** of the function and are determined by finding all points x for which $f(x) = 0$. A real zero of a function is

1. an ***x*-value** for which $f(x) = 0$.
2. a real **root** of the equation $f(x) = 0$.
3. an ***x*-intercept** for the graph of $y = f(x)$.

If zero is in the domain of f, then $f(0)$ is the ***y*-intercept** of the graph. Since each x in the domain of a function corresponds to one and only one $f(x)$ in the range, a function cannot have more than one y-intercept.

A way to determine the range of a function is to look at its graph.

Adding or subtracting a positive constant k to or from $f(x)$ is called a **vertical shift.** Adding or subtracting a positive constant h to or from x is called a **horizontal shift.** Vertical and horizontal shifts are summarized in the following table:

Type of change	Effect on y = f(x)
$y = f(x) + k$	vertical shift: k units up
$y = f(x) - k$	vertical shift: k units down
$y = f(x + h)$	horizontal shift: h units to left
$y = f(x - h)$	horizontal shift: h units to right

The graph of $y = -f(x)$ is a **reflection** of $y = f(x)$ in the x-axis. Multiplying $f(x)$ by $c > 1$ enlarges or **stretches** the graph of f. Multiplying $f(x)$ by $c < 1$ reduces or **shrinks** the graph of f. The graph of a function $y = f(x)$ can be modified by a combination of shifts, reflections, or a stretching or shrinking. It is customary to call $y = f(x)$ the **parent function** of the function that results from the modification(s).

A line $x = k$ is a **vertical asymptote** if the function approaches $-\infty$ or ∞ as x approaches k from the left or right. A line $y = h$ is a **horizontal asymptote** if the function approaches h as x increases or decreases without bound. A function $y = g(x)$ is an **oblique asymptote** or **slant asymptote** if the function approaches $g(x)$ as x increases or decreases without bound.

5.5 A function f is **one-to-one** if and only if $f(a) = f(b)$ implies that $a = b$; that is, if $(a, c) \in f$ and $(b, c) \in f$, then $a = b$. In a one-to-one function each first component is paired with *exactly one* second component *and* each second component is paired with *exactly one* first component. The **horizontal line test** for one-to-one functions states the following: If any horizontal line can be drawn so that it cuts the graph of a function in more than one point, the function is *not* a one-to-one function. If in an interval whenever $x_1 > x_2$, $f(x_1) < f(x_2)$, the function f is increasing in the interval; if $f(x_1) > f(x_2)$, the function is decreasing in the interval; if $f(x_1) = f(x_2)$, the function is constant in the interval. A function is **monotone** if it is increasing or decreasing on its entire domain. A monotone increasing or decreasing function is one-to-one.

A function is **even** if for every x in the domain of f, $-x$ is in the domain of f and $f(-x) = f(x)$. A function is **odd** if for every x in the domain of f, $-x$ is in the domain of f and $f(-x) = -f(x)$. The graph of an even function is **symmetric** with respect to the ***y*-axis.** The graph of an odd function is **symmetric** with respect to the **origin.**

5.6 Two functions f and g are **equal,** written $f = g$, if and only if their domains are equal and $f(x) = g(x)$ for all x in their common domain.

The following definitions for the arithmetic of functions hold:

If the ranges of f and g are real numbers,

- The **sum** of f and g, denoted $f + g$, is defined by

$$(f + g)(x) = f(x) + g(x)$$

- The **difference** of f and g, denoted $f - g$, is defined by

$$(f - g)(x) = f(x) - g(x)$$

- The **product** of f and g, denoted $f \cdot g$, is defined by

$$(f \cdot g)(x) = f(x)g(x)$$

- The **quotient** of f and g, denoted $(f/g)(x)$, is defined by

$$(f/g)(x) = \frac{f(x)}{g(x)} \quad (g(x) \neq 0)$$

These definitions hold for all real numbers x in the domain of *both* f and g.

5.7 The **composite** of f and g, denoted $f \circ g$, is the function defined by $(f \circ g)(x) = f(g(x))$, where $R_g \subseteq D_f$. Composition of functions is *not* a commutative process; that is, in general, $(f \circ g)(x) \neq (g \circ f)(x)$.

5.8 Two functions f and g are called **inverses** of each other if and only if $(f \circ g)(x) = (g \circ f)(x) = x$, where $R_g \subseteq D_f$ and $R_f \subseteq D_g$. The inverse of a function f is usually denoted f^{-1} (read "f inverse"). If a function is one-to-one, then f^{-1} may be found by interchanging the first and second components in the ordered pairs of f. If f is a one-to-one function, f^{-1} is the set of ordered pairs obtained from f by interchanging the first and second components of each of its ordered pairs. The graphs of f and f^{-1} are **reflections** of one another about the line $\boldsymbol{y = x}$. Two common methods for determining the equation that defines the inverse of a function follow:

Method 1: Set $(f \circ f^{-1})(x) = x$ and solve for $f^{-1}(x)$.

Method 2: First, interchange x and y in $y = f(x)$, then solve $x = f(y)$ for y. If a function is not one-to-one, we can restrict the domain so that the function is one-to-one in the **restricted domain.** The inverse of the function will be defined in the restricted domain of the function.

5.9 The concept of function is fundamental to all of mathematics. A **linear function** is defined by $y = f(x) = mx + b$ $(m \neq 0)$. The domain and range are both equal to the set of real numbers. The zeros are found by solving the equation $mx + b = 0$. Thus, $x = -\frac{b}{m}$ is the only zero of the function. The graph crosses the x-axis at the point $\left(-\frac{b}{m}, 0\right)$. The graph of $y = mx + b$ is a **straight line** with **slope** $\boldsymbol{m}$ and **$\boldsymbol{y}$-intercept** $\boldsymbol{b}$**.** This is called the **slope-intercept** form of a line. Every **linear equation** $Ax + By = C$ with $B \neq 0$ determines a linear function. The line determined by the equation has slope $-\frac{A}{B}$ and y-intercept $\frac{C}{B}$. The equation of a line can be determined using one of the following:

1. The **slope-intercept form:** $y = mx + b$, where m is the slope of the line and b is its y-intercept

2. The **standard form:** $Ax + By = C$, where A and are not both zero
3. The **point-slope form:** $y - y_1 = m(x - x_1)$, where m is the slope of the line and (x_1, y_1) is a point on the line

The following table summarizes linear equations:

Slope intercept form (functional form)	$y = mx + b$
Point-slope form	$y - y_1 = m(x - x_1)$
Standard form	$Ax + By = C$
Horizontal line	$y = k$ for any constant k
Vertical line (not a function)	$x = h$ for any constant h

A **constant function** is defined by $y = f(x) = b$. The domain is the set of real numbers and the range is the set $\{b\}$ consisting of the single element b. If $b \neq 0$, the constant function $f(x) = b$ has no zeros; if $b = 0$, every value of x is a zero of the function. The graph of a constant function is a horizontal line (has slope zero) that is $|b|$ units above or below the x-axis.

A **quadratic function** is defined by $y = f(x) = ax^2 + bx + c$, $(a \neq 0)$. The domain is the set of all real numbers and the range is a subset of the real numbers. The zeros are determined by finding the roots of the quadratic equation $ax^2 + bx + c = 0$. If $b^2 - 4ac > 0$, there are two real unequal zeros. If $b^2 - 4ac = 0$, there is one real zero (double root). If $b^2 - 4ac < 0$, there are no real zeros for the function. The graph of a quadratic function defined by $f(x) = ax^2 + bx + c$ is always a **parabola.** If $a > 0$, the parabola opens **upward** and has a **minimum.** If $a < 0$, the parabola opens **downward** and has a **maximum.** The parabola is **symmetric** about a vertical line through its vertex. This line is called its **axis of symmetry.** The graph of the function may or may not intersect the x-axis, depending on the solution set of $ax^2 + bx + c = 0$. The graph of the equation $f(x) = a(x - h)^2 + k$ $(a \neq 0)$ is a parabola with vertex (h, k). The parabola opens upward if $a > 0$ and downward if $a < 0$. This is the **standard form** for the equation of a parabola. By **completing the square,** any quadratic function can be put in standard form. The parent function for quadratic functions is $y = x^2$. By knowing its basic shape, sketches of quadratics that result when **transformations** such as shifts, reflections, stretching, and shrinking are applied to $y = x^2$ usually are more easily obtained than by plotting points.

A **polynomial function** is defined by $y = P(x) = a_n x^n + a_{n-1}x^{n-1} + a_{n-2}x^{n-2} + \ldots + a_1 x^1 + a_0$, where n is a nonnegative integer and $a_n \neq 0$. The domain is the set of real numbers. If the polynomial has odd degree, the range is also the set of real numbers. If the polynomial has even degree, the range is a subset of the real numbers. Linear, constant, and quadratic functions are special types of polynomial functions. The zeros are the solutions of the equation $P(x) = 0$. A number r is a **zero,** or **root,** of a polynomial $P(x)$ if $P(r) = 0$. If r is a real number, the graph of $P(x)$ intersects the x-axis at r.

Allowing complex roots, the **Fundamental Theorem of Algebra** assures us that $P(x) = 0$ has a solution. It states: Every polynomial of degree one or greater has at least one root.

The **Remainder Theorem** states: If a polynomial $P(x)$ is divided by $x - a$, the remainder is $P(a)$.

A root r of $P(x)$ has **multiplicity** k if $(x - r)^k$ is a factor of $P(x)$. Allowing complex roots and counting multiplicities, we have the following: **Every polynomial of positive degree *n* has exactly *n* roots.** Thus, every linear function has exactly one root; every quadratic function has exactly two roots; and so on.

The **Factor Theorem** states: $P(r) = 0$ if and only if $x - r$ is a factor of $P(x)$.

The **Rational Root Theorem** states: If $P(x) = a_n x^n + a_{n-1}x^{n-1} + a_{n-2}x^{n-2} + \ldots + a_1 x^1 + a_0$ is a polynomial function with integral coefficients and $\frac{p}{q}$ is a **rational root** of $P(x)$ in lowest terms, then p is a factor of a_0 and q is a factor of a_n.

Let $P(x)$ be a polynomial function with real coefficients. If $a + bi$ is a complex root of $P(x)$, then its **conjugate** $a - bi$ is also a root of $P(x)$.

Performing operations with **complex numbers** is essentially like working with binomials. The rules follow:

$i^2 = -1$
$\sqrt{-b^2} = bi \; (b \geq 0)$
The conjugate of $a + bi$ is $a - bi$.
$(a + bi)(a - bi) = a^2 + b^2$
$(a + bi) + (c + di) = (a + c) + (b + d)i$
$(a + bi)(c + di) = (ac - bd) + (ad + bc)i$

Let $P(x)$ be a polynomial function with real coefficients. If a and b are real numbers such that $P(a)$ and $P(b)$ have opposite signs, then $P(x)$ has at least one zero between a and b.

A **rational function** is defined by $y = f(x) = \frac{p(x)}{q(x)}$, where $p(x)$ and $q(x)$ are polynomials. The domain excludes any real number for which $q(x) = 0$. The range will be a subset of the real numbers. The zeros, if any, are the values of x for which $f(x) = 0$. To graph rational functions, you will usually need to look for **asymptotes** associated with the function. An **absolute value function** is defined by $y = f(x) = |ax + b|$. The domain is the set of all real numbers and the range is restricted to nonnegative values. The function has a zero at $x = -\frac{b}{a}$.

A **square root function** is defined by $y = f(x) = \sqrt{ax + b}$. The domain is the set of all real numbers such that $ax + b \geq 0$. The range is the set of nonnegative real numbers $[0, \infty)$. The function has a zero at $x = -\frac{b}{a}$. The graph of the function is to the right of $x = -\frac{b}{a}$ and above the x-axis.

A **greatest integer function** is defined by $y = f(x) = [x]$ = the greatest integer less than or equal to x. The domain is the set of all real numbers and the range is the set of integers. The zeros of the function lie in the interval $[0, 1)$.

An **exponential function** is defined by $y = f(x) = b^x$ $(b \neq 1,\ b > 0)$, where b is called the **base** of the exponential function. The domain is the set of real numbers, and the range is the set of positive real numbers. The function has no zeros. Exponential functions have the following properties:

For real numbers b $(b \neq 1,\ b > 0)$ and x, y:

$b^x > 0$ for all real numbers x.

$b^0 = 1$

$b^1 = b$

If $x = y$, then $b^x = b^y$.

$b^{-x} = \frac{1}{b^x}$.

$b^x b^y = b^{x+y}$

$\frac{b^x}{b^y} = b^{x-y}$

$(b^x)^y = b^{xy}$

The graph of $y = f(x) = b^x$

- passes through the points $(0, 1)$ and $(1, b)$ and is asymptotic to the x-axis;
- is an increasing function if $b > 1$;
- is a decreasing function if $0 < b < 1$.

The graph is located in the first and second quadrants only.

Two exponential functions that are of particular importance are $y = f(x) = 10^x$ and $y = f(x) = e^x$.

A **logarithmic function** is defined by $y = f(x) = \log_b x \Leftrightarrow b^y = x$ $(x > 0)$, where b is called the **base** of the logarithmic function $(b \neq 1,\ b > 0)$. The domain is the set of positive real numbers and the range is the set of real numbers. The function has a zero at $x = 1$. The graph of the function is located in the first and fourth quadrants only. Logarithms are **exponents.** If $y = \log_b x$, then y is the *exponent* that is used on b to get x; that is, $b^y = x$. The logarithm function is the **inverse** of the exponential function, and vice versa. We have the following properties for logarithmic functions:

For real numbers b ($b \neq 1$, $b > 0$), u, v, w, and p:

$\log_b b^x = x$

$b^{\log_b x} = x$

$\log_b b = 1$ (because $b^1 = b$)

$\log_b 1 = 0$ (because $b^0 = 1$)

If $\log_b u = \log_b v$, then $u = v$.

$\log_b \frac{1}{u} = -\log_b u$.

$\log_b uv = \log_b u + \log_b v$

$\log_b \frac{u}{v} = \log_b u - \log_b v$

$\log_b u^p = p \log_b u$

For real numbers b ($b \neq 1$, $b > 0$), x, and y, the graph of $y = f(x) = \log_b x$

- passes through (1, 0) and b, 1) and is asymptotic to the y-axis.
- is an increasing function if $b > 1$.
- is a decreasing function if $0 < b < 1$.

The logarithm function $y = f(x) = \log_{10} x$ (the **common logarithmic function**) is the inverse of the exponential function $y = f(x) = 10^x$. The logarithm function $y = f(x) = \ln x$ (the **natural logarithmic function)** is the inverse of the exponential function $y = f(x) = e^x$.

The variable y is said to **vary directly** as the nth power of the variable x if $y = f(x) = kx^n$ (where k is a positive constant and $n > 0$). The variable y is said to **vary indirectly** as the nth power of the variable x if $y = f(x) = \frac{k}{x^n}$ (where k is a positive constant). The constant k is called the **constant of variation.**

5.10 The four basic kinds of **conics** are the **parabola, circle, ellipse,** and **hyperbola.** The equation of a conic can be written as

$$Ax^2 + By^2 + Cx + Dy + F = 0 \qquad (A \neq 0 \text{ and } B \neq 0)$$

This equation defines a relation that is not a function and has different graphs depending on the values of the coefficients A, B, and C. The graph of the equation $Ax^2 + By^2 + Cx + Dy + F = 0$ ($A \neq 0$ and $B \neq 0$) takes the following form:

- a **circle** if $A = B$, with standard form $(x - h)^2 + (y - k)^2 = r^2$, where (h, k) is the **center** and the **radius** is $|r|$ units.
- an **ellipse** if $AB > 0$, with standard form $\frac{(x-h)^2}{a^2} + \frac{(y-k)^2}{b^2} = 1$. The **center** is at (h, k). The ellipse has **vertices** $(h - a, k)$, $(h + a, k)$, $(h, k - b)$, and $(h, k + b)$. The line segment joining the vertices $(h - a, k)$ and $(h + a, k)$ is a **horizontal axis** of symmetry and the line segment joining the vertices $(h, k - b)$ and $(h, k + b)$ is a **vertical axis** of symmetry. The longer axis is called the **major axis** and the shorter axis is called the **minor axis.** The length of the two axes are $2|a|$ and $2|b|$.
- a **hyperbola** if $AB < 0$. There are two standard forms:

Form (1) $$\frac{(x-h)^2}{a^2} - \frac{(y-k)^2}{b^2} = 1$$

The center is at (h, k). The hyperbola opens left and right along the line $y = k$, and it passes through the **vertices** $(h - a, k)$ and $(h + a, k)$. It has the intersecting lines $y = k + \frac{b}{a}(x - h)$ and $y = k - \frac{b}{a}(x - h)$ as (slanting) **asymptotes.** The asymptotes are the diagonals of a rectangle with dimensions $2|a|$ by $2|b|$ centered at (h, k).

Form (2) $$\frac{(y-k)^2}{b^2} - \frac{(x-h)^2}{a^2} = 1$$

The center is at (h, k). The hyperbola opens up and down along the line $x = h$, and it passes through the **vertices** $(h, k - b)$, and $(h, k + b)$. As in the previous case, it has the intersecting lines $y = k + (x - h)$ and $y = k - (x - h)$ as (slanting) **asymptotes.** If the equation of a conic section is not in standard form, it can be put in standard form by completing the squares on the x and y terms.

5.11 The solution set of an **inequality in two variables** is the relation consisting of all ordered pairs that make the inequality a true statement. The graph of a two-variable inequality is the set of all points in the solution set of the inequality.
The graph of a two-variable *linear* inequality is a **half-plane.**
To graph two-variable inequalities:
   1. Graph the equation that results when the inequality symbol is replaced with an equal sign. Use a dashed line for $<$ or $>$ inequalities and a solid line for $\leq$ or $\geq$ inequalities. This is the boundary line or curve.
   2. Select and shade the correct portion of the plane by testing a point that is *not* on the boundary (the origin is usually a good choice unless the boundary passes through it). If the coordinates of the point satisfy the inequality, shade the portion of the plane containing the test point; if not, shade the portion of the plane that does *not* contain the test point.

5.12 Functions have extensive application in real-world situations.

PRACTICE PROBLEMS FOR CHAPTER 5

Evaluate:

1. $f(5)$ if $f(x) = 7 - 2x$
2. $f(-5)$ if $f(x) = 7 - 2x$

3. $h(10)$ if $h(x) = \sqrt{8x + 1}$
4. $P(-1)$ if $P(x) = 2x^4 - 2x^3 + x^2 + 5x - 8$
5. $g(3) - g(0)$ if $g(x) = -5x + 1$
6. $\dfrac{f(y + 5)}{5}$ if $f(x) = 7 - 2x$
7. $h(5)$ if $h(x) = 2^x$
8. $\dfrac{f(x + h) - f(x)}{h}$ if $f(x) = 7 - 2x$

Write the equation of the line in slope-intercept form for each of the following:

9. slope 3, passes through $(4, -2)$
10. $(5, 0)$ and $(7, -1)$ lie on the line
11. y-intercept 10, parallel to $y = \dfrac{3}{7}x - 3$
12. passes through $(-4, 8)$ and $(2, 8)$
13. perpendicular to the line through $y = 2x$ and has y-intercept 3

Find the vertex and tell whether the y-value is a maximum or minimum for the function:

14. $y = -x^2 + 8x - 24$
15. $y = 3x^2 - 2x - 8$

State the domain and range for the following functions:

16. $f(x) = x^2 - 7x + 10$
17. $f(x) = \dfrac{1}{x + 2}$
18. $f(x) = \sqrt{x + 2}$

Find the zeros of each of the following functions:

19. $f(x) = 3x + 12$
20. $f(x) = x^2 - 7x + 10$
21. $f(x) = \sqrt{x + 2}$
22. $f(x) = \ln x$
23. $f(x) = x^3 - 8$

Solve:

24. A manufacturer has total revenue function given by

$$R(q) = 1.25q$$

where q is the number of items sold and the cost per item is \$1.25. The total cost function is given by

$$C(q) = 5{,}000 + 0.50q$$

The profit function is given by

$$P(q) = R(q) - C(q)$$

Determine the break-even point, which is defined as the level of production at which profit is zero, that is, $P(q) = 0$.

25. A ball is thrown into the air, following a path given by the function $f(t) = -16t^2 + 96t$, where $f(t)$ is the height in feet when time t is in seconds. Find the maximum height reached by the ball before it returns to earth.

SOLUTIONS TO PRACTICE PROBLEMS

1. $f(5) = 7 - 2(5) = -3$
2. $f(-5) = 7 - 2(-5) = 17$
3. $h(10) = \sqrt{8(10) + 1} = \sqrt{81} = 9$
4. $P(-1) = 2(-1)^4 - 2(-1)^3 + (-1)^2 + 5(-1) - 8 = -8$
5. $g(3) - g(0) = (-5(3) + 1) - (-5(0) + 1) = -14 - 1 = -15$
6. $\dfrac{f(y+5)}{5} = \dfrac{7 - 2(y+5)}{5} = \dfrac{7 - 2y - 10}{5} = -\dfrac{2y+3}{5}$
7. $h(5) = 2^5 = 32$
8. $$\frac{f(x+h) - f(x)}{h} = \frac{7 - 2(x+h) - (7 - 2x)}{h}$$
$$= \frac{7 - 2x - 2h - 7 + 2x}{h} = \frac{-2h}{h} = -2$$

Write the equation of the line in slope-intercept form for each of the following:

9. slope 3, passes through $(4, -2)$

$$y - (-2) = 3(x - 4)$$
$$y + 2 = 3x - 12$$
$$y = 3x - 14$$

10. $(5, 0)$ and $(7, -1)$ lie on the line

$$m = \frac{-1 - 0}{7 - 5} = -\frac{1}{2}$$

$$y - 0 = -\frac{1}{2}(x - 5)$$

$$y = -\frac{1}{2}x + \frac{5}{2}$$

11. y-intercept 10, parallel to $y = \dfrac{3}{7}x - 3$

$$m = \frac{3}{7} \quad \text{(parallel lines have equal slopes)}$$

$$y = \frac{3}{7}x + 10$$

12. passes through $(-4, 8)$ and $(2, 8)$

$$m = \frac{8 - 8}{2 - (-4)} = 0$$

$$y - 8 = 0(x - 2)$$

$$y = 8$$

13. perpendicular to the line through $y = 2x$ and has y-intercept 3

The slope of the line will be $-\frac{1}{2}$, the negative reciprocal of 2. Using the slope-intercept form, the equation is $y = -\frac{1}{2}x + 3$.

Find the vertex and tell whether the y-value is a maximum or minimum for the function:

14. $y = -x^2 + 8x - 24$

$$y = -x^2 + 8x - 24$$

$$y = -(x^2 - 8x) - 24$$

$$y = -(x^2 - 8x + 4^2) + 4^2 - 24$$

$$y = -(x^2 - 8x + 4^2) + 16 - 24$$

$$y = -(x - 4)^2 - 8$$

Vertex: $(4, -8)$; -8 is a maximum, since $a = -1 < 0$.

15. $y = 3x^2 - 2x - 8$

$$y = 3x^2 - 2x - 8$$

$$y = 3\left(x^2 - \frac{2}{3}x\right) - 8$$

$$y = 3\left(x^2 - \frac{2}{3}x + \left(\frac{1}{3}\right)^2\right) - 3\left(\frac{1}{3}\right)^2 - 8$$

$$y = 3\left(x - \frac{1}{3}\right)^2 - 3\left(\frac{1}{9}\right) - 8$$

$$y = 3\left(x - \frac{1}{3}\right)^2 - \frac{1}{3} - \frac{24}{3}$$

$$y = 3\left(x - \frac{1}{3}\right)^2 - \frac{25}{3}$$

Vertex: $(\frac{1}{3}, -\frac{25}{3})$; $-\frac{25}{3}$ is a minimum, since $a = 3 > 0$.

State the domain and range for the following functions:

16. $f(x) = x^2 - 7x + 10$; This is a quadratic function. The domain is the set of real numbers. The range can be found by finding the vertex of the parabola.

$$f(x) = x^2 - 7x + 10$$

$$f(x) = x^2 - 7x + \left(\frac{7}{2}\right)^2 + 10 - \left(\frac{7}{2}\right)^2$$

$$f(x) = x^2 - 7x + \left(\frac{7}{2}\right)^2 + 10 - \frac{49}{4}$$

$$f(x) = \left(x - \frac{7}{2}\right)^2 + \frac{40}{4} - \frac{49}{4}$$

$$f(x) = \left(x - \frac{7}{2}\right)^2 - \frac{9}{4}$$

The vertex is $\left(\frac{7}{2}, -\frac{9}{4}\right)$. Since $a > 0$, the parabola turns up and y has minimum value of $-\frac{9}{4}$. The range is $\left[-\frac{9}{4}, \infty\right)$.

17. $f(x) = \dfrac{1}{x + 2}$; The function is undefined when $x = -2$. The domain is the set of real numbers except -2. As x gets extremely large or extremely small, $f(x)$ approaches 0. The range of the function is all real numbers except 0.
18. $f(x) = \sqrt{x + 2}$; The function is undefined when $x + 2$ is negative. The domain is the set of real number $x \geq -2$. The range is $[0, \infty)$.

Find the zeros of each of the following functions:

19. $f(x) = 3x + 12$

Solve $3x + 12 = 0$:

$$3x = -12$$

$$x = -4$$

-4 is the only zero.

20. $f(x) = x^2 - 7x + 10$

Solve $x^2 - 7x + 10 = 0$:

$$(x - 5)(x - 2) = 0$$
$$x = 5 \text{ or } x = 2$$

5 and 2 are the only zeros.

21. $$f(x) = \sqrt{x + 2}$$

Solve $\sqrt{x + 2} = 0$:

Square both sides:

$$x + 2 = 0$$
$$x = -2$$

Check: $\sqrt{-2 + 2} \stackrel{?}{=} 0$

$$0 \stackrel{\checkmark}{=} 0$$

-2 is the only zero.

22. $f(x) = \ln x$. The logarithm function crosses the x-axis at the point $(1, 0)$, so 1 is the only zero.

23. $f(x) = x^3 - 8$

Solve $x^3 - 8 = 0$:

$(x - 2)(x^2 + 2x + 4) = 0$

$x = 2$ is the only real zero.

There are two complex zeros:

$$x = \frac{-2 \pm \sqrt{4 - 16}}{2} = \frac{-2 \pm \sqrt{-12}}{2} = \frac{-2 \pm i\sqrt{12}}{2} = \frac{-2 \pm i2\sqrt{3}}{2}$$
$$= -1 \pm i\sqrt{3}$$

(See Appendix E for techniques for simplifying radicals.)

Solve:

24. A manufacturer has total revenue function given by

$$R(q) = 1.25q$$

where q is the number of items sold and the cost per item is \$1.25. The total cost function is given by

$$C(q) = 5{,}000 + 0.50q$$

The profit function is given by

$$P(q) = R(q) - C(q)$$

Determine the break-even point, which is defined as the level of production at which profit is zero, that is, $P(q) = 0$.

Solution:

$$P(q) = R(q) - C(q)$$
$$P(q) = (1.25q) - (5{,}000 + 0.50q)$$
$$P(q) = 1.25q - 5{,}000 - 0.50q$$
$$P(q) = 0.75q - 5{,}000$$

To find the break-even point, we solve $0.75q - 5{,}000 = 0$:

$$0.75q - 5{,}000 = 0$$
$$0.75q = 5{,}000$$
$$q = 6{,}667 \text{ items (approximately)}$$

The break-even point occurs when about 6,667 items are sold.

25. A ball is thrown into the air, following a path given by the function $f(t) = -16t^2 + 96t$, where $f(t)$ is the height in feet when time t is in seconds. Find the maximum height reached by the ball before it returns to earth.

Solution: The graph of the function $f(t) = -16t^2 + 96t$ is a parabola that turns down. The maximum height will be reached at the vertex of the parabola. To find the coordinates of the vertex, complete the square on the right side:

$$f(t) = -16t^2 + 96t$$
$$f(t) = -16(t^2 - 6t + 9) + 16(9)$$
$$f(t) = -16(t - 3)^2 + 144$$

The vertex is (3, 144).

Solution: The maximum height of 144 feet is reached at time $t = 3$ seconds.

6
SYSTEMS OF EQUATIONS

6.1 SOLVING A SYSTEM OF TWO LINEAR EQUATIONS IN TWO VARIABLES

A **system of two linear equations in two variables** consists of a pair of linear equations in the same variables, such as the following:

$$\begin{cases} 3x - y = 1 \\ x + y = -5 \end{cases} \qquad \begin{cases} 2x - 3y = 12 \\ 5x + 2y = 11 \end{cases} \qquad \begin{cases} 0.08x + 0.05y = 270 \\ 0.07x + 0.06y = 285 \end{cases}$$

The equations in each system are to be considered simultaneously.

> To solve a system of linear equations in two variables means to find all ordered pairs of the variables that make both equations true simultaneously. There are three possibilities:
>
> 1. The system has exactly *one solution.*
> 2. The system has *no solution.*
> 3. The system has *infinitely many solutions.*

The system is **consistent** if it has at least *one solution:* at least one ordered pair makes both equations true simultaneously. The system is **inconsistent** if it has *no solution:* no ordered pair makes both equations true simultaneously. The system is **dependent** if it has *infinitely many solutions:* infinitely many ordered pairs make both equations true simultaneously. The **solution set** is the collection of all solutions.

To illustrate graphically (Figure 6.1), consider the three possibilities for the two lines that represent the equations in the system:

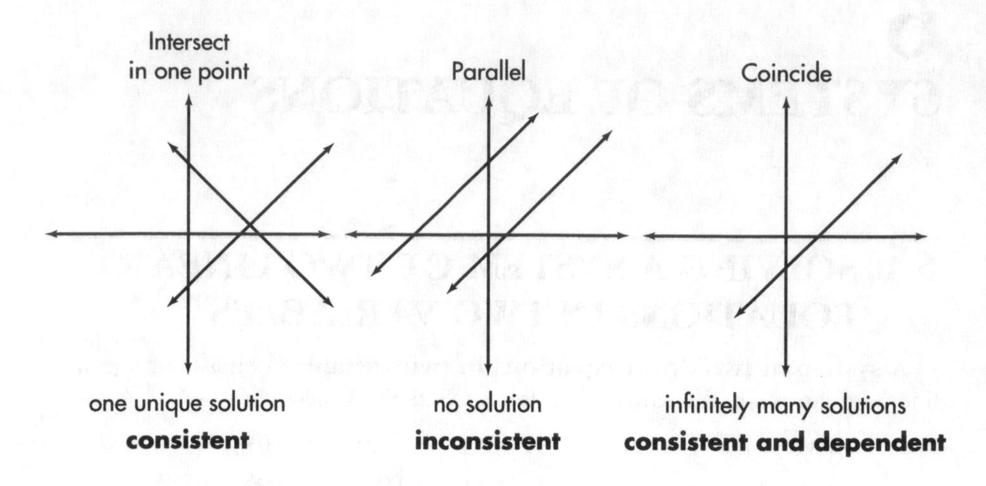

FIGURE 6.1

For example, the ordered pair $(-1, -4)$ is the unique solution to the system:

$$\begin{cases} 3x - y = 1 \\ x + y = -5 \end{cases}$$

This is so because if we substitute $x = -1$ and $y = -4$ into the two equations, we obtain two true statements:

$$3x - y = 1 \longrightarrow 3(-1) - (-4) \stackrel{?}{=} 1 \longrightarrow 1 \stackrel{\checkmark}{=} 1$$

$$x + y = -5 \longrightarrow -1 + (-4) \stackrel{?}{=} -5 \longrightarrow -5 \stackrel{\checkmark}{=} -5$$

Graphically, as shown in Figure 6.2, the two lines that represent the equations intersect in exactly one point, $(-1, -4)$.

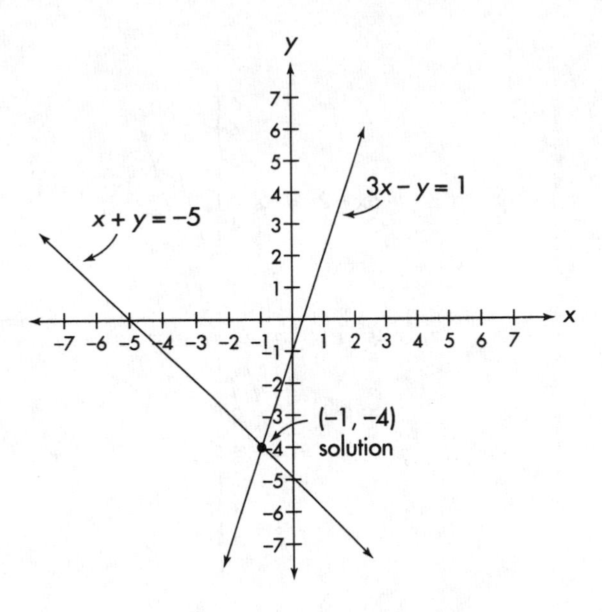

FIGURE 6.2

Thus, the system is consistent, with the common point of intersection, $(-1, -4)$, as the unique solution.

The system below has no solution:

$$\begin{cases} 3x - y = 1 \\ 6x - 2y = -2 \end{cases}$$

Graphically, as shown in Figure 6.3, the two lines that represent the equations are parallel to each other in the plane:

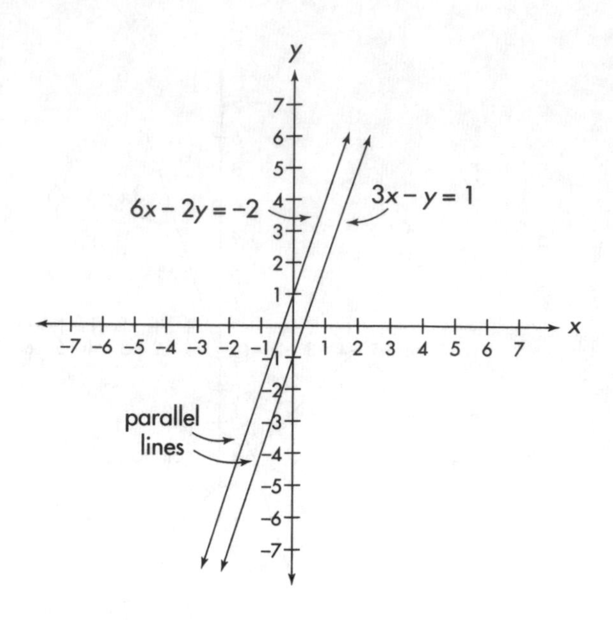

FIGURE 6.3

Thus, the system is inconsistent, having no solution, since the two lines have no common points.

The system below has infinitely many solutions:

$$\begin{cases} 3x - y = 1 \\ 6x - 2y = 2 \end{cases}$$

Graphically, as shown in Figure 6.4, the two lines that represent the equations coincide (lie on top of each other) in the plane:

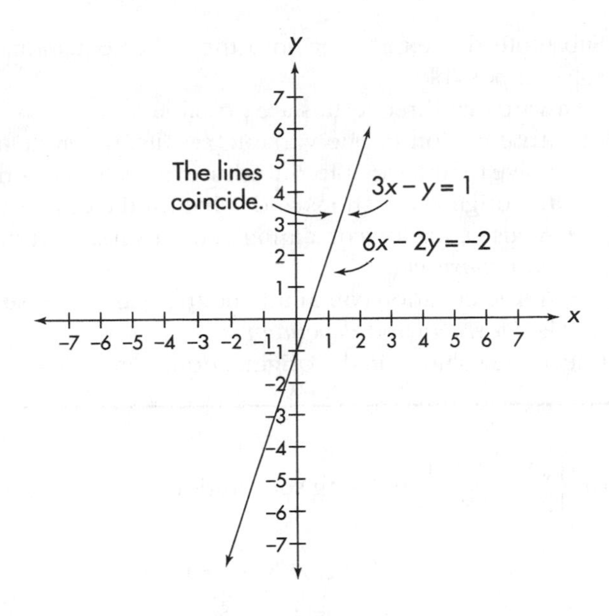

FIGURE 6.4

Thus, the system is dependent, since the two lines have infinitely many points in common.

Systems of linear equations are solved by going through a series of steps to transform the equations into equivalent equations from which the solution can be easily obtained.

> The following operations give equivalent equations, and may be used to solve a system of linear equations:
>
> 1. Interchanging two equations
> 2. Multiplying an equation by a nonzero constant
> 3. Multiplying an equation by a nonzero constant and adding the result to another equation

Three methods commonly used to solve a system of linear equations are **substitution, elimination,** and **transformation of the augmented matrix.**

> To solve a system of linear equations by using substitution:
>
> 1. Select the simpler equation and express one of the variables in terms of the other.

2. Substitute this expression into the other equation, simplify, and solve, if possible.
   Be aware that three results are possible:
   - An equation in one variable results: the system is *consistent.* Solve for the variable. Substitute the value obtained into one of the original equations and solve for the other variable.
   - A false equation containing neither variable results: the system is *inconsistent.*
   - A true equation containing neither variable results: the system is *consistent* and *dependent.*
3. Check the solution in the original equations.

To solve $\begin{cases} 3x - y = 1 \\ x + y = -5 \end{cases}$ by using substitution:

Solve $x + y = -5$ for x:

$$x + y = -5$$
$$x = -5 - y$$

Substitute into $3x - y = 1$ and solve for y:

$$3(-5 - y) - y = 1$$
$$-15 - 3y - y = 1$$
$$-4y = 16$$
$$y = -4$$

Substitute -4 for y in $x + y = -5$ and solve:

$$x + (-4) = -5$$
$$x = -1$$

Check $x = -1$ and $y = -4$ in each equation:

$$3x - y = 1 \longrightarrow 3(-1) - (-4) \stackrel{?}{=} 1 \longrightarrow 1 \stackrel{\checkmark}{=} 1$$
$$x + y = -5 \longrightarrow -1 + (-4) \stackrel{?}{=} -5 \longrightarrow -5 \stackrel{\checkmark}{=} -5$$

Thus, the solution is $x = -1$ and $y = -4$.

To solve a system of linear equations by using elimination:

1. Write both equations in standard form: $ax + by = c$.
2. If necessary, multiply one or both of the equations by a nonzero constant or constants to make the coefficients of one of the variables sum to zero.

3. Add the equations.
   Be aware that three results are possible:
   - An equation in one variable results: the system is *consistent.* Solve for the variable. Substitute the value obtained into one of the original equations and solve for the other variable.
   - A false equation containing neither variable results: the system is *inconsistent.*
   - A true equation containing neither variable results: the system is *consistent* and *dependent.*
4. Check the solution in the original equations.

- To solve $\begin{cases} 2x - 3y = 12 \\ 5x + 2y = 11 \end{cases}$ by using elimination:

  Multiply the first equation by 2 and the second equation by 3:

$$2x - 3y = 12 \xrightarrow{\text{2 times first}} 4x - 6y = 24$$

$$5x + 2y = 11 \xrightarrow{\text{3 times second}} 15x + 6y = 33$$

Add the equations and solve:

$$19x = 57$$
$$x = 3$$

Substitute 3 for x in $2x - 3y = 12$ and solve:

$$\begin{aligned} 2(3) - 3y &= 12 \\ 6 - 3y &= 12 \\ -3y &= 6 \\ y &= -2 \end{aligned}$$

Check $x = 3$ and $y = -2$ in each equation:

$$2x - 3y = 12 \longrightarrow 2(3) - 3(-2) \stackrel{?}{=} 12 \longrightarrow 12 \stackrel{\checkmark}{=} 12$$

$$5x + 2y = 11 \longrightarrow 5(3) + 2(-2) \stackrel{?}{=} 11 \longrightarrow 11 \stackrel{\checkmark}{=} 11$$

Thus, the solution is $x = 3$ and $y = -2$.

- To solve $\begin{cases} 3x - y = -1 \\ 6x - 2y = -2 \end{cases}$ by using elimination:

  Multiply the first by -2; do not change the second equation:

$$3x - y = -1 \xrightarrow{-2 \text{ times first}} -6x + 2y = 2$$

$$6x - 2y = -2 \xrightarrow{\text{no change to second}} 6x - 2y = -2$$

Add the equations:

$$0 = 0$$

The resulting equation contains no variable and is true. Therefore, the system has infinitely many solutions.

- To solve $\begin{cases} 0.08x + 0.05y = 270 \\ 0.07x + 0.06y = 285 \end{cases}$ by using elimination:

  Multiply the first equation by 0.07 and the second equation by -0.08:

$$0.08x + 0.05y = 270 \xrightarrow{0.07 \text{ times first}} 0.0056x + 0.0035y = 18.9$$

$$0.07x + 0.06y = 285 \xrightarrow{-0.08 \text{ times second}} -0.0056x - 0.0048y = -22.8$$

Add the equations and solve:

$$-0.0013y = -3.9$$
$$y = 3{,}000$$

Substitute 3,000 for y in $0.08x + 0.05y = 270$ and solve:

$$\begin{aligned} 0.08x + 0.05(3{,}000) &= 270 \\ 0.08x + 150 &= 270 \\ 0.08x &= 120 \\ x &= 1{,}500 \end{aligned}$$

Check $x = 1{,}500$ and $y = 3{,}000$ in each equation:

$$0.08x + 0.05y = 270 \rightarrow 0.08(1{,}500) + 0.05(3{,}000) \stackrel{?}{=} 270 \rightarrow 270 \stackrel{\checkmark}{=} 270$$

$$0.07x + 0.06y = 285 \rightarrow 0.07(1{,}500) + 0.06(3{,}000) \stackrel{?}{=} 285 \rightarrow 285 \stackrel{\checkmark}{=} 285$$

Thus, the solution is $x = 1{,}500$ and $y = 3{,}000$.

If you examine the examples for solving by elimination, the variables themselves are not manipulated in the procedure. Arithmetic operations are performed on the *coefficients* of the variables and the *constant* terms to reach the solution. With this in mind, consider the rectangular array associated with the system $\begin{cases} 2x - 3y = 12 \\ 5x + 2y = 11 \end{cases}$, composed of the numbers that appear in the system:

$$\left[\begin{array}{cc|c} 2 & -3 & 12 \\ 5 & 2 & 11 \end{array}\right]$$

This array is called the **augmented matrix** for the system. The horizontal lines of numbers are called **rows** and the vertical lines of numbers are called **columns.** When forming this matrix, first write both equations in standard form: $ax + by = c$. Place coefficients of corresponding variables underneath each other and use zero for the coefficient of a variable that does not appear in one of the equations. Use a vertical line to separate the constants on the right from the coefficients.

Since the rows of the augmented matrix correspond to the equations in the associated system, appropriate operations on the rows should transform the augmented matrix so that the solution to the system is easily obtainable.

The following *row operations* may be used to transform the augmented matrix:

1. Interchanging two rows
2. Multiplying a row by a nonzero constant
3. Multiplying a row by a nonzero constant and adding the result to another row

As you can see, the row operations correspond to the operations used on equations when solving a system.

To solve a system of linear equations by using the augmented matrix:

1. Write both equations in standard form: $ax + by = c$.
2. Write the augmented matrix.
3. Use row operations to transform the matrix into the form:

$$\left[\begin{array}{cc|c} 1 & 0 & x_0 \\ 0 & 1 & y_0 \end{array}\right]$$

   The solution is $x = x_0$ and $y = y_0$, when the system is consistent. Be aware that three results are possible:

   - $\left[\begin{array}{cc|c} 1 & 0 & x_0 \\ 0 & 1 & y_0 \end{array}\right]$: the system is *consistent* with solution (x_0, y_0);
   - $\left[\begin{array}{cc|c} 1 & k & k_1 \\ 0 & 0 & k_2 \end{array}\right]$, where $k_2 \neq 0$: the system is *inconsistent;*
   - $\left[\begin{array}{cc|c} 1 & k & k_1 \\ 0 & 0 & 0 \end{array}\right]$: the system is *consistent* and *dependent.*

4. Check the solution in the original equations.

To solve a system such as $\begin{cases} a_1x + b_1y = c_1 \\ a_2x + b_2y = c_2 \end{cases}$ using the augmented matrix, first write the augmented matrix:

$$\left[\begin{array}{cc|c} a_1 & b_1 & c_1 \\ a_2 & b_2 & c_2 \end{array}\right]$$

Then use row operations to transform the augmented matrix into the desired equivalent form:

$$\left[\begin{array}{cc|c} 1 & 0 & x_0 \\ 0 & 1 & y_0 \end{array}\right]$$

An efficient way to do this is to proceed systematically as follows:

$$\left[\begin{array}{cc|c} a_1 & b_1 & c_1 \\ a_2 & b_2 & c_2 \end{array}\right] \to \left[\begin{array}{cc|c} 1 & \cdots & \cdots \\ \cdots & \cdots & \cdots \end{array}\right] \to \left[\begin{array}{cc|c} 1 & \cdots & \cdots \\ 0 & \cdots & \cdots \end{array}\right] \to \left[\begin{array}{cc|c} 1 & \cdots & \cdots \\ 0 & 1 & \cdots \end{array}\right] \to \left[\begin{array}{cc|c} 1 & 0 & x_0 \\ 0 & 1 & y_0 \end{array}\right]$$

When the system is consistent, the solution is $x = x_0$ and $y = y_0$.

A helpful message: If you obtain a row of zeros except for the last element as you are transforming the augmented matrix, it is unnecessary to continue with the process because the system is inconsistent when this happens.

To solve the system $\begin{cases} 2x - 3y = 12 \\ 5x + 2y = 11 \end{cases}$ using the augmented matrix:

Write the augmented matrix:

$$\left[\begin{array}{cc|c} 2 & -3 & 12 \\ 5 & 2 & 11 \end{array}\right]$$

Perform appropriate elementary row operations:

$$\left[\begin{array}{cc|c} 2 & -3 & 12 \\ 5 & 2 & 11 \end{array}\right] \xrightarrow{\frac{1}{2}\text{ row 1}} \left[\begin{array}{cc|c} 1 & -1.5 & 6 \\ 5 & 2 & 11 \end{array}\right] \xrightarrow{-5\text{ row 1} + \text{row 2}} \left[\begin{array}{cc|c} 1 & -1.5 & 6 \\ 0 & 9.5 & -19 \end{array}\right] \xrightarrow{\frac{1}{9.5}\text{ row 2}}$$

$$\left[\begin{array}{cc|c} 1 & -1.5 & 6 \\ 0 & 1 & -2 \end{array}\right] \xrightarrow{1.5 \cdot \text{row 2} + \text{row 1}} \left[\begin{array}{cc|c} 1 & 0 & 3 \\ 0 & 1 & -2 \end{array}\right]$$

Thus, the solution is $(3, -2)$, which is the same as obtained using elimination.

The above procedure is called **Gauss-Jordan elimination.** You may find it easier to stop using elementary row operations at an earlier point. Many times, it is more convenient to stop when the augmented matrix is in a form like the following:

$$\left[\begin{array}{cc|c} 1 & b & k_1 \\ 0 & 1 & k_2 \end{array}\right]$$

This is called **triangular form** because when you convert it back to a system of linear equations, they have a triangular appearance:

$$\begin{aligned} x + by &= k_1 \\ y &= k_2 \end{aligned}$$

Then you can use **back-substitution** to solve for the variables.

For instance, in the above example, you would proceed as follows:

Write the augmented matrix:

$$\left[\begin{array}{cc|c} 2 & -3 & 12 \\ 5 & 2 & 11 \end{array}\right]$$

Perform appropriate elementary row operations:

$$\left[\begin{array}{cc|c} 2 & -3 & 12 \\ 5 & 2 & 11 \end{array}\right] \xrightarrow{\frac{1}{2}\text{ row 1}} \left[\begin{array}{cc|c} 1 & -1.5 & 6 \\ 5 & 2 & 11 \end{array}\right] \xrightarrow{-5\text{ row 1} + \text{row 2}} \left[\begin{array}{cc|c} 1 & -1.5 & 6 \\ 0 & 9.5 & -19 \end{array}\right] \xrightarrow{\frac{1}{9.5}\text{ row 2}} \left[\begin{array}{cc|c} 1 & -1.5 & 6 \\ 0 & 1 & -2 \end{array}\right]$$

Stop here and write the system of equations:

$$\begin{cases} x - 1.5y = 6 \\ y = -2 \end{cases}$$

Then substitute $y = -2$ back into the first equation, yielding

$$\begin{aligned} x - 1.5(-2) &= 6 \\ x + 3 &= 6 \\ x &= 3 \end{aligned}$$

Thus, the solution is $(3, -2)$, which is the same as already obtained.

6.2 LINEAR SYSTEMS IN THREE VARIABLES

The solution to a system of three equations with three variables can be solved using the methods previously shown. In general, it is efficient to solve such systems by using the augmented matrix.

- To solve the system $\begin{cases} 3x - 2y + 4z = 5 \\ 2x + 5y - 3z = 11 \\ -5x + 6y + 2x = 17 \end{cases}$ using Gauss-Jordan elimination:

Write the augmented matrix:

$$\left[\begin{array}{ccc|c} 3 & -2 & 4 & 5 \\ 2 & 5 & -3 & 11 \\ -5 & 6 & 2 & 17 \end{array}\right]$$

Perform appropriate elementary row operations:

$$\left[\begin{array}{ccc|c} 3 & -2 & 4 & 5 \\ 2 & 5 & -3 & 11 \\ -5 & 6 & 2 & 17 \end{array}\right] \xrightarrow{\frac{1}{3}\text{ row 1}} \left[\begin{array}{ccc|c} 1 & -\frac{2}{3} & \frac{4}{3} & \frac{5}{3} \\ 2 & 5 & -3 & 11 \\ -5 & 6 & 2 & 17 \end{array}\right] \xrightarrow[5\text{ row 1} + \text{row 3}]{-2\text{ row 1} + \text{row 2}} \left[\begin{array}{ccc|c} 1 & -\frac{2}{3} & \frac{4}{3} & \frac{5}{3} \\ 0 & \frac{19}{3} & -\frac{17}{3} & \frac{23}{3} \\ 0 & \frac{8}{3} & \frac{26}{3} & \frac{76}{3} \end{array}\right] \xrightarrow{\frac{3}{19}\text{ row 2}}$$

$$\begin{array}{c}\tfrac{2}{3}\text{ row 2 + row 1}\\ -\tfrac{8}{3}\text{ row 2 + row 3}\end{array} \qquad \tfrac{57}{630}\text{ row 3}$$

$$\left[\begin{array}{ccc|c} 1 & -\frac{2}{3} & \frac{4}{3} & \frac{5}{3} \\ 0 & 1 & -\frac{17}{19} & \frac{23}{19} \\ 0 & \frac{8}{3} & \frac{26}{3} & \frac{76}{3} \end{array}\right] \rightarrow \left[\begin{array}{ccc|c} 1 & 0 & \frac{42}{57} & \frac{141}{57} \\ 0 & 1 & -\frac{17}{19} & \frac{23}{19} \\ 0 & 0 & \frac{630}{57} & \frac{1,260}{57} \end{array}\right] \rightarrow$$

$$\begin{array}{c}-\tfrac{42}{57}\text{ row 3 + row 1}\\ \tfrac{17}{19}\text{ row 3 + row 2}\end{array}$$

$$\left[\begin{array}{ccc|c} 1 & 0 & \frac{42}{57} & \frac{141}{57} \\ 0 & 1 & -\frac{17}{19} & \frac{23}{19} \\ 0 & 0 & 1 & 2 \end{array}\right] \rightarrow \left[\begin{array}{ccc|c} 1 & 0 & 0 & 1 \\ 0 & 1 & 0 & 3 \\ 0 & 0 & 1 & 2 \end{array}\right]$$

Thus, the solution is $x = 1$, $y = 3$, $z = 2$.

Check the solution in each equation:

$$\begin{array}{lcrcr} 3x - 2y + 4z = 5 & \rightarrow & 3(1) - 2(3) + 4(2) \stackrel{?}{=} 5 & \rightarrow & 5 \stackrel{\checkmark}{=} 5 \\ 2x + 5y - 3z = 11 & \rightarrow & 2(1) + 5(3) - 3(2) \stackrel{?}{=} 11 & \rightarrow & 11 \stackrel{\checkmark}{=} 11 \\ -5x + 6y + 2z = 17 & \rightarrow & -5(1) + 6(3) + 2(2) \stackrel{?}{=} 17 & \rightarrow & 17 \stackrel{\checkmark}{=} 17 \end{array}$$

- To solve the system $\begin{cases} -2x + y - 3z = -8 \\ -x + 2y - z = 3 \\ 3x + 2y - 2z = -10 \end{cases}$ by reducing to triangular form:

Write the augmented matrix:

$$\left[\begin{array}{ccc|c} -2 & 1 & -3 & -8 \\ -1 & 2 & -1 & 3 \\ 3 & 2 & -2 & -10 \end{array}\right]$$

Perform appropriate elementary row operations:

$$\text{interchange row 1 and row 2} \qquad -1\text{ row 1} \qquad \begin{array}{c}2\text{ row 1 + row 2}\\ -3\text{ row 1 + row 3}\end{array}$$

$$\left[\begin{array}{ccc|c} -2 & 1 & -3 & -8 \\ -1 & 2 & -1 & 3 \\ 3 & 2 & -2 & -10 \end{array}\right] \rightarrow \left[\begin{array}{ccc|c} -1 & 2 & -1 & 3 \\ -2 & 1 & -3 & -8 \\ 3 & 2 & -2 & -10 \end{array}\right] \rightarrow \left[\begin{array}{ccc|c} 1 & -2 & 1 & -3 \\ -2 & 1 & -3 & -8 \\ 3 & 2 & -2 & -10 \end{array}\right] \rightarrow$$

$$\begin{bmatrix} 1 & -2 & 1 & | & -3 \\ 0 & -3 & -1 & | & -14 \\ 0 & 8 & -5 & | & -1 \end{bmatrix} \xrightarrow{-\frac{1}{3}\text{ row 2}} \begin{bmatrix} 1 & -2 & 1 & | & 3 \\ 0 & 1 & \frac{1}{3} & | & \frac{14}{3} \\ 0 & 8 & -5 & | & -1 \end{bmatrix} \xrightarrow{-8\text{ row 2} + \text{row 3}} \begin{bmatrix} 1 & -2 & 1 & | & -3 \\ 0 & 1 & \frac{1}{3} & | & \frac{14}{3} \\ 0 & 0 & -\frac{23}{3} & | & -\frac{115}{3} \end{bmatrix} \xrightarrow{\substack{3\text{ row 2} \\ -3\text{ row 3}}}$$

$$\begin{bmatrix} 1 & -2 & 1 & | & -3 \\ 0 & 3 & 1 & | & 14 \\ 0 & 0 & 23 & | & 115 \end{bmatrix}$$

Now, use the last matrix to return to the system of equations:

$$\begin{cases} x - 2y + z = -3 \\ \quad 3y + z = 14 \\ \qquad 23z = 115 \end{cases}$$ This is in triangular form.

From the last equation, we obtain:

$$23z = 115$$
$$z = 5$$

Substituting $z = 5$ into the second equation yields:

$$3y + 5 = 14$$
$$3y = 9$$
$$y = 3$$

Substituting $z = 5$ and $y = 3$ into the first equation yields:

$$x - 2(3) + 5 = -3$$
$$x - 6 + 5 = -3$$
$$x - 1 = -3$$
$$x = -2$$

Thus, the solution is $x = -2$, $y = 3$, $z = 5$.

Check the solution in each equation:

$$-2x + y - 3z = -8 \rightarrow -2(-2) + (3) - 3(5) \stackrel{?}{=} -8 \rightarrow -8 \stackrel{\checkmark}{=} -8$$
$$-x + 2y - z = 3 \rightarrow -(-2) + 2(3) - (5) \stackrel{?}{=} 3 \rightarrow 3 \stackrel{\checkmark}{=} 3$$
$$3x + 2y - 2z = -10 \rightarrow 3(-2) + 2(3) - 2(5) \stackrel{?}{=} -10 \rightarrow -10 \stackrel{\checkmark}{=} -10$$

A helpful message: Although the process of transforming the augmented matrix using row operations is efficient, the arithmetic involved is tedious and time-consuming. Most graphics calculators can perform elementary row operations. You will find this tool will greatly reduce the labor-intensive nature of the above procedures. As you work through the process, record the row operations as you perform them with the graphics calculator, so that if you make a mistake, you can check your work more easily.

6.3 SOLVING OTHER SYSTEMS OF EQUATIONS IN TWO VARIABLES

You can use the methods of this chapter to solve other systems of equations in two variables, such as

$$\begin{cases} 5x - y = 12 \\ y + 2x^2 = 0 \end{cases}$$

To solve this system:

Rearrange the second equation and observe that the coefficients of y sum to 0:

$$\begin{array}{r} 5x - y = 12 \\ 2x^2 + y = 0 \\ \hline \end{array}$$

Add the equations: $2x^2 + 5x = 12$

Solve the resulting quadratic equation:

$$2x^2 + 5x - 12 = 0$$

$$(2x - 3)(x + 4) = 0$$

$$2x - 3 = 0 \quad \text{or} \quad x + 4 = 0$$

$$2x = 3 \quad \text{or} \quad x = -4$$

$$x = 1.5$$

A helpful message: Do not make the mistake of stopping at this point and writing $(1.5, -4)$ as the solution. You have obtained two distinct values for x and now must obtain the corresponding y-value for each.

Substitute $x = 1.5$ and $x = -4$ into $5x - y = 12$ to find the corresponding y-value for each:

$$5(1.5) - y = 12 \quad \text{or} \quad 5(-4) - y = 12$$

$$-y = 4.5 \quad \text{or} \quad -y = 32$$

$$y = -4.5 \quad \text{or} \quad y = -32$$

Check both ordered pairs obtained:

Check $(1.5, -4.5)$:

$$5x - y = 12 \longrightarrow 5(1.5) - (-4.5) \stackrel{?}{=} 12 \longrightarrow 12 \stackrel{\checkmark}{=} 12$$

$$y + 2x^2 = 0 \longrightarrow -4.5 + 2(1.5)^2 \stackrel{?}{=} 0 \longrightarrow 0 \stackrel{\checkmark}{=} 0$$

Check $(-4, -32)$:

$$5x - y = 12 \longrightarrow 5(-4) - (-32) \stackrel{?}{=} 12 \longrightarrow 12 \stackrel{\checkmark}{=} 12$$

$$y + 2x^2 = 0 \longrightarrow -32 + 2(-4)^2 \stackrel{?}{=} 0 \longrightarrow 0 \stackrel{\checkmark}{=} 0$$

Thus, the solution set contains exactly two ordered pairs: $(1.5, -4.5)$ and $(-4, -32)$.

To help you recognize that the solution set contains two points, you should sketch a rough graph of the two equations. Rewrite them in functional form:

$$\begin{cases} y = 5x - 12 \\ y = -2x^2 \end{cases}$$

Since $y = 5x - 12$ is a linear function, its graph is a line with slope 5 and intercept -12. Since $y = -2x^2$ is a quadratic function, its graph is a parabola with vertex at the origin, opening downward. A sketch (Figure 6.5) will show that the line cuts the parabola in two points:

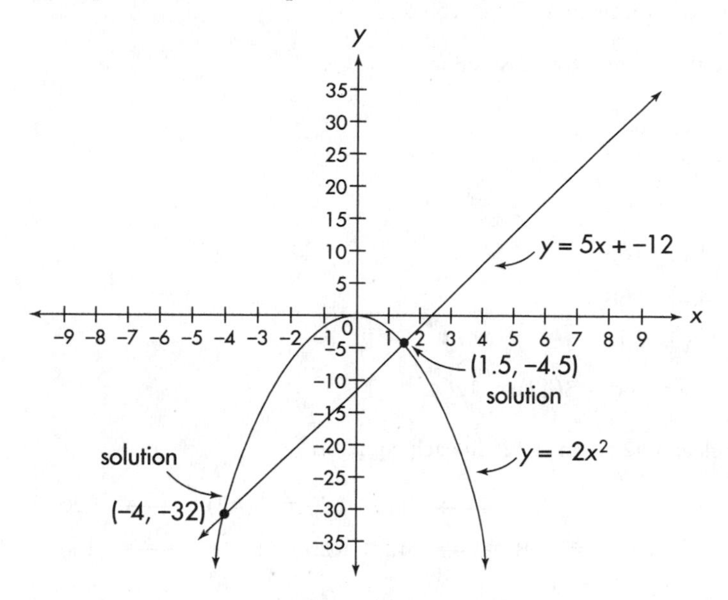

FIGURE 6.5

6.4 APPLICATIONS OF SYSTEMS OF EQUATIONS

Systems of equations can be used in problem solving when more than one unknown is involved. In order to determine a solution, you will need to have as many equations as you have variables.

Solve: One number is three times another number. The sum of the two numbers is 168. Find the numbers.

Solution: Represent all unknowns with variable expressions:

Let x = the smaller number.

Let y = the larger number.

Write the two equations:

$$\underbrace{\text{The larger number}}_{y}\ \underbrace{\text{is}}_{=}\ \underbrace{\text{three}}_{3}\ \underbrace{\text{times}}_{\cdot}\ \text{the}\ \underbrace{\text{smaller number}}_{x}.$$

First equation: $y = 3x$

$$\text{The}\ \underbrace{\text{sum of the two numbers}}_{x+y}\ \text{is 168.} \quad = 168$$

Second equation: $x + y = 168$

Solve the system (by substitution):

$$\begin{cases} y = 3x \\ x + y = 168 \end{cases}$$

$$\begin{aligned} x + y &= 168 \\ x + 3x &= 168 \\ 4x &= 168 \\ x &= 42 \\ y &= 3x = 3(42) = 126 \end{aligned}$$

Check $x = 42$ and $y = 126$ in each equation:

$$\begin{aligned} y = 3x &\longrightarrow 126 \stackrel{?}{=} 3(42) \longrightarrow 126 \stackrel{\checkmark}{=} 126 \\ x + y = 168 &\longrightarrow 42 + 126 \stackrel{?}{=} 168 \longrightarrow 168 \stackrel{\checkmark}{=} 168 \end{aligned}$$

Solution: The two numbers are 42 and 126.

Check the solution in the context of the problem:

126 is three times 42. ✓ The sum of 126 and 42 is 168. ✓

Solve: A collection of quarters and dimes amounts to $7.50. If there are 48 coins in all, find the number of each coin.

Solution:

Represent all unknowns with variable expressions:

Let $x =$ number of quarters in the collection.

Let $y =$ number of dimes in the collection.

Make a table to organize the information:

Denomination	Quarters	Dimes	Total
Value per coin	\$0.25	\$0.10	not applicable
Number of coins	x	y	48
Value (in dollars)	\$0.25(no. of quarters) = $0.25x$	\$0.10(no. of dimes) = $0.10y$	\$7.50

Write two equations based on the information in the table:

$$\underbrace{\text{Number of quarters}}_{x} + \underbrace{\text{Number of dimes}}_{y} = \underbrace{\text{Total number of coins}}_{48}$$

First equation: $x + y = 48$

$$\underbrace{\text{Total value of quarters}}_{\$0.25x} + \underbrace{\text{Total value of dimes}}_{\$0.10y} = \underbrace{\text{Total value of all the coins}}_{\$7.50}$$

Second equation: $\$0.25x + \$0.10y = \$7.50$

Solve the system (by elimination), omitting the units for convenience:

$$\begin{cases} x + y = 48 \\ 0.25x + 0.10y = 7.50 \end{cases}$$

Multiply first equation by -0.10 and add to second equation:

$$\begin{array}{rcrcl} x + y = 48 & \longrightarrow & -0.10x + -0.10y & = & -4.80 \\ 0.25x + 0.10y = 7.50 & \longrightarrow & 0.25x + \quad 0.10y & = & 7.50 \\ \hline & & 0.15x & = & 2.70 \end{array}$$

Solve for x: $x = 18$ (the no. of quarters)

Solve for y:

$x + y = 48 \rightarrow 18 + y = 48 \rightarrow y = 48 - 18 \rightarrow y = 30$ (the no. of dimes)

Check $x = 18$ and $y = 30$ in each equation:

$x + y = 48 \rightarrow 18 + 30 \stackrel{?}{=} 48 \rightarrow 48 \stackrel{\checkmark}{=} 48$

$\$0.25x + \$0.10y = \$7.50 \rightarrow \$0.25(18) + \$0.10(30) \stackrel{?}{=} \$7.50 \rightarrow \$7.50 \stackrel{\checkmark}{=} 7.50$

Solution: There are 18 quarters and 30 dimes in the collection.

Check the solution in the context of the problem:

The value of 18 quarters is \$0.25 times 18. This is \$4.50;

the value of 30 dimes is \$0.10 times 30. This is \$3.00.

The collection of quarters and dimes amounts to \$4.50 plus \$3.00. This is \$7.50. ✓

18 quarters and 30 dimes equals 48 coins in all. ✓

Or check the solution in the table:

Denomination	Quarters	Dimes	Total
Value per coin	\$0.25	\$0.10	not applicable
Number of coins	18	30	18 + 30 = 48 ✓
Value	\$0.25(18) = \$4.50	\$0.10(30) = \$3.00	\$4.50 + \$3.00 = \$7.50 ✓

Solve: The length of a rectangle is 3 meters more than its width. If the area is 10 square meters, find the dimensions of the rectangle.

(Note: See Chapter 4 for a solution using one variable).

Solution:

Formula needed:

The area of a rectangle with length l and width w is $A = lw$.

Represent all unknowns with variable expressions:

Let l = the length of the rectangle (in meters).

Let w = the width of the rectangle (in meters).

Make a diagram:

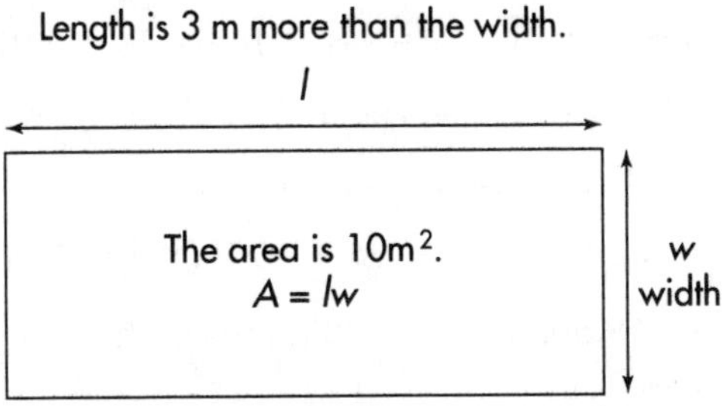

Write two equations using the information in the diagram:

$$\underbrace{\text{The length}}_{l} \ \underbrace{\text{is}}_{=} \ \underbrace{\text{3 meters}}_{\text{3 meters}} \ \underbrace{\text{more than}}_{+} \ \underbrace{\text{the width}}_{w}.$$

First equation: $l = 3 + w$ meters

$$\underbrace{\text{The area}}_{A} \text{ is } 10 \text{ m}^2. \quad A = 10 \text{ m}^2$$

Second equation: $lw = 10 \text{ m}^2$

Solve the system (by substitution), omitting the units for convenience:

$$\begin{cases} l = 3 + w \\ lw = 10 \end{cases}$$

$$lw = 10$$
$$(3 + w)w = 10$$
$$3w + w^2 = 10$$
$$w^2 + 3w - 10 = 0$$
$$(w - 2)(w + 5) = 10$$

$w = 2$ meters or $w = -5$ meters (reject because width should not be negative)

Solve for l: $l = 3 + w$ meters $\longrightarrow$ 3 meters + 2 meters = 5 meters

Check $l = 5$ m and $y = 2$ m in each equation:

$l = 3 \text{ m} + w \longrightarrow 5 \text{ m} \stackrel{?}{=} 3 \text{ m} + 2 \text{ m} \longrightarrow 5 \text{ m} \stackrel{\checkmark}{=} 5 \text{ m}$

$lw = 10 \text{ m}^2 \longrightarrow (5 \text{ m})(2 \text{ m}) \stackrel{?}{=} 10 \text{ m}^2 \longrightarrow 10 \text{ m}^2 \stackrel{\checkmark}{=} 10 \text{ m}^2$

Solution: The rectangle has dimensions 5 meters by 2 meters.

Check the solution in the context of the problem:

Three meters more than the width is 3 meters plus 2 meters, which is 5 meters, the same as the length of the rectangle. ✓

The area of the rectangle is 5 meters times 2 meters, which is 10 m^2. ✓

Or check the diagram:

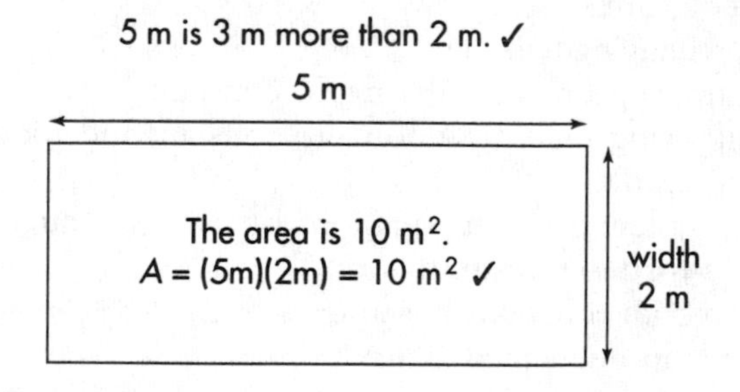

A helpful message: The methods in this chapter can be used to solve any system of linear equations: four equations with four variables, five equations with five variables, and so forth. However, the arithmetic becomes rather tedious, so you should probably use a programmable calculator or a computer when confronted with such a problem.

CHAPTER 6 SUMMARY

6.1 A **system of two linear equations in two variables** consists of a pair of linear equations in the same variables. To **solve a system** of linear equations in two variables means to find all ordered pairs of the variables that make both equations true simultaneously. The system is **consistent** if it has *one or more solutions;* the system is **inconsistent** if it has *no solution;* and the system is **consistent** and **dependent** if it has *infinitely many solutions.* The **solution set** is the collection of all solutions.

Three methods commonly used to solve a system of linear equations are **substitution, elimination,** and **transformation** of the augmented matrix.

To solve a system of linear equations by using **substitution:**

1. Select the simpler equation and express one of the variables in terms of the other.
2. Substitute this expression into the other equation, simplify, and solve, if possible.
3. Check the solution in the original equations.

To solve a system of linear equations by using **elimination:**

1. Write both equations in standard form: $ax + by = c$.
2. If necessary, multiply one or both of the equations by a nonzero constant or constants to make the coefficients of one of the variables sum to zero.
3. Add the equations.
4. Check the solution in the original equations.

The rectangular array composed of the numbers that appear in a system is called the **augmented matrix** for the system. The horizontal lines of numbers are called **rows** and the vertical lines of numbers are called **columns.** The following row operations may be used to transform the augmented matrix:

1. Interchanging two rows
2. Multiplying a row by a nonzero constant
3. Multiplying a row by a nonzero constant and adding the result to another row

To solve a system of linear equations by using the **augmented matrix** and **Gauss-Jordan elimination:**

1. Write both equations in standard form: $ax + by = c$.
2. Write the augmented matrix.

3. Use row operations to transform the matrix into the form:

$$\left[\begin{array}{cc|c} 1 & 0 & x_0 \\ 0 & 1 & y_0 \end{array}\right]$$

The solution is $x = x_0$ and $y = y_0$, when the system is consistent.

4. Check the solution in the original equations.

You may find it easier to stop using elementary row operations on the augmented matrix when it is in **triangular form.** The corresponding system of equations can be solved using **back-substitution.**

6.2 The solution to a **system of three equations with three variables** can be solved using substitution, elimination, or transformation of the augmented matrix. In general, it is efficient to solve such systems by using the augmented matrix.

6.3 The methods of this chapter can be used to solve other systems of equations in two variables.

6.4 Systems of equations can be used in problem solving when more than one unknown is involved. In order to determine a solution, you will need to have as many equations as you have variables.

PRACTICE PROBLEMS FOR CHAPTER 6

Solve the following systems using any method:

1. $\begin{cases} 2x + y = 6 \\ x - y = -3 \end{cases}$

2. $\begin{cases} 4x - 8y = 4 \\ 3x - 6y = 3 \end{cases}$

3. $\begin{cases} -2t + 3u = 1 \\ 3t - 7u = 6 \end{cases}$

4. $\begin{cases} 2x + 5y = 1 \\ 3x + 4y = -9 \end{cases}$

5. $\begin{cases} x + 3y = 4 \\ 2x + 6y = 10 \end{cases}$

6. $\begin{cases} 0.08x + 0.06y = 678 \\ y = x + 1{,}500 \end{cases}$

7. $\begin{cases} 2l + 2w = 20 \\ lw = 24 \end{cases}$

8. $\begin{cases} 2x + 3y - 4z = 3 \\ 3x - 4y + z = -7 \\ -5x + 2y + 4z = 9 \end{cases}$

9. Mrs. Lazarine has two investments, one yielding 8% annual simple interest and the other yielding 6% annual simple interest. The total annual yield from the two investments is $678. If she invested $1,500 more at 6% than at 8%, how much is invested at each rate?

10. The perimeter of a rectangular walk-in closet is 20 feet. Its area is 24 square feet. Find the dimensions of the closet.

SOLUTIONS TO PRACTICE PROBLEMS

1. $\begin{cases} 2x + y = 6 \\ x - y = -3 \end{cases}$

To solve $\begin{cases} 2x + y = 6 \\ x - y = -3 \end{cases}$ by using elimination:

Add the equations:

$$\begin{array}{r} 2x + y = 6 \\ \underline{x - y = -3} \\ 3x = 3 \\ x = 1 \end{array}$$

Solve for x: $3x = 3$, $x = 1$

Solve for y: $x - y = -3 \longrightarrow 1 - y = -3 \longrightarrow -y = -4 \longrightarrow y = 4$

Check:

$$2x + y = 6 \longrightarrow 2(1) + 4 \stackrel{?}{=} 6 \longrightarrow 6 \stackrel{\checkmark}{=} 6$$

$$x - y = -3 \longrightarrow 1 - 4 \stackrel{?}{=} -3 \longrightarrow -3 \stackrel{\checkmark}{=} -3$$

Thus, the solution is $x = 1$ and $y = 4$.

2. $\begin{cases} 4x - 8y = 4 \\ 3x - 6y = 3 \end{cases}$

To solve $\begin{cases} 4x - 8y = 4 \\ 3x - 6y = 3 \end{cases}$ by using the augmented matrix:

Write the augmented matrix:

$$\left[\begin{array}{cc|c} 4 & -8 & 4 \\ 3 & -6 & 3 \end{array}\right]$$

Perform appropriate elementary row operations:

$$\left[\begin{array}{cc|c} 4 & -8 & 4 \\ 3 & -6 & 3 \end{array}\right] \xrightarrow{\frac{1}{4}\text{ row 1}} \left[\begin{array}{cc|c} 1 & -2 & 1 \\ 3 & -6 & 3 \end{array}\right] \xrightarrow{-3\text{ row 1} + \text{row 2}} \left[\begin{array}{cc|c} 1 & -2 & 1 \\ 0 & 0 & 0 \end{array}\right]$$

all zeros

Since row 2 is all zeros, the system is dependent; that is, it has infinitely many solutions.

3. $\begin{cases} -2t + 3u = 1 \\ 3t - 7u = 6 \end{cases}$

To solve $\begin{cases} -2t + 3u = 1 \\ 3t - 7u = 6 \end{cases}$ by using the augmented matrix:

Write the augmented matrix:

$$\left[\begin{array}{rr|r} -2 & 3 & 1 \\ 3 & -7 & 6 \end{array}\right]$$

Perform appropriate elementary row operations:

$$\left[\begin{array}{rr|r} -2 & 3 & 1 \\ 3 & -7 & 6 \end{array}\right] \xrightarrow{-\frac{1}{2}\text{ row 1}} \left[\begin{array}{rr|r} 1 & -1.5 & -0.5 \\ 3 & -7 & 6 \end{array}\right] \xrightarrow{-3\text{ row 1} + \text{row 2}} \left[\begin{array}{rr|r} 1 & -1.5 & -0.5 \\ 0 & -2.5 & 7.5 \end{array}\right] \xrightarrow{-\frac{1}{2.5}\text{ row 2}}$$

$$\left[\begin{array}{rr|r} 1 & -1.5 & -0.5 \\ 0 & 1 & -3 \end{array}\right] \xrightarrow{1.5\text{ row 2} + \text{row 1}} \left[\begin{array}{rr|r} 1 & 0 & -5 \\ 0 & 1 & -3 \end{array}\right]$$

Thus, the solution is $t = -5$, $u = -3$.

Check:

$$-2t + 3u = 1 \longrightarrow -2(-5) + 3(-3) \stackrel{?}{=} 1 \longrightarrow 1 \stackrel{\checkmark}{=} 1$$

$$3t - 7u = 6 \longrightarrow 3(-5) - 7(-3) \stackrel{?}{=} 6 \longrightarrow 6 \stackrel{\checkmark}{=} 6$$

4. $\begin{cases} 2x + 5y = 1 \\ 3x + 4y = -9 \end{cases}$

To solve $\begin{cases} 2x + 5y = 1 \\ 3x + 4y = -9 \end{cases}$ by using the augmented matrix:

Write the augmented matrix:

$$\left[\begin{array}{rr|r} 2 & 5 & 1 \\ 3 & 4 & -9 \end{array}\right]$$

Perform appropriate elementary row operations:

$$\left[\begin{array}{rr|r} 2 & 5 & 1 \\ 3 & 4 & -9 \end{array}\right] \xrightarrow{\frac{1}{2}\text{ row 1}} \left[\begin{array}{rr|r} 1 & 2.5 & 0.5 \\ 3 & 4 & -9 \end{array}\right] \xrightarrow{-3\text{ row 1} + \text{row 2}} \left[\begin{array}{rr|r} 1 & 2.5 & 0.5 \\ 0 & -3.5 & -10.5 \end{array}\right] \xrightarrow{-\frac{1}{3.5}\text{ row 2}}$$

$$\xrightarrow{-2.5 \text{ row } 2 + \text{row } 1}$$

$$\left[\begin{array}{cc|c} 1 & 2.5 & 0.5 \\ 0 & 1 & 3 \end{array}\right] \longrightarrow \left[\begin{array}{cc|c} 1 & 0 & -7 \\ 0 & 1 & 3 \end{array}\right]$$

Thus, the solution is $x = -7$, $y = 3$.

Check:

$$2x + 5y = 1 \longrightarrow 2(-7) + 5(3) \stackrel{?}{=} 1 \longrightarrow 1 \stackrel{\checkmark}{=} 1$$

$$3x + 4y = -9 \longrightarrow 3(-7) + 4(3) \stackrel{?}{=} -9 \longrightarrow -9 \stackrel{\checkmark}{=} -9$$

5. $\begin{cases} x + 3y = 4 \\ 2x + 6y = 10 \end{cases}$

To solve $\begin{cases} x + 3y = 4 \\ 2x + 6y = 10 \end{cases}$ by using the augmented matrix:

Write the augmented matrix:

$$\left[\begin{array}{cc|c} 1 & 3 & 4 \\ 2 & 6 & 10 \end{array}\right]$$

Perform appropriate elementary row operations:

$$-2 \text{ row } 1 + \text{row } 2$$

$$\left[\begin{array}{cc|c} 1 & 3 & 4 \\ 2 & 6 & 10 \end{array}\right] \longrightarrow \left[\begin{array}{cc|c} 1 & 3 & 4 \\ 0 & 0 & 2 \end{array}\right]$$

zeros nonzero constant

Since row 2 is all zeros except for the last number, the system is inconsistent; that is, it has no solution.

6. $\begin{cases} 0.08x + 0.06y = 678 \\ y = x + 1{,}500 \end{cases}$

To solve $\begin{cases} 0.08x + 0.06y = 678 \\ y = x + 1{,}500 \end{cases}$ by using substitution:

Substitute $y = x + 1{,}500$ into $0.08x + 0.06y = 678$ and solve:

$$0.08x + 0.06y = 678$$

$$0.08x + 0.06(x + 1{,}500) = 678$$

$$0.08x + 0.06x + 90 = 678$$

$$0.14x = 588$$

$$x = 4{,}200$$

Solve for y: $y = x + 1{,}500 \longrightarrow y = 4{,}200 + 1500 \longrightarrow y = 5{,}700$

Check:

$$0.08x + 0.06y = 678 \longrightarrow 0.08(4{,}200) + 0.06(5{,}700) \stackrel{?}{=} 678 \longrightarrow 678 \stackrel{\surd}{=} 678$$

$$y = x + 1{,}500 \longrightarrow 5{,}700 \stackrel{?}{=} 4{,}200 + 1{,}500 \longrightarrow 5{,}700 \stackrel{\surd}{=} 5{,}700$$

Thus, the solution is $x = 1{,}500$ and $y = 3{,}000$.

7. $\begin{cases} 2l + 2w = 20 \\ lw = 24 \end{cases}$

To solve $\begin{cases} 2l + 2w = 20 \\ lw = 24 \end{cases}$ by using substitution:

Solve the 1st equation for w:

$$2l + 2w = 20$$
$$w = 10 - l$$

Substitute $w = 10 - l$ into $lw = 24$ and solve:

$$lw = 24$$
$$l(10 - l) = 24$$
$$10l - l^2 = 24$$
$$l^2 - 10l + 24 = 0$$
$$(l - 4)(l - 6) = 0$$

Solve for l: $l = 4$ or $l = 6$

Solve for the corresponding w values:

When $l = 4 \longrightarrow w = 10 - l = 10 - 4 = 6$

When $l = 6 \longrightarrow w = 10 - l = 10 - 6 = 4$

Check both ordered pairs obtained:

Check (4, 6):

$$w = 10 - l \longrightarrow 6 \stackrel{?}{=} 10 - 4 \longrightarrow 6 \stackrel{\surd}{=} 6$$

$$lw = 24 \longrightarrow (4)(6) \stackrel{?}{=} 24 \longrightarrow 24 \stackrel{\surd}{=} 24$$

Check (6, 4):

$$w = 10 - l \longrightarrow 4 \stackrel{?}{=} 10 - 6 \longrightarrow 4 \stackrel{\surd}{=} 4$$

$$lw = 24 \longrightarrow (6)(4) \stackrel{?}{=} 24 \longrightarrow 24 \stackrel{\surd}{=} 24$$

Thus, the solution set contains exactly two ordered pairs: (4, 6) and (6, 4).

8. $\begin{cases} 2x + 3y - 4z = 3 \\ 3x - 4y + z = -7 \\ -5x + 2y + 4z = 9 \end{cases}$

$$\text{To solve} \begin{cases} 2x + 3y - 4z = 3 \\ 3x - 4y + z = -7 \\ -5x + 2y + 4z = 9 \end{cases}$$

Write the augmented matrix:

$$\begin{bmatrix} 2 & 3 & -4 & 3 \\ 3 & -4 & 1 & -7 \\ -5 & 2 & 4 & 9 \end{bmatrix}$$

Perform appropriate elementary row operations:

$\frac{1}{2}$ row 1 -3 row 1 + row 2, 5 row 1 + row 3 $-\frac{1}{8.5}$ row 2

$$\left[\begin{array}{ccc|c} 2 & 3 & -4 & 3 \\ 3 & -4 & 1 & -7 \\ -5 & 2 & 4 & 9 \end{array}\right] \rightarrow \left[\begin{array}{ccc|c} 1 & 1.5 & -2 & 1.5 \\ 3 & -4 & 1 & -7 \\ -5 & 2 & 4 & 9 \end{array}\right] \rightarrow \left[\begin{array}{ccc|c} 1 & 1.5 & -2 & 1.5 \\ 0 & -8.5 & 7 & -11.5 \\ 0 & 9.5 & -6 & 16.5 \end{array}\right] \rightarrow$$

-1.5 row 2 + row 1, -9.5 row 2 + row 3 $\frac{17}{31}$ row 3 $\frac{13}{17}$ row 3 + row 1, $\frac{14}{17}$ row 3 + row 2

$$\left[\begin{array}{ccc|c} 1 & 1.5 & -2 & 1.5 \\ 0 & 1 & -\frac{14}{17} & \frac{23}{17} \\ 0 & 9.5 & -6 & 16.5 \end{array}\right] \rightarrow \left[\begin{array}{ccc|c} 1 & 0 & -\frac{13}{17} & -\frac{9}{17} \\ 0 & 1 & -\frac{14}{17} & \frac{23}{17} \\ 0 & 0 & \frac{31}{17} & \frac{62}{17} \end{array}\right] \rightarrow \left[\begin{array}{ccc|c} 1 & 0 & -\frac{13}{17} & -\frac{9}{17} \\ 0 & 1 & -\frac{14}{17} & \frac{23}{17} \\ 0 & 0 & 1 & 2 \end{array}\right] \rightarrow$$

$$\left[\begin{array}{ccc|c} 1 & 0 & 0 & 1 \\ 0 & 1 & 0 & 3 \\ 0 & 0 & 1 & 2 \end{array}\right]$$

Thus, the solution is $x = 1$, $y = 3$, $z = 2$.

Check:

$$2x + 3y - 4z = 3 \rightarrow 2(1) + 3(3) - 4(2) \stackrel{?}{=} 3 \rightarrow 3 \stackrel{\surd}{=} 3$$
$$3x - 4y + z = -7 \rightarrow 3(1) - 4(3) + 2 \stackrel{?}{=} -7 \rightarrow -7 \stackrel{\surd}{=} -7$$
$$-5x + 2y + 4z = 9 \rightarrow -5(1) + 2(3) + 4(2) \stackrel{?}{=} 9 \rightarrow 9 \stackrel{\surd}{=} 9$$

9. Mrs. Lazarine has two investments, one yielding 8% annual simple interest and the other yielding 6% annual simple interest. The total annual yield from the two investments is $678. If she invested $1,500 more at 6% than at 8%, how much is invested at each rate?

Let x = amount invested at 8%.
Let y = amount invested at 6%.

Make a table to organize the facts:

	8% Investment	**6% Investment**	**Total**
Amount of investment	x	$y = \$1,500 + x$	$x + y$
Simple interest rate	8% = 0.08	6% = 0.06	not needed
Time invested	1 year	1 year	1 year
Annual interest earned	$0.08x$	$0.06y$	$678

Write two equations using the information in the table:

First equation: $0.08x + 0.06y = \$678$
Second equation: $y = \$1,500 + x$

Solve the system, omitting the units for convenience:

$$\begin{cases} 0.08x + 0.06y = 678 \\ y = 1,500 + x \end{cases}$$

From Problem 6, the solution to the system yields $x = \$4,200$ and $y = \$5,700$.

Solution: The amount invested at 8% is $4,200, and the amount invested at 6% is $5,700.

Check the solution in the context of the problem:

Mrs. Lazarine has two investments, one yielding 8% annual simple interest and the other yielding 6% annual simple interest. The total annual yield from the two investments is $678. If she invested $1,500 more at 6% than at 8%, how much is invested at each rate?

The $4,200 invested for 1 year earns 8% of $4,200, which is $336. The $5,700 invested for 1 year earns 6% of $5,700, which is $342. The total annual return on the two investments is $336 plus $342, which is $678. ✓ If you add $1,500 to $4,200, the amount invested at 8%, you obtain $5,700, which is the amount invested at 6%. ✓

Check the table:

	8% Investment	**6% Investment**	**Total**
Amount of investment	\$4,200	\$5,700 = \$1,500 + \$4,200 ✓	not needed
Simple interest rate	8% = 0.08	6% = 0.06	Do not add rates unless they have the *same* base.
Time invested	1 year	1 year	1 year
Annual interest earned	0.08(\$4,200) = \$336	0.06(\$5,700) = \$342	\$336 + \$342 = \$678 ✓

10. The perimeter of a rectangular walk-in closet is 20 feet. Its area is 24 square feet. Find the dimensions of the closet.

Formula needed:

The perimeter of a rectangle with length l and width w is $P = 2l + 2w$.

Represent all unknowns with variables expressions:

Let l = the length of the closet (in feet).
Let w = the width of the closet (in feet).

Draw and label a diagram using the information in the problem:

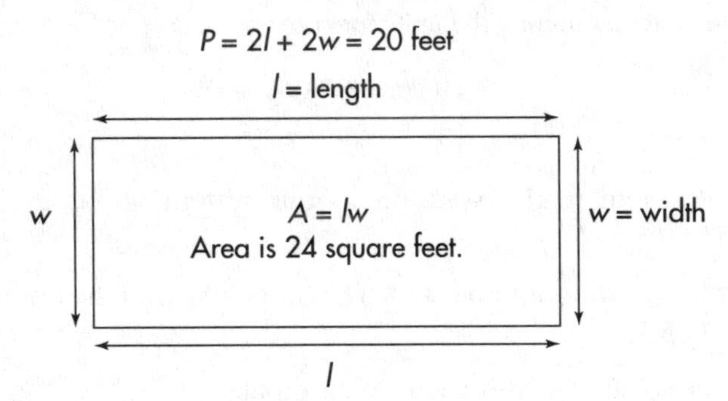

Write two equations using the diagram:

First equation: $2l + 2w = 20$ ft
Second equation: $lw = 24$ ft^2

Solve the system, omitting the units for convenience:

$$\begin{cases} 2l + 2w = 20 \\ lw = 24 \end{cases}$$

From Problem 7, the solution to this system of equations yields

$$l = 4 \text{ ft},\ w = 6 \text{ ft} \quad \text{or} \quad l = 6 \text{ ft},\ w = 4 \text{ ft}$$

Solution: The closet has dimensions 6 feet by 4 feet.

Check the solution in the context of the problem:

The perimeter of a rectangular walk-in closet is 20 feet.
Its area is 24 square feet. Find the dimensions of the closet.
The closet has dimensions 6 feet by 4 feet. The perimeter is 2 times 6 feet plus 2 times 4 feet, which is 20 feet. ✓ Its area is 6 feet times 4 feet, which is 24 square feet. ✓

Check the solution in the diagram:

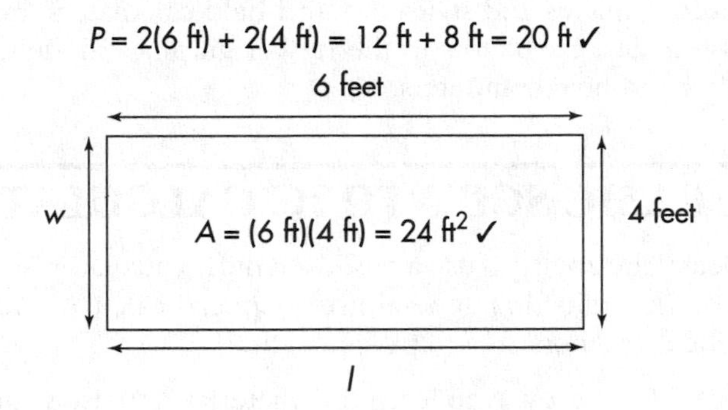

APPENDIX A CALCULATOR USE

There are various makes and styles of hand-held calculators, ranging from nonscientific 4-function calculators to the newer graphics models, which are really like small, hand-held computers.

USING A BASIC SCIENTIFIC CALCULATOR

At the very least, the ability to use a basic scientific calculator is essential for work in algebra. The following general list of special calculator features and keys for scientific calculators should prove useful.

A helpful message: The best way to learn this material is to use your calculator as you read through this appendix. Also, since key symbols may vary from calculator to calculator, use the manufacturer's booklet provided with your calculator to ensure that you are using the functions correctly.

Algebraic Logic

One of the main differences in hand-held calculators is the order in which operations are performed. Most use either algebraic logic or chain logic. When you key a sequence such as

$$40 - 2 \times 9 =$$

into an algebraic-logic calculator, the calculations are *not* necessarily executed in the order in which you entered them. What the calculator does is execute the operations in order, according to the **algebraic hierarchy of operations.** For the above example:

$$\text{Multiplication first: } 2 \times 9 = 18$$

$$\text{then subtraction: } 40 - 18 = 22$$

The same sequence entered into a chain-logic calculator would be executed as follows:

$$\text{Subtraction first: } 40 - 2 = 38;$$

$$\text{then multiplication: } 38 \times 9 = 342$$

In chain-logic calculators, the operations are executed in the order in which they were entered. As you can see, there is quite a difference in the results! Many students are unaware that calculators differ this way. Business majors may be surprised to learn that, frequently, business analyst calculators are designed to have chain logic. What is important is that you know how your calculator works, so you can use it effectively. *For an algebra course you should use a calculator that follows the algebraic hierarchy of operations.*

Scientific Notation

When the result of a calculation is too large or too small to be displayed, the calculator will show the result as a number with a small raised exponent to the upper right. For example, if you multiply 1,234,567 times 7,654,321, the calculator displays 9.449772^{12}. This represents scientific notation; it means 9.449772 times 10^{12}. If you multiply 0.000012 times 0.000021, the calculator displays 2.52^{-10}, which means 2.52 times 10^{-10}.

Special Keys and Their Functions

Key	Function
[+], [−], [×], [÷] Add, Subtract, Multiply, and Divide keys	Each performs the indicated operation and follows the algebraic hierarchy of operations when more than one operation is used in a calculation.
[+/−] Change Sign key	Changes the sign of the nonzero value in the display.
[(] and [)] Parentheses keys	Used to group calculations, usually for the purpose of overriding the order of operations that would occur otherwise. Multiple sets of parentheses may be used, but *must always occur in left-right pairs*.
[2nd] Shift key	Allows keys to have more than one purpose. The function printed on each key is its primary function, which you can use by simply pressing the key. If you want the secondary function, you must press [2nd], followed immediately by the function key.
[INV] Inverse key	Also allows keys to have two purposes. Pressing [INV] before certain function keys results in the inverse of the function.
[CLR] Clear key	The [CLR] key clears the calculator for a new set of operations. Pressing [CLR] returns the display to zero.
[CE] Clear Entry key	Enables you to correct any error you might have made in the current entry *without* affecting any result you have already calculated as well as any operations you may have started.

STO or MEM Store to Memory key	Instructs the calculator to put the value that is currently in display into the memory location *in place of any value that might be there already*. Some calculators have multiple memory locations, so you may have to press STO followed by a number to indicate which memory location you want to use.
RCL or MR Recall Memory key	Instructs the calculator to display the value currently in memory. If your calculator has multiple memories, you will have to press RCL followed by the number that corresponds to the memory location you want to access.
SUM or M+ Add to Memory key	Instructs the calculator to add the number in display to the contents in the memory location. Again you may have to specify a specific memory location by keying a number after pressing SUM.
$1/x$ Reciprocal key	Calculates the reciprocal of the number in the display.
x^2 Squaring key	Calculates the square of the number in the display. May be the secondary function on the $\sqrt{x}$ key, in which case x^2 is accessed using 2nd $\sqrt{x}$ or INV $\sqrt{x}$.
$\sqrt{x}$ Square Root key	Calculates the *principal* square root of the number in the display. May be the secondary function on the x^2 key, in which case $\sqrt{x}$ is accessed using 2nd x^2 or INV x^2.
y^x Exponentiation and Root key	Raises a number y to the x power or root. Key the number to be raised to the power, then y^x, followed by the power or root desired. Examples: • 3 y^x 4 = 81 • 81 y^x .25 = 3
LOG Common Logarithm key	Calculates the logarithm to the base 10 of the number in display. May be the secondary function on the 10^x key, in which case LOG is accessed using 2nd 10^x or INV 10^x.

10^x Powers of 10 key	Raises 10 to the power of the number in the display. May be the secondary function on the LOG key, in which case 10^x is accessed using 2nd LOG or INV LOG.
LN Common Logarithm key	Calculates the logarithm to the base *e* of the number in display. May be the secondary function on the e^x key, in which case LN is accessed using 2nd e^x or INV e^x.
e^x Powers of *e* key	Raises *e* to the power of the number in the display. May be the secondary function on the LN key, in which case e^x is accessed using 2nd LN or INV LN.
EE Scientific Notation key	Used to enter a number times a power of 10, up to a power of 99. Enter the number, press EE, then enter the power of 10 desired.
MODE Mode key	Use varies from calculator to calculator. Usually, will allow you to change from degree to radian mode when working with trigonometric functions.
SIN, COS, TAN Trigonometric Sine, Cosine, and Tangent keys	Each calculates the indicated trigonometric function of the number (in degrees or radians) in display. You can set your calculator to accept degrees or radians using the MODE key.
SIN^{-1}, COS^{-1}, TAN^{-1} Inverse Trigonometric Sine, Cosine, and Tangent keys	Each determines the angle (in degrees or radians) whose indicated trigonometric function results in the number in display. Usually, these are the secondary functions on the corresponding trigonometric function keys, so they are accessed using the 2nd or INV key.

USING A GRAPHING CALCULATOR

A graphing calculator, although not necessary (unless required by the instructor) in most college algebra classes, can be used to help you visualize concepts and problem situations and can also facilitate arithmetic computations. It can be used for observing the effects various parameters have on graphs of relationships and functions, for determining the zeros of higher-order polynomials and to make general statements about their graphs, for performing operations with matrices, and for doing various other algebraic tasks.

USING OTHER TECHNOLOGY

Other tools that can be used for algebraic manipulations and problem solving are computer algebra and calculus systems such as *Mathematica, Derive, Maple,* and *The Mathematics Exploration Toolkit (MET)*.

A helpful message: Calculators and computers can do computations for you, but they cannot *think* for you. They cannot tell you how to put the numbers in the right place in an equation or formula. *You* have to do the thinking. Algebra involves thinking carefully and deliberately about what you are doing and why you are doing it.

APPENDIX B MEASUREMENT UNITS AND CONVERSIONS

LENGTH

English System

1 foot (ft) = 12 inches (in.)
1 yard (yd) = 36 in.
1 mile (mi) = 1,760 yd
= 5,280 ft

Metric System

1 centimeter (cm) = 10 millimeters (mm)
1 decimeter (dm) = 10 cm
1 meter (m) = 10 dm
= 100 cm
= 1,000 mm
1 kilometer (km) = 1,000 m

AREA

English System

1 ft^2 = 144 $in.^2$
1 yd^2 = 9 ft^2
= 1,296 $in.^2$
1 acre = 4,840 yd^2
= 43,560 ft^2
1 mi^2 = 640 acres

Metric System

1 cm^2 = 100 mm^2
1 dm^2 = 100 cm^2
1 m^2 = 100 dm^2
1 m^2 = 10,000 cm^2
1 m^2 = 1,000,000 mm^2
1 km^2 = 1,000,000 m^2

VOLUME AND CAPACITY

English System

1 ft^3 = 1,728 $in.^3$
1 yd^3 = 27 ft^3
= 46,656 $in.^3$
1 cup = 8 fluid ounces (oz)
1 pint (pt) = 2 cups
= 16 oz
1 qt = 2 pt
= 32 oz
1 gallon (gal) = 4 qt
= 8 pt
= 128 oz

Metric System

1 cm^3 (cc) = 1,000 m^3
1 dm^3 = 1,000 cc
1 m^3 = 1,000 dm^3
= 1,000,000 cc
= 1,000,000,000 mm^3
1 km^3 = 1,000,000,000 m^3
1 liter (L) = 1,000 cc
= 1,000 milliliters (ml)
= 100 centiliters (cl)
1 ml = 1 cc
1 kiloliter (kl) = 1,000 L

WEIGHT

English System

1 pound (lb) = 16 ounces (oz)
1 ton (T) = 2,000 lb

Metric System

1 gram (g) = 1,000 milligrams (mg)
= 100 centigrams (cg)

= 10 decigrams (dg)
1 kilogram (kg) = 1,000 grams
1 g = weight of 1 cc (1 ml) of water at 4° C

TIME

1 minute (min) = 60 seconds (s)
1 hour (hr) = 60 min
= 3,600 s
1 day = 24 hr
1 week = 7 days
1 month = 30 days (for ordinary accounting)
1 year (yr) = 12 months
= 52 weeks
= 365 days
1 leap year = 366 days
1 decade = 10 yr
1 century = 10 decades
= 100 yr

APPROXIMATE EQUIVALENTS

English to Metric	*Metric to English*
1 in. = 2.54 cm (exactly)	1 cm ≅ 0.3937 in.
1 ft = 30.48 cm	1 m ≅ 39.37 in.
= 0.3048 m	≅ 3.281 ft
1 yd = 91.44 cm	≅ 1.094 yd
= 0.9144 m	
1 mi ≅ 1.609 km	1 km ≅ 0.6214 mi
1 oz ≅ 28.349 g	1 g ≅ 0.035 oz
1 lb ≅ 454 g	
≅ 0.454 kg	1 kg ≅ 2.205 lb
1 ton ≅ 907.18 kg	1,000 kg ≅ 1.1 tons
1 oz ≅ 29.573 ml (cc)	
1 cup ≅ 237 ml (cc)	
1 qt ≅ 0.946 L	1 L ≅ 1.057 qt
1 gal ≅ 3.785 L	

Other Approximate Equivalents

1 ft^3 ≅ 7.48 gal	0° C = 32° F freezing point of water
62.4 lb ≅ weight of 1 ft^3 of water	100° C = 212° F boiling point of water

CONVERSION FACTORS FOR METRIC SYSTEM

*Kilo*unit = 1,000 units
*Hecto*unit = 100 units
*Deca*unit = 10 units
*Deci*unit = 0.1 unit
*Centi*unit = 0.01 unit
*Milli*unit = 0.001 unit

APPENDIX C
COMMON FORMULAS

Temperature

$F = \frac{9}{5}C + 32$ F = degrees Fahrenheit

$C = \frac{5}{9}(F - 32)$ C = degrees Centigrade

Percentage

$P = RB$

P = percentage
R = rate
B = base

Business Formulas

Simple Interest

$I = Prt$

I = simple interest earned
P = principal invested or present value of S
r = annual simple interest rate
t = time in years

$S = P(1 + rt)$ S = maturity value of P

Compound Interest

$S = P\left(1 + \frac{r}{m}\right)^{mt}$

S = maturity value, or the compound amount of P

P = original principal, or the present value of S

r = stated annual percentage rate

m = number of compoundings per year

t = time in years

$S - P$ = compound interest annuity

Ordinary Simple Annuity

$$S = R\left[\frac{(1+i)^n - 1}{i}\right]$$

S = amount of the annuity

R = periodic payment

r = stated annual percentage rate

m = number of payments (compoundings) per year

$i = r/m$ = interest rate per compounding period

n = total number of payments

Rn = total of payments

$S - Rn$ = interest earned

Amortization

$$A = R\left[\frac{1 - (1+i)^{-n}}{i}\right]$$

A = amount financed

R = periodic payment

r = stated annual percentage rate

m = number of payments (compoundings) per year

$i = r/m$ = interest rate per compounding period

n = total number of payments

Rn = total of payments

$Rn - A$ = interest paid

Distance

$d = rt$

d = distance traveled

r = (uniform) rate of speed

t = time

Geometry Formulas

Triangle

$$\text{area} = \frac{1}{2}bh$$

$$\text{area} = \sqrt{s(s-a)(s-b)(s-c)},$$

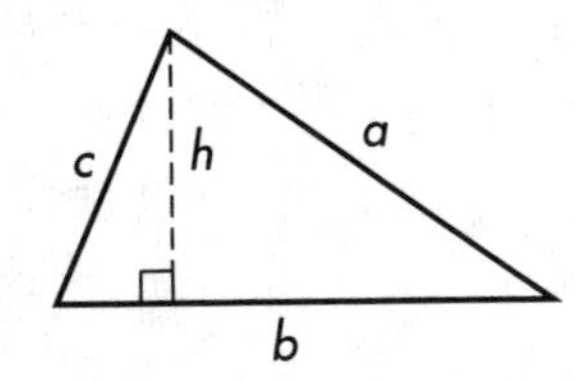

where $s = \dfrac{a + b + c}{2}$

perimeter $= a + b + c$

Equilateral Triangle

area $= \dfrac{\sqrt{3}}{4} s^2$

Perimeter $= 3s$

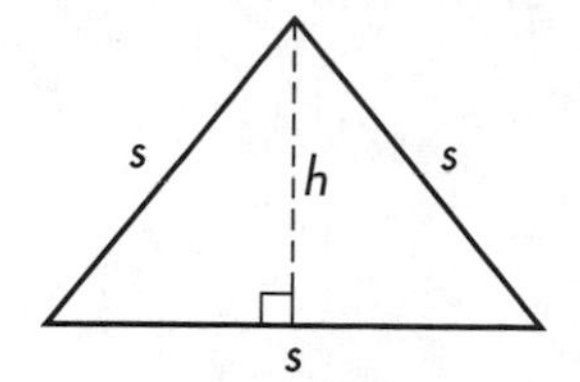

Isosceles Triangle

area $= \dfrac{1}{2} b \sqrt{a^2 - \dfrac{b^2}{4}}$

Perimeter $= 2a + b$

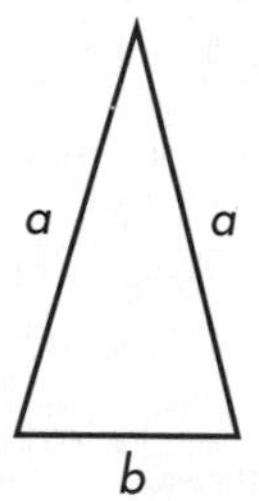

Square

area $= s^2$

perimeter $= 4s$

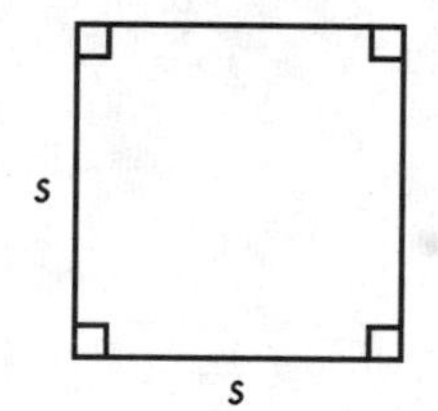

Rcctangle

area $= lw$

perimeter $= 2l + 2w$

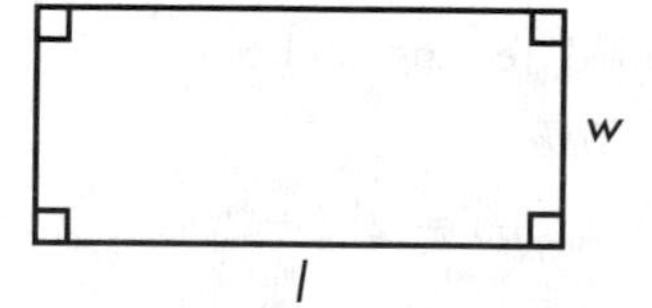

Parallelogram

area $= bh$

perimeter $= 2b + 2a$

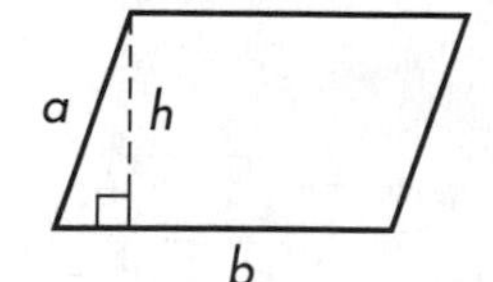

Trapezoid

area $= \dfrac{1}{2} (a + b)h$

perimeter $= a + b + c + d$

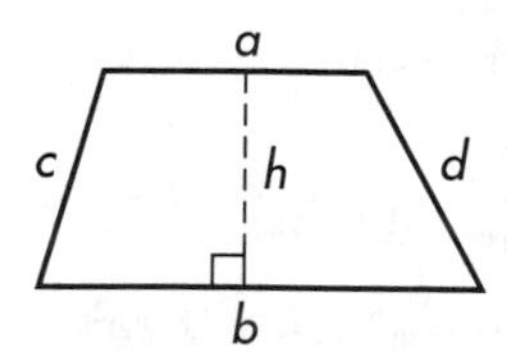

Circle

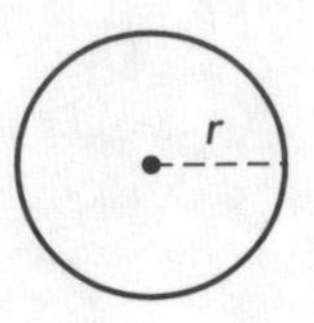

area = πr^2

circumference = $2\pi r = \pi d$

diameter $d = 2r$ (radius)

Sector of Circle

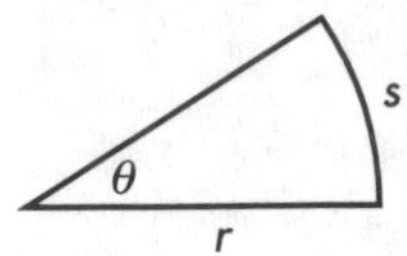

area = $\dfrac{\theta r^2}{2}$

$s = r\theta$

Sphere

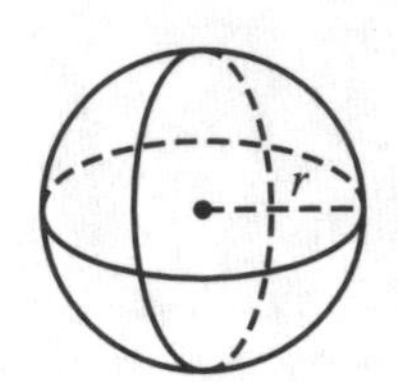

volume = $\dfrac{4}{3}\pi r^3$

surface area = $4\pi r^2$

Ellipse

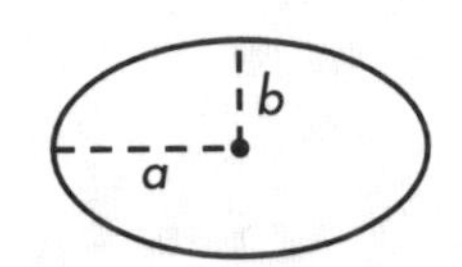

area = πab

circumference = $2\pi\sqrt{\dfrac{a^2 + b^2}{2}}$

Cube

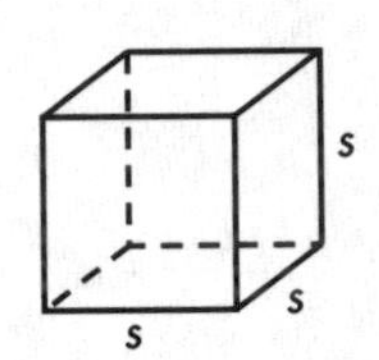

volume = s^3

total surface area = $6s^2$

Parallelepiped (rectangular box)

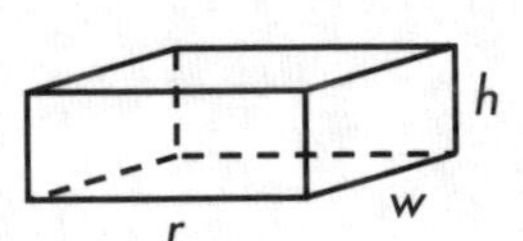

volume = lwh

total surface area = $2hl + 2hw + 2lw$

Right Circular Cone

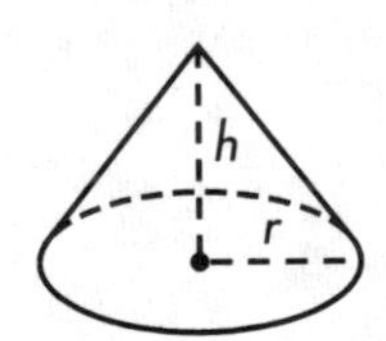

volume = $\dfrac{\pi r^2 h}{3}$

lateral surface area = $\pi r\sqrt{r^2 + h^2}$

total surface area = $\pi r\sqrt{r^2 + h^2} + \pi r^2$

Right Circular Cylinder

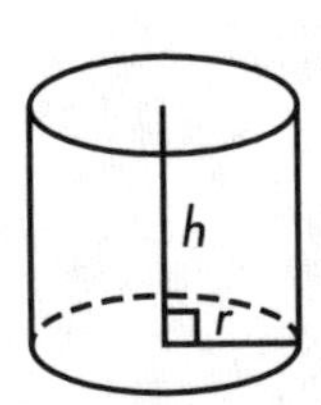

volume = $\pi r^2 h$

lateral surface area = $2\pi rh$

total surface area = $2\pi rh + 2\pi r^2$

Other Geometric Concepts

The sum of the measures of the three angles of a triangle is 180°.
Two angles are complementary when their sum is 90°.
Two angles are supplementary when their sum is 180°.
A 90° angle is called a right angle.
An acute angle measures less than 90°.
An obtuse angle measures more than 90° but less than 180°.
When two lines intersect, they form four vertical angles. Opposite vertical angles are equal.
The height of a triangle is the perpendicular distance from a vertex to the opposite side.
In an isosceles triangle, the angles opposite the equal sides are equal.
In any triangle, the sum of the lengths of any two sides is greater than the length of the third side.
If two triangles are similar, their corresponding angles are equal and corresponding sides are proportional.

Trigonometry Formulas

A = measure of $\angle A$

B = measure of $\angle B$

C = measure of $\angle C = 90°$

a = side opposite $\angle A$

b = side adjacent to $\angle A$

c = hypotenuse

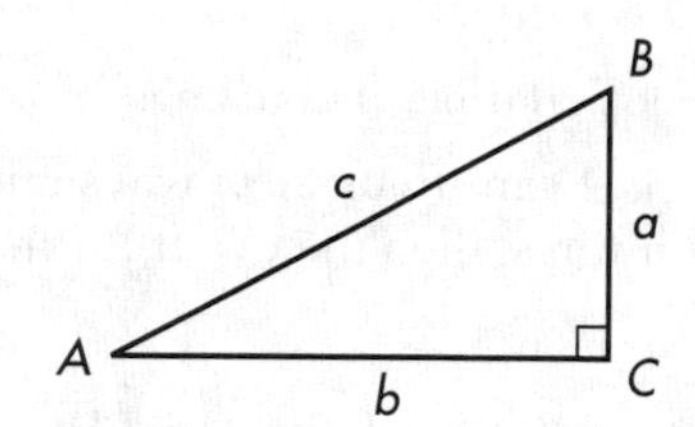

Basic Formulas

$$\sin A = \frac{\text{side opposite}}{\text{hypotenuse}} = \frac{a}{c} \qquad \csc A = \frac{\text{hypotenuse}}{\text{side opposite}} = \frac{c}{a}$$

$$\cos A = \frac{\text{side adjacent}}{\text{hypotenuse}} = \frac{b}{c} \qquad \sec A = \frac{\text{hypotenuse}}{\text{side adjacent}} = \frac{c}{b}$$

$$\tan A = \frac{\text{side opposite}}{\text{side adjacent}} = \frac{a}{b} \qquad \cot A = \frac{\text{side adjacent}}{\text{side opposite}} = \frac{b}{a}$$

$A = 90° - B$

$B = 90° - A$

Pythagorean Theorem: $c^2 = a^2 + b^2$

Radian-Degree Conversions

$$1 \text{ degree} = \frac{\pi}{180} \text{ radians} \quad 1 \text{ radian} = \frac{180}{\pi} \text{ degrees}$$

TABLE OF VALUES FOR TRIGONOMETRIC FUNCTIONS OF SPECIAL ANGLES

angle (degrees)	angle (radians)	sine	cosine	tangent	cotangent	secant	cosecant
0	0	0	1	0	undefined	1	undefined
30°	$\frac{\pi}{6}$	$\frac{1}{2}$	$\frac{\sqrt{3}}{2}$	$\frac{1}{\sqrt{3}}$	$\sqrt{3}$	$\frac{2}{\sqrt{3}}$	2
45°	$\frac{\pi}{4}$	$\frac{1}{\sqrt{2}}$	$\frac{1}{\sqrt{2}}$	1	1	$\sqrt{2}$	$\sqrt{2}$
60°	$\frac{\pi}{3}$	$\frac{\sqrt{3}}{2}$	$\frac{1}{2}$	$\sqrt{3}$	$\frac{1}{\sqrt{3}}$	2	$\frac{2}{\sqrt{3}}$
90°	$\frac{\pi}{2}$	1	0	undefined	0	undefined	1

Counting Formulas

Multiplication Principle of Counting:

If an event A can occur in m ways and, after A has occurred, an event B can occur in n ways, then the event A followed by B can occur in mn ways.

Factorial Notation

$n! = n(n-1)(n-2)(n-3)\cdots(3)(2)(1),\ n \geq 1 \quad 0! = 1$

Permutations

$$P(n, r) = \frac{n!}{(n-r)!} = \text{number of permutations of } n \text{ things taken } r \text{ at a time}$$

Combinations

$$C(n, r) = \binom{n}{r} = \frac{n!}{r!(n-r)!} = \text{number of combinations of } n \text{ things taken } r \text{ at a time}$$

Probability and Statistical Concepts

Probability Formulas

Probability of an event A:

Assuming equal likelihood of outcomes,

$$P(A) = \frac{\text{number of outcomes favorable to event } A}{\text{total number of possible outcomes}}$$

Probability of the complement of an event A, denoted $\overline{A}$:

$P(\overline{A}) = 1 - P(A)$

The rule of unions:

$P(A \cup B) = P(A) + P(B) - P(A \cap B)$

For mutually exclusive events A and B (that is, for two events A and B that cannot occur simultaneously):

$P(A \cap B) = 0$

and therefore,

$P(A \cup B) = P(A) + P(B)$

The conditional probability of event A given the occurrence of event B:

$$P(A|B) = \frac{P(A \cap B)}{P(B)}$$

assuming $P(B) \neq 0$

The probability of the intersection of events A and B:

$P(A \cap B) = P(A)P(B|A)$

Events A and B are independent if:

both $P(A|B) = P(A)$ and $P(B|A) = P(B)$

and, most usefully:

$P(A \cap B) = P(A)P(B)$

Statistical Concepts

$$\text{mean} = \frac{\text{sum of the numbers}}{\text{how many numbers you have}}$$

median = the middlemost value in a set of numbers that have been ordered according to size. For an even number of values, the median is the arithmetic average of the two middle values. For an odd number of values, the median is the middle value.

mode = the most frequently occurring value in a set of numbers.

range = the largest number minus the smallest number in the set of numbers.

Formulas from Science

Gas Laws

$\frac{p_1 v_1}{T_1} = \frac{p_2 v_2}{T_2}$ (General)

v_1 = volume at pressure p_1 and temperature T_1

v_2 = volume at pressure p_2 and temperature T_2

$\frac{v_1}{T_1} = \frac{v_2}{T_2}$ (Charles' Law)

$$\frac{p_1}{p_2} = \frac{v_2}{v_1} \text{ (Boyle's Law)}$$

Specific Gravity

$$\text{specific gravity of substance} = \frac{\text{weight of given volume of substance}}{\text{weight of equal volume of water}}$$

Lever

$$\frac{W_1}{W_2} = \frac{L_2}{L_1}$$

W_1 = force at a distance L_1 from the fulcrum
W_2 = force at a distance L_2 from the fulcrum

Pulley

$$\frac{R_1}{R_2} = \frac{d_2}{d_1}$$

R_1 = revolutions per minute of pulley of diameter d_1
R_2 = revolutions per minute of pulley of diameter d_2

Other Mathematical Formulas and Terminology

Binomial Theorem

$$(x+y)^n = x^n + \binom{n}{1}x^{n-1}y + \binom{n}{2}x^{n-2}y^2 + \binom{n}{3}x^{n-3}y^3 + \cdots + \binom{n}{n-1}xy^{n-1} + y^n$$

Arithmetic Progression

*n*th term: $a_n = a_1 + (n-1)d$

sum of first *n* terms: $\sum_{k=1}^{n} a_k = \frac{n}{2}(a_1 + a_n)$

Geometric Progression

*n*th term: $a_n = a_1 r^{n-1}$

sum of first *n* terms: $\sum_{k=1}^{n} a_k = \frac{a(1-r^n)}{1-r}$

sum of infinite geometric series: $\sum_{k=1}^{\infty} a_k = \frac{a_1}{1-r}$, provided $|r| < 1$

Principle of Mathematical Induction

Let S be a subset of the natural numbers N such that

(*i*) $1 \in S$,

(*ii*) if $k \in S$, then $k + 1 \in S$.

Then $S = N$.

Complex Numbers

$i^2 = -1$

$\sqrt{-b^2} = ib$ (b nonnegative)

$(a + bi) + (c + di) = (a + c) + (b + d)i$

$(a + bi)(c + di) = (ac - bd) + (ad + bc)i$

conjugate of $a + bi = a - bi$

$(a + bi)(a - bi) = a^2 + b^2$

modulus of $a + bi = \sqrt{a^2 + b^2}$

APPENDIX D SYNTHETIC DIVISION

Long division can be very tedious. **Synthetic division** is a shortcut method you can use to divide a polynomial by a binomial of the form $x - r$. To see how it works, divide $4x^3 - 3x + 1$ by $x - 2$:

$$\begin{array}{r|l} & 4x^2 + 8x \;\; + 13 \\ x - 2 & 4x^3 + 0x^2 - \;\; 3x + 1 \\ & \underline{4x^3 - 8x^2} \\ & \qquad 8x^2 - \;\; 3x \\ & \qquad \underline{8x^2 - 16x} \\ & \qquad\qquad 13x + \;\; 1 \\ & \qquad\qquad \underline{13x - 26} \\ & \text{remainder} \rightarrow \qquad 27 \end{array} \qquad \begin{array}{r|l} & 4 + 8 + 13 \\ 1 - 2 & 4 + 0 - \;\; 3 + 1 \\ & \underline{4 - 8} \\ & \quad 8 - \;\; 3 \\ & \quad \underline{8 - 16} \\ & \qquad 13 + \;\; 1 \\ & \qquad \underline{13 - 26} \\ & \text{remainder} \rightarrow \quad 27 \end{array}$$

On the left is the usual long division process; on the right is the same process using only the coefficients of the term. Thus, we can simplify the long division by working with the coefficients only—as long as we are careful to use 0 as a coefficient for missing powers of x. By condensing vertically and omitting unnecessary coefficients in the process, we obtain the following procedure:

$$\text{Example: Divide } P(x) = \underbrace{4x^3 - 3x + 1}_{\text{dividend}} \text{ by } \underbrace{x - 2}_{\text{divisor}}$$

A helpful message: Notice since $x - r = x - 2$, in this example, $r = 2$.

Step 1: Write both dividend and divisor in descending powers of x, using a coefficient of 0 when a power of x is missing, then write the coefficients as shown:

$$4 + 0 - 3 + 1$$

Step 2: Write $r = 2$, as shown:

$$2 \rfloor 4 + 0 - 3 + 1$$

Step 3: Bring down the first coefficient:

$$\begin{array}{r|l} 2 & \underline{4 + 0 - 3 + 1} \\ & 4 \end{array}$$

Step 4: Multiply the first coefficient by $r = 2$, write the product under the second coefficient, and add:

$$\begin{array}{r|l} 2 & 4 + 0 - 3 + 1 \\ & \underline{\quad + 8 \qquad\qquad} \\ & 4 + 8 \end{array}$$

Step 5: Multiply the sum by $r = 2$, write the product under the third coefficient, and add:

$$\begin{array}{r|l} 2 & 4 + 0 - 3 + 1 \\ & \underline{\quad + 8 + 16 \quad} \\ & 4 + 8 + 13 \end{array}$$

Step 6: Repeat Step 5 until all the coefficients in the dividend have been used. Separate the final sum, which is the remainder, as shown:

$$\begin{array}{r|l} 2 & 4 + 0 - 3 + 1 \\ & \underline{\quad + 8 + 16 + 26} \\ & 4 + 8 + 13 \;\lfloor\; 27 \leftarrow \text{remainder} \end{array}$$

Step 7: Write the quotient and remainder using the coefficients:

Solution: $P(x) = 4x^3 - 3x + 1$ divided by $x - 2$ is $4x^2 + 8x + 13$, with remainder 27.

Applying the remainder theorem, you now know that:

$$P(2) = 4(2)^3 - 3(2) + 1 = 27$$

Thus, you can use synthetic division as a shortcut to evaluate a polynomial at a given value.

Synthetic division is also a quick way to determine if a given value is a zero of a polynomial. If the remainder is zero when you complete the division, the number is a zero of the polynomial.

Example: Is -3 a zero of the polynomial $P(x) = x^3 + 2x^2 - 15x - 36$?

$$\begin{array}{r|l} -3 & 1 + 2 - 15 - 36 \\ & \underline{\quad - 3 + 3 + 36} \\ & 1 - 1 - 12 \;\lfloor\; 0 \leftarrow \text{remainder} \end{array}$$

Solution: Since the remainder is zero, we know $P(-3) = 0$; thus, -3 is a zero of $P(x) = x^3 + 2x^2 - 15x - 36$.

Applying the factor theorem to the above example, we can say that since $P(-3) = 0$, then $x - (-3) = x + 3$ is a factor of $P(x) = x^3 + 2x^2 - 15x - 36$.

In summary:

USES FOR SYNTHETIC DIVISION

(1) To evaluate a polynomial $P(x)$ for a given value r:
Divide $P(x)$ by $x - r$; $P(r)$ equals the remainder.

(2) To determine whether a given value r is a zero of a polynomial $P(x)$:
Divide $P(x)$ by $x - r$; if the remainder is 0, r is a zero of $P(x)$.

(3) To determine whether $x - r$ is a factor of a polynomial $P(x)$:
Divide $P(x)$ by $x - r$; if the remainder is 0, $x - r$ is a factor of $P(x)$.

APPENDIX E
SIMPLIFYING RADICALS

If $a^n = x$, a is called an ***n*th root** of x, where n is a natural number. This relationship can be expressed as

$$a = x^{\frac{1}{n}} = \sqrt[n]{x}$$

The $\sqrt{\ }$ is called a **radical,** x is called the **radicand,** and n is called the **index** and indicates which root is desired. If no index is written, it is understood to be 2 and the radical expression indicates the *principal* square root of the radicand. We have the following rules for radicals when x, y are real numbers m, n positive integers and the radical expression denotes a real number:

1. $\sqrt[n]{x^n} = x$ if n is odd
2. $\sqrt[n]{x^n} = |x|$ if n is even
3. $\sqrt[n]{x^m} = (\sqrt[n]{x})^m$
4. $(\sqrt[n]{x})(\sqrt[n]{y}) = \sqrt[n]{xy}$
5. $\dfrac{\sqrt[n]{x}}{\sqrt[n]{y}} = \sqrt[n]{\dfrac{x}{y}} \qquad (y \neq 0)$
6. $\sqrt[m]{\sqrt[n]{x}} = \sqrt[mn]{x}$
7. $\sqrt[pn]{x^{pm}} = \sqrt[n]{x^m}$
8. $a(\sqrt[n]{x}) + b(\sqrt[n]{x}) = (a + b)(\sqrt[n]{x})$

These rules form the basis for simplifying radical expressions. A radical is **simplified** when

a. the radicand contains no variable factor raised to a power equal to or greater than the index of the radical;
b. the radicand contains no constant factor that can be expressed as a power equal to or greater than the index of the radical;
c. the radicand contains no fractions;
d. no fractions contain radicals in the denominator;
e. the index of the radical is reduced to its lowest value.

For example:

- $\sqrt{12} = (\sqrt{4})(\sqrt{3}) = 2(\sqrt{3})$ is simplified.
- $\sqrt[3]{24a^5b^6} = (\sqrt[3]{8a^3b^6})(\sqrt[3]{3a^2}) = 2ab^2\sqrt[3]{3a^2}$ is simplified.

- $\frac{\sqrt{54}}{\sqrt{6}} = \sqrt{\frac{54}{6}} = \sqrt{9} = 3$ is simplified.
- $\frac{1}{\sqrt{2}} = \left(\frac{1}{\sqrt{2}}\right)\frac{\sqrt{2}}{\sqrt{2}} = \frac{\sqrt{2}}{2}$ is simplified.

Since square roots occur so frequently, the remainder of the examples will use only square root radicals.

Radicals that have the same index and the same radicand are **like radicals.** To add or subtract like radicals, combine their coefficients and write the result as the coefficient of the common radical factor. Indicate the sum or difference of unlike radicals.

$$5\sqrt{3} + 2\sqrt{3} = 7\sqrt{3}$$

You may have to simplify the radical expressions before combining them:

$$5\sqrt{3} + \sqrt{12} = 5\sqrt{3} + (\sqrt{4})(\sqrt{3}) = 5\sqrt{3} + 2\sqrt{3} = 7\sqrt{3}$$

To multiply radicals that have the same index, multiply their coefficients to find the coefficient of the product. Multiply the radicands to find the radicand of the product. Simplify the results. For example:

$$(5\sqrt{3})(2\sqrt{3}) = 10\sqrt{9} = 10(3) = 30$$

For a sum or difference, treat the factors as you would binomials, being sure to simplify radicals after you multiply. For example:

- $(2\sqrt{3} + 5\sqrt{7})(\sqrt{3} - 3\sqrt{6}) = 2\sqrt{9} - 6\sqrt{18} + 5\sqrt{21} - 15\sqrt{42} =$
  $2(3) - 6\sqrt{9}\,\sqrt{2} + 5\sqrt{21} - 15\sqrt{42} = 6 - 18\sqrt{2} + 5\sqrt{21} - 15\sqrt{42}$
- $(1 - \sqrt{3})(1 + \sqrt{3}) = 1 + \sqrt{3} - \sqrt{3} - \sqrt{9} = 1 - 3 = -2$

A technique called **rationalizing** is used to remove radicals from the denominator of a fraction. For square root radicals, if the denominator contains a single term, multiply the numerator and denominator by the smallest radical that will produce a perfect square in the denominator. For example:

$$\frac{5}{\sqrt{3}} = \left(\frac{5}{\sqrt{3}}\right)\left(\frac{\sqrt{3}}{\sqrt{3}}\right) = \frac{5\sqrt{3}}{3}$$

If the denominator contains a sum or difference of square roots, multiply the numerator and denominator by a difference or sum that will cause the middle term to sum to zero when you multiply. For example:

$$\frac{5}{1 - \sqrt{3}} = \left(\frac{5}{1 - \sqrt{3}}\right)\left(\frac{1 + \sqrt{3}}{1 + \sqrt{3}}\right) = \left(\frac{5(1 + \sqrt{3})}{1 - 3}\right) = -\frac{5 + 5\sqrt{3}}{2}$$

GLOSSARY

Abscissa. The first number in an ordered pair; the *x*-coordinate.

Absolute value. The absolute value of a number x, represented by $|x|$, is its distance from zero on the number line. For any number x, $|x| = x$ if x is positive or zero and $|x| = -x$ if x is negative.

Absolute value function. A function defined by an equation of the form $f(x) = |x|$.

Acute angle. Any angle that is less than a right angle.

Addend. Any one of a set of numbers to be added. In the sum $3 + 4 + 10$, 3, 4, and 10 are the addends.

Addition. A binary operation performed on a pair of numbers called *addends* to obtain a unique *sum.* To add signed numbers: (1) if two numbers have like signs, find the sum of their absolute values and use the same sign; (2) if two numbers have opposite signs, find the difference of their absolute values and use the sign of the number having the greater absolute value.

Addition of matrices. The sum of two $m \times n$ matrices is an $m \times n$ matrix in which the elements are the sums of the corresponding elements of the given matrices.

Addition of rational expressions. To add rational expressions: (1) find a common denominator; (2) express each term as an equivalent rational expression having the common denominator; (3) add or subtract numerators; (4) write the result over the common denominator; and (5) simplify.

Addition property of equality. For any numbers a, b, and c, if $a = b$, then $a + c = b + c$.

Addition property of inequality. For any numbers a, b, and c, if $a < b$, then $a + c < b + c$. If $a > b$, then $a + c > b + c$. Similar statements hold for $\leq$ and $\geq$.

Additive identity. The number zero is the additive identity. When we add zero to any number, the sum is that number.

Additive inverse. If the sum of two numbers is zero, they are additive inverses of each other.

Algebraic expression. Any symbol or combination of symbols that represents a number. In an expression containing + or − signs, the quantities connected by the + or − signs are the *terms* of the expression.

Algebraic hierarchy of operations. See *order of operations.*

Algorithm. A systematic procedure for performing a computation.

Altitude. In a plane geometric figure, the perpendicular distance from a vertex to the opposite side, called the *base.* In general, the perpendicular distance that measures the height of a figure.

Angle. The union of two rays that have a common endpoint.

Angle of depression or elevation. The angle from the horizontal upward or downward to a line of sight.

Antilogarithm. If $\log_b x = a$, x is the antilogarithm of a.

Area. The measure of a surface expressed in square units.

Argument of a function. In the functional notation $y = f(x)$, x is the argument of the function f.

Arithmetic sequence. A sequence in which the difference between successive terms is a constant.

Arithmetic series. The indicated sum of an arithmetic sequence.

Arrow (or mapping) diagram. A pictorial representation of a function that shows the correspondence between the domain elements and the range elements.

Associative property. For any real numbers a, b, and c,
Addition: $(a+b)+c=a+(b+c)$
Multiplication: $(a \cdot b) \cdot c=a \cdot (b \cdot c)$

Asymptote. A line to which a curve gets closer and closer as the distance from the origin increases, e.g.,

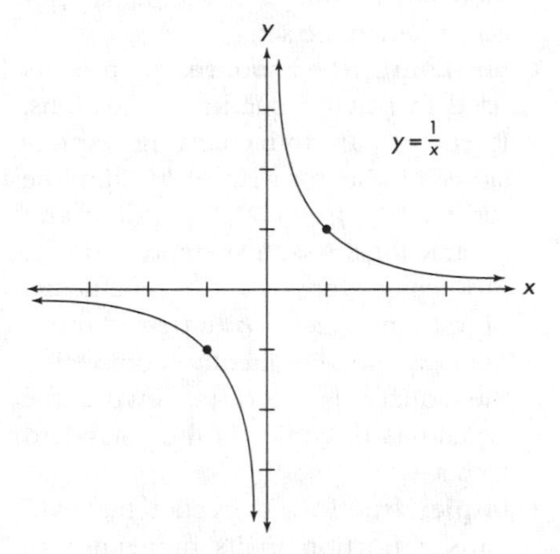

The asymptotes for $y=\frac{1}{x}$ are the x- and y-axes.

Asymptote, horizontal. A line $y=b$ is a horizontal asymptote if the function approaches b as x increases or decreases without bound.

Asymptote, oblique (or slant). A function $y=g(x)$ is an oblique asymptote if the function approaches $g(x)$ as x increases or decreases without bound.

Asymptote, vertical. A line $x=k$ is a vertical asymptote if the function increases or decreases without bound as x approaches k from the left or right.

Augmented matrix. A matrix containing the coefficients and constants of a system of linear equations.

Average. See *mean.*

Axes. See *coordinate axes.*

Axiom. A principle assumed or accepted without proof.

Axis of symmetry. If the graph of a function is folded so that its two sides coincide, the line on which the fold occurs is the axis of symmetry.

Base. In the exponential expression b^n, b is the base, which is used as a factor n times.

Base *e*. The base used in the system of natural logarithms. Its value to five decimal places is 2.71828. The expression $\log_e x$ is usually written as $\ln x$.

Base of a logarithm. The arbitrarily selected number that is raised to a power (the logarithm) to obtain a given number. In the logarithmic expression $\log_b x$, b is the base. Any positive number except one can be used as a base in a system of logarithms.

Binary operation. An operation that is performed on only two numbers at a time.

Binomial. A polynomial consisting of two unlike terms.

Binomial coefficient. The binomial coefficient $\binom{n}{r}$ means $\frac{n!}{r!(n-r)!}$.

Binomial expansion. The sum of terms that is the result of multiplying out a power of a binomial.

Binomial theorem. A theorem that tells how to expand a power of a binomial. Symbolically, $(x+y)^n=\sum_{r=0}^{n}\binom{n}{r}x^{n-r}y^r$.

Bisect. To divide in half.

Boundary. A line or curve separating a plane into two regions.

Canceling. The act of dividing like factors out of the numerator and denominator of a fraction.

Cartesian coordinates. The system of rectangular coordinates named in honor of René Descartes, French philosopher and mathematician, who conceived many of the basic ideas of analytic geometry.

Cartesian product. The Cartesian product of two sets A and B, denoted

$A \times B$, is the set of all ordered pairs with first component from A and second component from B.

Circle. The set of points in a plane at a constant distance from a fixed point in that plane. The fixed point is the *center* of the circle and the constant distance is the *radius* of the circle. The standard form of the equation of a circle centered at the origin with radius r is $x^2 + y^2 = r^2$.

Circumference. The distance around the boundary of a circle.

Closed interval. A segment of the number line between two points a and b that includes the endpoints. It consists of all x such that $a \leq x \leq b$ and is written as $[a, b]$. If the interval includes only one endpoint, it is called either half-open or half-closed, and is written as either $(a, b]$ or $[a, b)$, with the bracket at the closed end.

Closure property. For any real numbers a, b, and c,
Addition: $a + b$ is a real number.
Multiplication: $a \cdot b$ is a real number.

Coefficient. In a term with two or more factors, the coefficient of each factor is the product of all the other factors in that term.

Coefficient matrix. A matrix containing the coefficients of the variable in a system of linear equations.

Coefficient, numerical. The numerical factor of a term. If no numerical coefficient is written, it is understood to be one; e.g., 6 is the numerical coefficient in $6x$; 1 (understood) is the numerical coefficient in xy.

Coefficients of a polynomial. The numbers $a_n, a_{n-1}, a_{n-2}, \ldots, a_0$ in the polynomial $a_nx^n + a_{n-1}x^{n-1} + a_{n-2}x^{n-2} + \ldots + a_1x^1 + a_0$.

Collect like terms. The process of using the distributive property to simplify expressions containing like terms.

Combination. An arrangement of objects where the order is not a consideration.

Common difference. The constant difference between successive terms of an arithmetic sequence.

Common logarithms. Logarithms that use 10 as the base.

Common ratio. The constant ratio between successive terms of a geometric sequence.

Complement of a set. The set of elements in the universal set that are not contained in the set.

Completing the square. A process used in solving quadratic equations. It centers around changing expressions of the form $x^2 + bx$ into the form $(x + \frac{b}{2})^2 - (\frac{b}{2})^2$ by adding and subtracting the same quantity, namely, the square of half the coefficient of x, to make $x^2 + bx$ a perfect trinomial square. Frequently, completing the square is used to rewrite the equations of conics in their standard forms.

Complex fraction. A fraction that contains a fraction in its numerator or denominator or both.

Complex number. Any number that can be written in the form $a + bi$, where $i^2 = -1$. In the complex number $a + bi$, a is the real part and b is the imaginary part.

Composite number. A number greater than one that has factors other than one and itself. The term composite number refers only to integers, not to fractions or irrational numbers.

Composite of functions. The composite of functions f and g, denoted $f \circ g$, maps x into $f(g(x))$.

Composition of functions. The process of forming the composite of functions, denoted by the symbol $\circ$.

Compound sentence. A sentence in which two independent clauses are connected by the word "and" or the word "or."

Conditional probability. The conditional probability of event A, given event B, is given by the following formula: $P(A|B) = \frac{P(A \text{ and } B)}{P(B)}$, where $(B) \neq 0$.

Congruent. Figures are congruent if they can be placed one upon the other so that all parts coincide.

Conic section. A curve formed by the intersection of a plane with a conical surface.

Conjugate of a complex number. The conjugate of the complex number $a + bi$ is $a - bi$.

Conjunction. A conjunction of two statements is formed by connecting them with "and" and is true only when *both* statements are true.

Consecutive even integers. Integers that follow each other when counting by twos, beginning with an even integer.

Consecutive integers. Integers that follow each other when counting by ones.

Consecutive odd integers. Integers that follow each other when counting by twos, beginning with an odd integer.

Consistent system. A system of equations that has one or more solutions is consistent.

Constant. A quantity whose value does not change, or is regarded as fixed, during a mathematical process or discussion.

Constant function. The function defined by $y = b$. The graph is a horizontal line.

Constant in an interval. A function f is constant in an interval (x_1, x_2) if $f(x_1) = f(x_2)$ for every point in the interval.

Constant matrix. A matrix consisting of the constants in a system of linear equations.

Constant of variation. The constant k in equations such as $y = kx$ or $y = \frac{k}{x}$.

Constant term. A term with no variable.

Continuous graph. One that can be drawn without picking up the pencil from the paper.

Coordinate. A number, or one of a set of numbers, used to locate a point. To locate a point on a line, only one coordinate is needed; in the plane, two coordinates are needed; in three-dimensional space, three coordinates are needed; and so on.

Coordinate axes. The x- and y- axes in the coordinate plane.

Coordinates of a point on the plane. The ordered pair (x, y) of real numbers used to locate the point on the coordinate plane.

Coordinate plane. A plane in which a system has been set up for graphing ordered pairs of numbers as points in the plane using two perpendicular lines, called the coordinate axes, which intersect at the zero point for each.

Counting numbers. See *natural numbers.*

Cramer's rule. A rule used to solve a system of linear equations in two or more unknowns. It states that each unknown can be expressed as the ratio of two determinants.

Cube. The results of using a quantity as a factor three times; e.g., the cube of 2 is (2)(2)(2), or 8.

Cube root. If x^3 *is* a, then x is a cube root of a, as 2 is a cube root of 8.

Cubic polynomial. A polynomial of the third degree, such as $3x^3 - 4x^2 + 6$.

Decimal fraction. A proper fraction whose denominator is a power of 10. The denominators are not written, but are indicated by use of a dot called the *decimal point,* the power of 10 for the denominator being the number of digits on the right side of the dot. Examples are $\frac{2}{10} = 0.2$; $\frac{75}{100} = 0.75$; and $\frac{35}{100} = 0.035$.

Decimal, repeating (or nonterminating). A decimal in which after a certain digit, a set of one or more digits is repeated continuously. Any repeating decimal can be shown to be equal to a fraction (that is, a quotient of integers), and is therefore a rational number.

Decimal system. The number system that uses ten as its base.

Decimal, terminating. A decimal that can be written using a finite number of digits. Any terminating decimal can be shown to be equal to a fraction (that is, a quotient of integers), and is therefore a rational number. The denominator of the fraction will be divisible by either two, five, or both.

Decreasing. A function f is decreasing in an interval (x_1, x_2) if $f(x_1) > f(x_2)$ for every point in the interval.

Deductive reasoning. The process of applying general rules to specific situations.

Degenerate conic. The intersection of a conical surface and a plane containing the vertex of the conical surface. The result is a point, a single line, or two intersecting lines.

Degree of a monomial. The sum of the exponents of its variables.

Degree of a polynomial. The greatest degree of any of its terms after it has been simplified.

Degree of a term. The degree of a term is the sum of the exponents of its variables.

Denominate number. A quantity that includes both a number and a unit of measurement.

Density property. The density property for real numbers states that between any two different real numbers, there exists another real number.

Dependent. A system of equations that has an infinite number of solutions is dependent.

Dependent variable. In $y = f(x)$, y is the dependent variable; its value depends on the value of x.

Descartes' Rule of Signs. Suppose $P(x)$ is a polynomial whose terms are arranged in descending order. The number of positive real zeros of $y = P(x)$ is the same as the number of changes in sign (as viewed from left to right) of the coefficients of the terms, or is less by an even number. The number of negative real zeros is the same as the number of changes in sign (as viewed from left to right) of the coefficients of the terms of $P(-x)$, or is less by an even number.

Descending order. A polynomial is in descending order for a variable if the term with the greatest exponent for that variable is first, the term with the next greatest exponent for that variable is second, and so on.

Determinant. A numerical value assigned to a matrix that is a sum of certain products of its elements. For a 2×2 matrix $\begin{bmatrix} a & b \\ c & d \end{bmatrix}$, the determinant is the difference of the product of its diagonals: $\det\begin{bmatrix} a & b \\ c & d \end{bmatrix} = \begin{vmatrix} a & b \\ c & d \end{vmatrix} = ad - bc$.

Diagonal of a polygon. A line connecting two nonadjacent vertices.

Diameter. The length of a line segment that passes through the center of a circle and has endpoints on the circle.

Difference. The results obtained when one number is subtracted from another.

Difference of two cubes. A special case of factoring in which one factor is a binomial and one a trinomial, e.g., $x^3 - y^3 = (x - y)(x^2 + xy + y^2)$.

Difference of two squares. A special case of factoring in which the factors are the sum and difference of the square roots of two squares, e.g., $x^2 - y^2 = (x + y)(x - y)$.

Dimensions of a matrix. A matrix with m rows and n columns has dimensions $m \times n$.

Directed numbers. Numbers that have both magnitude and direction, such as the integers.

Direct variation. A function defined by an equation of the form $f(x) = kx^n$, where k is a nonzero constant.

Discriminant. The quantity $b^2 - 4ac$ is called the discriminant of the quadratic equation $ax^2 + bx + c = 0$. If the discriminant equals zero, the roots are real and equal. If it is positive, the roots are real and unequal. If it is negative, the roots are complex numbers.

Disjunction. A disjunction of two statements is formed by connecting them with "or" and is true when one or both statements are true.

Distance formula. The distance d between two points (x_1, y_1) and (x_2, y_2) is given by $d = \sqrt{(x_2 - x_1)^2 + (y_2 - y_1)^2}$.

Distance from a point to a line. The distance d from a point (x_0, y_0) to a line $Ax + By + C = 0$ is given by $d = \frac{|Ax_0 + By_0 + C|}{\sqrt{A^2 + B^2}}$.

Distributive property. For any real numbers a, b, and c, $a(b + c) = ab + ac$ and $(b + c)a = ba + ca$.

Divisible. A number is divisible by another number if, when it is divided by that number, a remainder of zero is obtained.

Division. The inverse of multiplication, defined as $a \div b = a \cdot \frac{1}{b}$.

Division property for radicals. For any real numbers a and b, $\sqrt{\frac{a}{b}} = \frac{\sqrt{a}}{\sqrt{b}}$ and $\frac{\sqrt{a}}{\sqrt{b}} = \sqrt{\frac{a}{b}}$, where $b \neq 0$.

Domain. (1) In a relation, the set of first coordinates. (2) In a function, the set of values for which the function is defined.

Domain of definition. The largest possible subset of the real numbers for which a function is defined.

Element of a set. Any object in a set.

Elimination (or addition) method. A method for solving systems of equations in which the equations are added together to eliminate variable(s). One or both equations may be multiplied by a constant.

Ellipse. The set of points in a plane such that the sum of their distances from two fixed points, called the *foci*, is constant. The standard form for an ellipse centered at the origin is $\frac{x^2}{a^2} + \frac{y^2}{b^2} = 1$.

Empty set. The set with no elements, symbolized by $\varnothing$. Also called the null set.

Equality. A statement that two quantities are equal.

Equation. A mathematical statement of equality.

Equiangular. Having all angles equal.

Equilateral. Having all sides equal.

Equilateral triangle. A triangle with all three sides congruent.

Equivalence relation. A relation "*" defined on a set A that has the following properties for elements a, b, and c in A:

1. $a * a$ (reflexive property)
2. If $a * b$, then $b * a$. (symmetric property)
3. If $a * b$ and $b * c$, then $a * c$. (transitive property)

The equality relation is an equivalence relation.

Equivalent equations. Equations that have the same solution set.

Equivalent systems. Systems of equations that have the same solution set.

Evaluate an expression. Substitute for the variable (or variables) in an expression and calculate the result.

Even function. A function f for which $f(-x) = f(x)$ for every x in its domain. If a function is even, its graph has symmetry about the y-axis.

Even number. A number that is divisible by two.

Exponent. A small raised number to the right of a quantity. For example, n is an exponent in the expression b^n. If the exponent is a natural number, it indicates the quantity it is attached to is to be used as a factor that many times.

Exponential equation. An equation in which the variable (or variables) occurs in an exponent.

Exponential function. A function defined by an equation of the form

$y = b^x$, where b is a fixed positive number other than one and x varies.

Exponential notation. In exponential notation, one number, the *power,* is represented in terms of another number, the *base,* raised to a specific *exponent.*

Exponentiation. Raising a quantity to a power.

Extraneous solution. A solution that is obtained but that does not check in the original equation.

Factor. When two or more quantities are multiplied, each is a factor of the product.

Factorial. The factorial of a positive integer is the product of all the positive integers less than or equal to that integer. Factorial n is denoted $n!$

Factor theorem. If $P(x)$ is a polynomial and $P(r) = 0$, then $x - r$ is a factor of $P(x)$.

Field properties. The properties the real numbers are assumed to possess under the binary operations of addition and multiplication.

FOIL method. A method for finding the product of two binomials. F.O.I.L. is a mnemonic that reminds you to (1) multiply the two *first* terms, (2) multiply the two *outer* terms, (3) multiply the two *inner* terms, and finally (4) multiply the two *last* terms; then simplify the results, if possible.

Formula. An equation that expresses the relation between two or more variables.

Fraction. A number written as the indicated quotient of two quantities.

Function. A relation in which each first component is paired with *one and only one* second component.

Fundamental theorem of algebra. Every polynomial of degree one or greater has at least one root.

Gauss-Jordan elimination. A method used to solve a system of linear equations.

Geometric sequence. A sequence in which the ratio of successive terms is the same number.

Geometric series. The indicated sum of a geometric sequence.

Graph of an equation. The set of points whose coordinates satisfy the equation.

Graph of a function. The graph of the ordered pairs that make up the function.

Greater than (>). A number a is greater than a number b if it is to the right of b on the number line.

Greatest common factor. The greatest common factor of two or more quantities is the greatest product that will divide evenly into each of the quantities.

Greatest integer function. A function defined by an equation of the form $f(x) = [x]$, which denotes the greatest integer less than or equal to x.

Grouping symbols. Symbols such as parentheses ( ), brackets [ ], braces { }, fraction bars, and so on, which indicate that the operation within them should be done first.

Half-plane. One of the two regions into which a line separates a plane. The line itself is called the *boundary* and might or might not be included.

Horizontal line. A line parallel to the x-axis. It has slope zero.

Horizontal line test. If any horizontal line can be drawn so that it cuts the graph of a function in more than one point, the function is *not* a one-to-one function.

Horizontal shift. A shift of the graph of a function right or left accomplished by adding or subtracting a positive constant to x.

Hyperbola. The set of points in a plane whose distances from two fixed points, called the *foci,* have a constant difference. The standard form for a hyperbola centered at the origin is $\frac{x^2}{a^2} - \frac{y^2}{b^2} = 1$.

Hypotenuse. The side opposite the right angle in a right triangle.

Identity. An equation that is true for all replacements for the variable (for which the equation has meaning.)

Imaginary number. Any complex number for which the imaginary part is not zero.

Inconsistent system. A system of equations that has no solution.

Increasing. A function f is increasing in an interval (x_1, x_2) if $f(x_1) < f(x_2)$ for every point in the interval.

Independent events. Two events are independent if the outcome of one does not affect the outcome of the other.

Independent variable. In $y = f(x)$, x is called the independent variable.

Index. The small number above a radical sign that indicates the root to be found; e.g, in $\sqrt[3]{64}$, 3 is the index and indicates the cube root is to be found.

Indirect (or inverse) variation. A function defined by an equation of the form $f(x) = \frac{k}{x^n}$.

Inductive reasoning. The process of forming general conclusions from specific cases.

Inequality. A mathematical statement containing an inequality symbol ($<$, $>$, $\leq$, $\geq$).

Inner (or dot) product. The inner product of (x_1, x_2, x_3) and (y_1, y_2, y_3) is $x_1y_1 + x_2y_2 + x_3y_3$.

Integers. The numbers in the set $\{\ldots, -3, -2, -1, 0, 1, 2, 3, \ldots\}$.

Intercept. The x-, y-, or z-coordinate of a point at which a graph crosses the corresponding axis.

Intersection of two sets. The set containing only the elements common to both sets.

Interval. A portion of the number line.

Interval, closed. An interval that contains its endpoints.

Interval, half-open (half-closed). An interval that contains exactly one of its endpoints.

Interval, open. An interval that contains neither of its endpoints.

Inverse function. For a given function $f(x)$, a function, designated $f^{-1}(x)$, such that if $f(a) = b$, then $f^{-1}(b) = a$. The inverse function "undoes" the given function. It can be strictly defined for multiple-valued functions only if they are limited to principal values.

Irrational numbers. The real numbers that cannot be written as the ratio of two integers, e.g., π, e, $\sqrt{2}$.

Isosceles triangle. A triangle that has at least two congruent sides.

Leading coefficient. The coefficient of the leading term.

Leading term. The term of highest degree in a polynomial.

Least common denominator. For a set of fractions, arithmetic or algebraic, the least common denominator is the least common multiple of their denominators. It can be obtained by factoring each denominator and taking the smallest product that contains all the factors of each denominator.

Least common multiple. The least common multiple of a set of quantities is the least product that is a multiple of each of the quantities. It can be obtained by factoring each quantity and taking the smallest product that contains all the factors of each quantity.

Less than (<). A number a is less than a number b if it is to the left of b on the number line.

Like terms. Monomials that differ only in their numerical coefficients; that is, their variable factors are the same (same letters with the same respective exponents).

Linear combination. A constant multiple of one equation added to a constant multiple of another equation.

Linear equation. An equation whose graph is a straight line. In a linear equation, the variables occur to first powers only, there are no products of variables, and no variable occurs in a denominator.

Linear function. A function defined by an equation of the form $y = mx + b$, where m and b are real numbers and $m \neq 0$.

Line segment. A part of the number line between two points that may or may not belong to the line segment.

Line of symmetry. If a figure is folded so that its two sides coincide, the line on which the fold occurs is the line of symmetry.

Locus. A set of points that satisfy a given condition.

Logarithmic function. A function defined by an equation of the form $f(x) = \log_b x$ $(x > 0)$, where b is called the *base* of the logarithmic function, $(b \neq 1, b > 0)$. The logarithm function is the inverse of the exponential function, and reciprocally.

Major axis. In an ellipse, the major axis is the segment with endpoints at the vertices of the ellipse.

Matrix. A rectangular array of elements. An $m \times n$ matrix has m rows and n columns.

Mean (or arithmetic mean). The arithmetic average of a set of numbers; that is, the sum of the numbers divided by how many numbers are in the set.

Measure of central tendency. A single number typifying or representing a set of numbers.

Median. The "middlemost" value in a set of numbers that have been ordered according to size. For an even number of values, the median is the arithmetic average of the two middle values. For an odd number of values, the median is the middle value.

Midpoint. The point that is halfway between the endpoints of a line. In a plane, its x-coordinate is the average of the x-coordinates of the endpoints and its y-coordinate is the average of the y-coordinates of the endpoints.

Minor axis. In an ellipse, the minor axis is the segment perpendicular to the major axis at the center, with endpoints on the ellipse.

Mode. The most frequently occurring value in a set of numbers.

Modulus. The modulus of a complex number $a + bi$ is $\sqrt{a^2 + b^2}$, the distance from the origin to the point (a, b) in the complex plane.

Monomial. An expression consisting of a numerical coefficient times one or more variables each raised to a nonnegative power.

Monotone. A function is monotone if it is either increasing or decreasing on its entire domain.

Multiple. In arithmetic, a product obtained by multiplying a given factor by a whole number. In general, any product can be considered to be a multiple of any of its factors.

Multiplication. A binary operation performed on a pair of numbers called *factors* to obtain a unique *product.* To multiply numbers: (1) if two numbers have like signs, find the product of their absolute values and use a positive sign; (2) if two numbers have opposite signs, find the product of their absolute values and use a negative sign.

Multiplication of matrices. The product of an $m \times n$ matrix, A, and an $n \times p$ matrix, B, is an $m \times p$ matrix, AB. The ijth element of AB is the inner product of the ith row of A and the jth column of B.

Multiplication principle of counting. If an event A can occur in m ways and, after A has occurred, an event B can occur in n ways, then the event A followed by B can occur in mn ways.

Multiplication property of equality. For any numbers a, b, and c, if $a = b$, then $ac = bc$.

Multiplication property of inequality. For any numbers a, b, and $c > 0$, if $a < b$, then $ac < bc$. If $a > b$, then $ac > bc$. For any numbers a, b, and $c < 0$, if $a < b$, then $ac > bc$. If $a > b$, then $ac < bc$. Similar statements hold for $\leq$ and $\geq$.

Multiplicative identity. The number one is the multiplicative identity. When we multiply any number by one, the product is that number.

Multiplicative inverse. If the product of two numbers is one, they are multiplicative inverses of each other.

Multiplicity. A root r of $P(x)$ has multiplicity k if $(x - r)^k$ is a factor of $P(x)$.

Mutually exclusive events. Events that cannot happen at the same time.

Mutually exclusive sets. Sets that do not intersect.

Natural logarithms. Logarithms that use the number e as the base.

Natural numbers. The numbers in the set $\{1, 2, 3, \ldots\}$. Also called the *counting numbers*.

Negative number. Any number less than zero.

Null set. See *empty set*.

Oblique line. A line that is neither horizontal nor vertical.

Odd function. A function f for which $f(-x) = -f(x)$ for every x in its domain. If a function is odd, its graph has symmetry about the origin.

Odd number. A number that is not divisible by two.

One-to-one function. A function in which each first component is paired with exactly one second component *and* each second component is paired with exactly one first component.

Open interval. A segment of the number line that does not include its endpoints.

Open sentence. A sentence containing a variable.

Opposites. Pairs of numbers that sum to zero. Also called *additive inverses*.

Order of operations. The order in which computations are to be performed when several operations are involved. Mathematicians have agreed upon the following order of operations: (1) operations enclosed within grouping symbols (parentheses, brackets, braces, etc.), starting within the innermost grouping symbol; (2) exponentiation—powers and roots—from left to right; (3) multiplication and division from left to right, *whichever comes first;* (4) addition and subtraction from left to right, *whichever comes first*.

Ordinate. The second number in an ordered pair; the y-coordinate.

Origin. On the number line, the point 0. In the plane, the point (0, 0).

Parabola. The set of points that are equidistant from a fixed point, called the *focus*, and a fixed line, called the *directrix*.

Parallel. Equidistant apart.

Parallelogram. A quadrilateral that has opposite sides congruent and parallel.

Percent. Hundredths; denoted by %; e.g., 5% of a quantity is $\frac{5}{100}$ of it.

Perfect square. A quantity that is the exact square of another quantity.

Perimeter. The distance around a closed figure.

Permutation. An arrangement of objects in a certain order.

Perpendicular. Intersecting at a right angle (90°).

Pi (π). The ratio of the circumference of a circle to its diameter. Pi is an irrational number approximately equal to 3.14159.

Polygon. A plane closed figure formed by line segments, called the *sides*, that meet at points, called the *vertices*. For example, a triangle is a three-sided polygon.

Polyhedron. A solid bounded by plane polygons.

Polynomial. A monomial or the sum of monomials.

Polynomial function. A function defined by an equation of the form $P(x) = a_nx^n + a_{n-1}x^{n-1} + a_{n-2}x^{n-2} + \ldots + a_1x^1 + a_0$.

Positive number. Any number greater than zero.

Power. The product obtained when a quantity is used as a factor a given number of times.

Prime factorization. The process of expressing a quantity as a product of prime factors only.

Prime number. A number greater than one whose only factors are one and itself. The term prime number refers only to integers, not to fractions or irrational numbers.

Prime polynomial. A polynomial that has no polynomial factors except itself and constants.

Principal root. For odd or even roots of nonnegative numbers, the nonnegative real root; for odd roots of negative numbers, the negative real root.

Principal square root. The nonnegative square root of a number.

Probability. The numerical value that expresses the chance that an event will occur.

Product. The result obtained from the operation of multiplication.

Property of zero products. If the product of two quantities is zero, at least one of the quantities is zero.

Proportion. A mathematical sentence stating that two ratios are equal, e.g., $\frac{a}{b} = \frac{c}{d}$. The quantities a and d are called the *means* and b and c are called the *extremes*, of the proportion. The product ab of the means equals the product bc of the extremes.

Pythagorean theorem. In a right triangle, the square of the length of the hypotenuse is equal to the sum of the squares of the lengths of the other two sides.

Quadrant. One of the four regions into which the coordinate axes divide the plane.

Quadratic equation. An equation that can be written in the form $ax^2 + bx + c = 0$, $(a \neq 0)$. The graph of a quadratic equation is a parabola.

Quadratic formula. For the quadratic equation $ax^2 + bx + c = 0$, $x = \frac{-b \pm \sqrt{b^2 - 4ac}}{2a}$.

Quadratic function. A function defined by an equation of the form $y = ax^2 + bx + c$, $a \neq 0$.

Quadrilateral. A four-sided polygon.

Quotient. The result obtained when two numbers are divided.

Radical. The indicated root of a quantity; e.g., $\sqrt{2}$ is a radical.

Radicand. The quantity under a radical sign, as 2 in $\sqrt{2}$.

Range. (1) In algebra, the set of second coordinates in a relation or function. (2) In statistics, the largest number minus the smallest number in a set of numbers.

Ratio. A comparison between two numbers; e.g., the ratio of three to four is written as $3:4$ or $\frac{3}{4}$.

Rational expression. An algebraic fraction in which both the numerator and denominator are polynomials. Values for which the denominator is zero are excluded.

Rational function. A function defined by an equation of the form $f(x) = \frac{p(x)}{q(x)}$, where $p(x)$ and $q(x)$ are polynomials $(q(x) \neq 0)$.

Rationalize. To remove radicals without altering the value of an expression.

Rational numbers. All numbers that can be expressed as the quotient of two integers.

Rational root theorem. If $P(x) = a_nx^n + a_{n-1}x^{n-1} + a_{n-2}x^{n-2} +$

$\ldots + a_1x^1 + a_0$ is a polynomial function with integral coefficients and $\frac{p}{q}$ is a rational root of $P(x)$ in lowest terms, then p is a factor of a_0 and q is a factor of a_n.

Ray. A subset of the number line that extends indefinitely in one direction only.

Real numbers. Any rational or irrational number.

Reciprocal. The multiplicative inverse of a nonzero number. It is the number whose product with the given number is one.

Rectangle. A quadrilateral that has four right angles.

Reflexive property for equality. For a number a, $a = a$.

Relation. A set of ordered pairs.

Restricted domain. The domain for which a function that is not one-to-one has an inverse.

Rhombus. A parallelogram with all four sides congruent.

Right angle. A 90° angle.

Root of an equation. A solution to an equation.

Row operation. Any of the following operations on the rows of a matrix used to transform it into a specific form: interchanging two rows, multiplying a row by a nonzero constant, and multiplying a row by a nonzero constant and adding the result to another row.

Scalar. A constant that is a real number.

Scalene triangle. A triangle in which no two sides are equal.

Scientific notation. A number that is written as a number greater than or equal to 1 but less than 10 multiplied times an integral power of 10; e.g., 64,000 is expressed in scientific notation as $6.4 \cdot 10^4$.

Sequence. A function whose domain is the natural numbers.

Series. The indicated sum of the terms of a sequence.

Set. A well-defined collection of objects.

Set, finite. A set that contains a finite number of elements.

Set, infinite. A set that is not finite.

Similar polygons. Polygons having corresponding angles equal and corresponding sides proportional to each other.

Slope of a line. If (x_1, y_1) and (x_2, y_2) are the coordinates of two points on a line, the slope of the line is given by $m = \frac{y_2 - y_1}{x_2 - x_1}$ $(x_1 \neq x_2)$. When a line slopes upward to the right, its slope is positive, and when a line slopes downward to the right, its slope is negative.

Solution of an equation. A value that makes the equation true.

Solution set for an equation. The collection of all solutions for an equation.

Solution set for a system of equations. A set of values for the unknowns that satisfies all the equations.

Sphere. The set of points in space that are equidistant from a fixed point.

Square. (1) In arithmetic or algebra, the product of a quantity multiplied times itself. (2) In geometry, a rectangle with all four sides congruent.

Square root. If x^2 is a, then x is a square root of a, as 3 is a square root of 9.

Symmetric property for equality. If $a = b$, then $b = a$.

Synthetic division. A method of dividing a polynomial in one variable by $x - r$ in which only the coefficients are written.

System of equations. A set of two or more equations.

Substitution property for equality. If $a = b$, then a can be replaced by b, or b replaced by a, in any statement without changing the meaning of the statement.

Subtraction. The inverse of addition, defined as $a - b = a + (-b)$.

Sum. The result obtained when numbers are added.

Tangent line. A line that touches a curve in only one point.

Term. A constant, variable, or any product of constants or variables.

Transitive property for equality. If $a = b$ and $b = c$, then $a = c$.

Transitive property for inequality. If $a < b$ and $b < c$, then $a < c$. If $a > b$ and $b > c$, then $a > c$. Similar statements hold for $\leq$ and $\geq$.

Translation. See *horizontal shift* and *vertical shift*.

Trapezoid. A quadrilateral that has exactly two parallel sides.

Triangle. A three-sided polygon.

Trichotomy property. For any two real numbers, exactly one of the following is true: $a < b$, $a = b$, or $a > b$.

Trinomial. A polynomial consisting of three unlike terms.

Union of two sets. The set containing all the elements in either or both of the sets.

Universal set. The set of all objects under consideration.

Variable. A quantity that represents any unspecified element from a set.

Vertical line. A line parallel to the y-axis. Its slope is undefined.

Vertical line test. If any vertical line can be drawn so that it cuts the graph of a relation in more than one point, the relation is *not* a function.

Vertical shift. A shift up or down of the graph of a function accomplished by adding or subtracting a positive constant to $f(x)$.

Volume. The measure of the capacity of a three-dimensional figure expressed in cubic units.

Whole numbers. The numbers in the set $\{0, 1, 2, 3, \ldots\}$.

***X*-intercept.** The x-coordinate at the point at which the graph of a function crosses the x-axis.

***Y*-intercept.** The y-coordinate at the point at which the graph of a function crosses the y-axis.

Zero of a function. A zero of a function f is a value x for which $f(x) = 0$.

INDEX

A
Abscissa, 169, 246
Absolute value, 17–21, 36
  equations, solving, 82–83
  function, graphs, 226, 252
  inequalities, solving, 83–87
Addition, 10 *See also:* Sums
Additive:
  identity property, 12, 36
  inverse property, 13–14, 36
Age problems, 132–136, 159
Algebraic:
  expression, 43, 65
  terminology, 43–44
Arrow diagram, functions, 180, 247
Associative property, 11–12, 36
Asymptotes, 193–195, 215, 225, 251–252
  hyperbola, 237–238, 254
Augmented matrix, transformation of, system of equations, 265, 268–273, 280–281
Axes, 169, 246
Axis of symmetry:
  ellipse, 235–236
  parabola, 215, 250

B
Base, 24
Binomial, 44
  factoring, 53–54
  FOIL, 46–47
Bounded intervals, 9, 36

C
Calculator use, 290–294
Cartesian:
  coordinate plane, 169–178, 246
  product, 178–179, 246
Changing the subject, 79–80, 115
Circle, 234–235, 253
Closed intervals, 9, 36
Closure property, 11, 36
Coefficients, 43, 65, 89, 115
Coin problems, 136–142, 159
Common logarithmic function, 232, 253
Commutative property, 11, 36
Complement, of a set, 2, 35
Completing the square, quadratic equations, 92–98
Complex numbers, 5–6, 35, 223–224, 251
Composite numbers, 2
Composition of functions, 202–204, 249
Conic sections, relations defined by, 234–239, 253
Conjugate pairs, 223, 251
Conjunction, 20, 37, 115
Consistent system of equations, 261–263, 267, 280
Constant function, 197, 214, 250
Constants, 6, 36
Coordinates, 169, 246
Cubes, 25
  difference of two, 47
    factoring, 53–54
  perfect, 3, 48
  root, 32
  sum of two, 47
    factoring, 53–54

D
Decimals:
  repeating, 4
  terminating, 4
Decreasing function, 197
Degree:
  monomial, 43
  polynomial, 44
Dependent variable, 182, 247
Differences, 21–22
  functions, 200–201, 248
  polynomials, 44–45
  rational expressions, 62–64, 67
  two cubes, 47
    factoring, 53–54
  two squares, 47
    factoring, 53–54
Direct variation, 234, 253
Discriminant, quadratic equations, 100, 116
Disjunction, 21, 37, 115
Distance:
  formula, 171–174, 246
  problems, 145–147, 159
Distributive property, 12, 36
  factoring polynomials, 52
Division, 16–17, 36
  by zero, 4
  long, polynomials, 50–51, 66
  synthetic, 306–307
Domain, 178, 246
  of definition, 184–185, 247
Double inequality, 86–87, 115

E
Elements, 1, 35
  first, 169
  second, 169
Elimination, systems of equation, 265–268, 280
Ellipse, 234–236, 253
Empty set, 1, 35
Equality, 6–7
  ordered pairs, 171
  properties of, 6–7
Equations:
  absolute value, solving, 82–83
  function, 181, 247
  linear, 213
  literal, 79, 115
  one-variable:
    linear, 73–78, 114
    solving, 73–124
  systems of, 261–289
  two-variable, solving for one variable, 80–81
Equivalent, 74, 114
Even:
  function, 199, 248
  integers, 3
Exponential:
  expression, 24
    product rule, 25–26
    quotient rule, 30–31
  functions, 229–230, 252
Exponents, 24–31, 37
  negative integer, 28–30
  rational, 33–34
  zero, 27–28
  See also: Logarithmic functions
Extraneous root, 108, 116–117

F
Factor theorem, 221–222, 251
Factoring:
  binomials, 53–54

completely, 58–59
four term polynomials, 57–59
grouping, 57–58
monomial factors, common, 52–53
polynomials, 51–59, 66
quadratic equations, 90–92, 115–116
trinomials, 55–57
Factors, 52
Field properties, 10–17
Finite set, 2
First element, 169
FOIL method, 46
Formulas:
common, 297–305
solving, 78–80, 115
Fraction, complex, simplifying, 64–65, 67
Fractional exponents, 32–33, 38
Functional notation, 179–185
Functions, 169–260
absolute value, 226, 252
application, 243–246, 254
arithmetic, 200–201
arrow diagram, 180
composition, 202–204
constant, 197, 214
decreasing, 197
definition, 179, 246–247
by variation, 234
domain of definition, 184–185, 247
equality, 200, 248
equation, 181
even, 199, 248
exponential, 229–230, 252
graphs, 207–234
horizontal shifts, 188–191, 247
reflection, 191, 248
shrinking, 192, 248
sketching, 185–195
stretching, 192, 248
vertical shifts, 188–191, 247
greatest integer, 228, 252
increasing, 197
inverse, 204–207, 249
reflection property, 206, 249
linear, 207–213, 249
logarithmic, 230–233, 252–253
monotone, 197–199, 248
number lines, 180
odd, 199–200, 248
ordered pairs, 180
parent, 190, 248
polynomials, 220–225, 251
quadratic, 215–220
rational, 225
special characteristics, 195–200
step, 228
table, 180
variation formulas, 234, 253
zeros of, 186
Fundamental theorem of algebra, 220, 251

G
Gauss-Jordan elimination, systems of equations, 270–272
Geometry problems, 149–152
Graphs:
absolute value function, 226
conic sections, relations defined by, 234–239
constant functions, 214, 250
exponential functions, 229–230, 252
function, 181, 247
horizontal shifts, 188–191, 247
reflection, 191, 248
shrinking, 192, 248
stretching, 192, 248
vertical shifts, 188–191, 247
sketching, 185–195
greatest integer function, 228, 252
inequalities, 84, 115
intervals, 10
linear functions, 207–213
logarithmic functions, 230–233
parabolas, 215–220, 250
polynomial functions, 220–225, 251
quadratic functions, 215–220, 250
rational functions, 225
square root functions, 226–227, 252
step functions, 228
Greater than, 8–9, 36
Greatest:
common monomial factor, 52–53
integer function, graphs, 228, 252

H
Half plane, 240
Horizontal:
asymptote, 194–195, 248
axis, 169, 246
ellipse, 235, 253
line:
slope, 176
test, 196–197, 248
shift, 188–191, 247
Hyperbola, 234, 237–239, 253–254

I
Identity, 74, 114
Imaginary numbers, 5–6
Increasing function, 197
Independent variable, 182, 247
Index, 33
Indirect variation, 234, 253
Inequalities, 8–9
double, 86–87, 115
intervals of, 10
linear, 239–241
one-variable linear, 83–87
quadratic, one variable, 111–114
relations defined by, 239–242
two variable:
linear, 239–241
quadratic, 241–242
Infinite set, 2
Integers, 3, 35
exponents, 24–31
problems, 132–136
Interest, simple, problems, 142–145, 159
Intersection, of sets, 2, 35
Intervals, 9–10, 36
notation, 9–10
Inverse of functions, 204–207, 249
reflection property, 206, 249
Irrational numbers, 4–5, 35

L
Language, mathematical, translating, 125–126
Less than, 8–9, 36
Like terms, 44
Lines:
parallel, slope, 177, 246
point-slope form, 209, 250
slope of, 175–178, 246
slope-intercept form, 208–209, 249–250
standard form, 209, 250

Linear:
- equations:
  - three variables, system, 271–273, 281
  - two variables, system, 261–271, 280
- functions, graphs, 207–213, 249

Literal equation, 79, 115
Logarithmic functions, 230–233, 252–253

M

Major axis, ellipse, 235–236
Matrix, augmented, transformation of, systems of equations, 265, 268–273, 280–281
Maximum, parabola, 215, 250
Measurement, units of, 126, 158, 295–296
Midpoint formula, 174–175, 246
Minimum, parabola, 215, 250
Minor axis, ellipse, 235–236
Mixture problems, 136–142, 159
Monomial, 43, 65
Monotone function, 197–199, 248
Multiplication, 10–11
Multiplicative:
- identity property, 12–13, 36
- inverse property, 14–16, 36

Multiplicity, 221, 251

N

Natural:
- exponential function, 229
- logarithmic function, 232, 253
- numbers, 2, 35

Negatives, 13–14
- integer exponents, 28–30, 37
- numbers, 4, 7, 36
- slope, 175

Nonnegative, 9
Nonpositive, 9
Null set, 1
Number:
- lines, functions, 180, 247
- set of, 2–6
- system, real, field properties of, 10–17

Numerical coefficient, 43, 65

O

Oblique asymptote, 194, 248
Odd:
- function, 199–200, 248
- integers, 3

One-to-one function, 195–197, 248
One-variable:
- absolute value:
  - equations, 82–83
  - inequalities, 83–87
- equations:
  - fractional exponents, 108–110
  - fractions, 104–107
  - radicals, 107–108
  - solving, 73–124
- linear inequalities, 83–87
- quadratic equations, 89–104, 115
  - factoring, 90–92, 115–116
  - square, completing, 92–98, 115–116
  - quadratic formula, 98–104, 115–116
- quadratic inequalities, 111–114

Open intervals, 9, 36
Operations, order of, 34–35, 38
Order relationships, 8–9
Ordered pair, 169, 246
- equality, 171
- function, 18, 2470

Ordinate, 169, 246
Origin, 7, 36, 269, 246

P

Parabola, 215–220, 234, 250, 253
- axis of symmetry, 215, 250
- maximum, 215, 250
- minimum, 215, 250
- vertex, 215

Parent function, 190, 248
Percent problems, 129–132, 158
Perfect:
- cubes, 3, 48
- square, 3
- trinomial squares, 47, 55–57

Point-slope form of a line, 209, 250
Polynomials, 43–76
- differences, 44–45
- expressions, simplifying, 48–49
- factoring, 51–59
  - four terms, 57–59
- functions, graphs, 220– 225, 251
- long division, 50–51, 66
- products, 45–48
- quotients by monomials, 49–50, 66
- rational roots, 222–223, 251
- root, 220, 251
- simplification, 44, 66
- sums, 44–45
- zero, 220, 251

Positive:
- integer exponents, 24–25, 37
- numbers, 7, 36
- slope, 175

Power, 24
- of a power, 27
- of product, 26
- of quotient, 26

Prime:
- factors, 52
- numbers, 2

Principal:
- cube root, 32
- square root, 32, 37–38

Problem solving, 125–168
- steps, 127–158
- strategy, 127, 158
- types of problems:
  - coin and mixture problems, 136–142
  - distance-rate-time, 145–147
  - geometry, 149–152
  - integer and age, 132–136
  - ratio-proportion-percent, 129–132
  - simple interest, 142–145
  - statistical, 157–158
  - trigonometry, 152–157
  - work, 147–149

Product, functions, 200–201, 248
Product rule for exponential expressions, 25–26
Products, 22–24, 37
- polynomials, 45–48
- power of, 26
- rational expressions, 61–62, 67
- zero, property of, 91, 116

Proper subset, 2
Property of zero products, 91, 116
Proportion, 107, 116
- problems, 129–132

Pythagorean theorem, 173

Q
Quadrants, 171, 246
Quadratic:
  factoring, 90–92, 115–116
  formula, 98–104, 115–116
  functions, graphs, 215–220, 250
  inequalities, 111–114
  square, completing, 92–98, 115–116
  trinomials, factoring, 55–57
Quadratic equations:
  discriminant, 100
  one-variable, 89–104, 115
Quotients, 22–24, 37
  functions, 200–201, 248
  polynomials by monomials, 49–50, 66
  power of, 26
  rational expressions, 61–62, 67
  rule for exponential expressions, 30–31

R
Radicals, 32–33, 38
  simplifying, 308–309
Radicand, 33, 38
Range, 178
Rate problems, 145–147, 159
Ratio problems, 129–132
Rational:
  exponents, 33–34, 38
  expressions, 43–76
    differences, 62–64, 67
    products, 61–62, 67
    quotients, 61–62, 67
    reducing, 59–61
    sums, 62–64, 67
  functions, graph, 225, 251
  numbers, 3, 35
  roots of a polynomial, 222–223, 251
Real numbers, 2, 5–6, 35
  line, 7–8, 169
  system, field properties of, 10–17
Reciprocals, 14–16
Reflection of a function, 191, 248
Reflexive property, 6, 36
Relations, 169–260
Remainder theorem, 221–222, 251
Repeating decimal, 4
Root, 4, 32–33
  extraneous, 108
  of an equation, 73, 114
  quadratic equation, 90, 115
  rational, polynomials, 222–223, 251
Rule, functions, 180, 247

S
Scientific notation, 31–32, 37
Second element, 169
Sets, 1–2, 35
  finite, 2
  infinite, 2
  numbers, 2–6
  solution, 74, 114
Shrinking function graphs, 192, 248
Sign, 14
Simple interest problems, 142–145, 159
Simplification;
  complex fractions, 64–65, 67
  polynomial expressions, 48–49, 66
Slant asymptote, 194, 248
Slope:
  -intercept form of a line, 208–209, 249–250
  of a line, 175–178, 246
  zero, 176, 246
Solution, 73–74
  checking, 73–74
  set, 74, 114
    quadratic inequalities, 112–114
    system of equations, 261–262
Solving:
  equations:
    one-variable, 73–124
      absolute value, 82–83
    two-variable, solving for one variable, 80–81
  formulas, 78–80, 115
  inequalities, one–variable linear, 83–87
Square roots, 32–33
  functions, graphs, 226–227, 252
Squares, 25
  completing, quadratic equations, 92–98
  difference of two, 47
    factoring, 53–54
  perfect, 3
    trinomial, 47
  sum of two, factoring, 53–54
Statistical problems, 157–158
Step functions, graphs, 228
Stretching function graphs, 192
Subject, changing, 79–80, 115
Subset, 1–2, 35
  proper, 2
Substitution, 36
  systems of equations, 265–266, 280
Subtraction, 16, 36 *See also:* Differences
Sum, 21–22, 37
  functions, 200–201, 248
  of two cubes, 47
    factoring, 53–54
  of two squares, factoring, 53–54
  polynomials, 44–45
  rational expressions, 62–64, 67
Symmetric property, 6, 36
Systems of equations, 261–289
  applications, 275–280
  back substitution, 270–271
  consistent, 261–263, 267, 280
  dependent, 261–262, 265, 267, 280
  elimination, 265–268, 280
  Gauss-Jordan elimination, 270–272, 280
  inconsistent, 261–262, 264, 267, 280
  linear:
    three variables, 271–273, 281
    two variables, 261–271, 280
  matrix, augmented, transformation of, 265, 268–273, 280–281
  solution set, 261, 280
  solving, 265–271
  substitution, 265–266, 280
  triangular form, 270, 272–273, 281
  two variables, other, 274–275

T
Tables, functions, 180, 247